Optimization and Inverse Problems in Electromagnetism

Optimization and Inverse Problems in Electromagnetism

Edited by

Marek Rudnicki

Academy of Humanities and Economics,
Łódź, Poland

and

Sławomir Wiak

Academy of Humanities and Economics,
Łódź, Poland

SPRINGER-SCIENCE+BUSINESS MEDIA, B.V.

A C.I.P. Catalogue record for this book is available from the Library of Congress.

ISBN 978-90-481-6375-5 ISBN 978-94-017-2494-4 (eBook)
DOI 10.1007/978-94-017-2494-4

Printed on acid-free paper

Table of Contents

3. Applications

Editor's Preface

The Workshop "Optimization and Inverse Problems in Electromagnetism" is a bi-annual event. The history dates back to 1989, Pavia, Italy. The following meetings were held in Warsaw, Poland (1992), Geneva, Switzerland (1994), Brno, Czech republic (1996), Jyväskylä, Finland (1998) and Turin, Italy (2000). The next Workshop will be held in Grenoble, France in 2004. The aim of the Workshop is to discuss recent developments and application in optimization and inverse methodologies for electromagnetic fields. The meeting is intended to be a forum for applied mathematicians, computer scientists and electrical engineers to exchange ideas, experience on the new developments, trends and applications from industrial and academic viewpoints on the topic. Particular attention is paid to optimization of electromagnetic devices and micro-systems along with software development. An important goal of the workshop is also stimulating personal contacts and co-operation, especially between industrial and academic institutions.

In the year 2002, from 12 to 14 September, the Workshop was jointly organised by The Academy of Humanities and Economics (AHE) and the Institute of Mechatronics and Information Systems (previously the Institute of Electrical Machines and Transformers) of the Technical University of Łódź.

The hosting institution, Academy of Humanities and Economics, is the oldest and the largest private university in Łódź, with over 18.000 students. It employs numerous outstanding scientists, academics and expert-practitioners. The Academy of Humanities and Economics participates in the majority of educational programs of the European Union such as Socrates, Leonardo da Vinci and Youth for Europe. AHE is a unique institution offering as well the exceptional course of creativity to all the students, which completes the traditional academic knowledge with additional skills that stimulate individual self-development.

The Institute of Mechatronics and Information Systems conducts teaching on the topic of electrical machines, mechatronics and computer science. It organises numerous scientific conferences, among which the International Symposium on Electromagnetic Fields in Electrical Engineering. The Institute cooperates with Polish Academy of Sciences (PAN), Institute of Electrical and Electronics Engineers (IEEE), International Compumag Society, etc. In the last fifty years the staff of the Institute published approximately 60 monographs and textbooks, about 800 papers, more than 300 scientific-technical expertise.

The contributions selected for the present volume cover a wide spectrum of inverse and optimal electromagnetic methodologies ranging from theoretical to practical applications. This is an outstanding feature of the meeting, which, to our knowledge is the only one focussed on that subject.

A number of new optimal and inverse methodologies were proposed. There are contributions related to dedicated software. The book consists of three thematic chapters covering:

- General Papers (survey of specific aspects of optimization and inverse problems in electromagnetism),
- Methodologies,
- Industrial Applications.

The book can be useful for students of electrical and electronic engineering, computer science, applied mathematics (PhD level) and to researchers interested in this topic.

Prof. Marek Rudnicki — Chairman

Prof. Sławomir Wiak — Scientific Secretary

PART 1

GENERAL PAPERS

Introductory remarks

There are three survey papers in this chapter. The first one entitled :"How to find new and fresh ideas" is written by Tapani Jokinen of Helsinki University of Technology, Finland, The paper deals with how people get ideas and how we can help the team members to co-operate in a group and see the positive sides of other people's ideas. Logical thinking is essentially ineffective in discovering new ideas, and sometimes it even prevents the birth of ideas. When the problem is left out of mind, it is all of a sudden worked out, although we have not given it a thought. Various methods have been developed for idea generation. The problem analysing and solving method, AIR-OPERA, is presented in the paper. Although not related directly to electromagnetic optimization, it provides valuable advice on creative thinking in electrical engineering.

The second paper by Antonio Savini of University of Pavia, Italy, entitled "Mimicking nature. How nature inspires methodologies in computational electromagnetics" may be considered as a survey of methodologies in computational electromagnetics inspired by nature. A brief overview on the attempts to apply mathematical models to investigate nature and, how to deduce mathematical tools from the study of nature is given, Specifically, we observe is a great deal of research work with the ambitious ultimate goal of getting "intelligent" computers, capable of solving problems without explicitly programming them. Moreover, evolutionary computation seems the natural way for programming and implementing bio-computers. From the software viewpoint, new biological processes like that of molecules of the immune systems combating the invading pathogenic micro-organisms are going to stimulate new search methodologies like artificial immune systems. This way, the circle of inspiration from nature is closed. The final author's conclusion is that the book of nature is inherently written in mathematical terms.

The last paper in this chapter entitled "Comparing Stochastic Methods on SMES Optimization" by O. Hajji, S. Brisset and P. Brochet is concerned with the optimization of a superconducting magnetic energy storage device by means of three stochastic methods: genetic algorithms, simulated annealing and tabu search (GA, SA, TS). Three different optimization strategies have been compared on the superconducting magnetic energy storage (SMES) benchmark

On one hand, stochastic methods, explore the search space better than the deterministic ones, but on the other hand, a large number of objective function evaluations is required. To decrease it, a new stop rule for GA, based on the response surface methodology, is proposed. A table of step vectors is applied to accelerate the convergence of SA. For the optimization of the SMES, the improved versions of stochastic algorithm have led to better solutions using significantly less time. The parameter sensitivity is studied and improvements are proposed. The paper may be considered as a comparative study of meta-heuristic optimization algorithms enlightened on a practical engineering design problem.

PART 2

METHODOLOGIES

Introductory remarks

There is still a widespread and remarkable interest within the electromagnetic community for the refinement of methodologies for field analysis and optimization technique. It is even possible to observe that the implementation of field and optimization theory in the computer codes has contributed to the clarification of the fundamental questions in such discipline like optimization algorithms. The integration of field analysis with optimization techniques has made it possible to move from computer-aided analysis to automated optimal design of the family of electromagnetic devices, electromechanical converters, micro-mechanical systems (MEMS), etc. Inverse problems actually could be solved, at least in principle, by means of the numerical optimization of an objective function, which depends on many variables and subject to prescribed constraints. Nevertheless, methodologies for optimal design are still complex; moreover the number of fundamental problems are still open. Problems like multiple local minima, large number of unknowns, multiple design criteria, etc. represent the main challenge to the development of computational electromagnetism.

Sixteen papers are included in this chapter; they may be divided into the following subgroups: Genetic Algorithms (1,7), Probability Theory (14,15), Optimization Techniques (2,3,4,5,6,8,9,10,12,13) and Applications (11,16).

The *first group of papers* is focused on applying genetic algorithms and advanced software to the analysis of:

- *Benchmark problems.* Studies have been conducted for the use of quasi-random sequences in the initialisation of a genetic algorithm. Sample points in a quasi-random sequence are designed to have very good distribution properties. Solving a large number of continuous benchmark problems from the literature has tested the modified genetic algorithms using quasi-random sequences in the initial population. The numerical results of three genetic algorithm implementations using different quasi-random sequences have been compared to those of a traditional implementation of using pseudo random numbers.
- *Grounding system.* A new method for quick determining the maximum touch voltage generated by a grounding system leaking a known current is proposed. The touch voltages in the points of the soil surface are calculated by the Maxwell's subarea method and the search for the maximum touch voltage is carried out by an ad hoc genetic program which is based both on the "one-point crossover" technique and on a mutation having a pre-arranged probability rate.

The *second group of papers* is devoted to general and specific inverse problems with a structure called Exponential Family:

- *Inverse Problems with a structure called Exponential Family.* This structure is pivotal in many fundamental theories (Bayesian Probability Theory, Maximum Entropy Principle, Fourier-Laplace Transforms, Generating Functionology, Sufficient Statistics, Lie Groups, ...), and offers extensive applications to Physical Theories and Complex Systems.

- Topology is manifest in Inverse Problems, Electromagnetic Theory and many other fields. Cartan's exterior calculus of differential forms, Lie groups and Exponential Family are central tools for investigating phenomena where some topological changes take place. A fundamental law of topological evolution is equivalent to the First Law of Statistical Thermodynamics, involving a topological interpretation of heat. We link Bayes' Theorem with Triality, D4-group, and Octonions. The proposed inner symmetries of Probability Theory itself correspond to an exceptional A-D-E sequence of Lie groups, with vital implications for fundamental physical laws.

The *next group of papers* deals with optimization techniques applied to the specific model and technical problems, while the numerical and analytical methods are successfully applied to:

- *Symbolic treatment of analytical models.* A symbolic treatment of analytical models is used to describe systems to be size-using methods with optimization under constraints techniques. Such models are equations based. In many cases, the models have to be arranged by the designer, to get a well organised and oriented description for their valuing in the optimization process. Indeed, a complete physical description for a system is often made of equations, algorithms and functions. To optimise on many criteria, the computer code that values them has to be performed. To provide an automatic help to organise the model elements, and to create a computer science code that carries out the model valuing. The description of these models is stored in a format that will allow easier code generation, for their valuing in different kind of tools, especially in optimization one.
- *Pareto criterion for electromechanical converters.* The shape synthesis of a linear actuator and a permanent-magnet motor is revisited in terms of multiobjective optimization using Pareto criterion of optimality. Best-compromise solutions are approximated by means of sample-and-rank method. The two case studies are proposed as model problems for inverse electromechanics.
- *Choice of automatic grounding grid.* A method for choosing, in a quite automatic way, the various grids and so for speeding up the designing process is proposed. The method is based on the determination of the maximum touch voltages, these being calculated by resorting to a genetic algorithm. Moreover, the building cost of the grounding grid is optimized thanks to the "Travelling Salesman" algorithm.
- *Gradient optimization algorithms in sizing of electrical devices.* The sizing of some electrical devices, mainly electrical circuits, using optimization under constraints, while . more accurately gradient optimization algorithms are proposed.
- *Practical improvement of stochastic algorithms with numerical models.* Combined use of solution space discretisation and solution history is implemented. discretisation gives the possibility to use a solution history where all the solutions evaluated are stored. Use of the solution history guarantees that a single solution candidate is calculated only once. This approach is useful for the stochastic algorithms that typically evaluate many solution candidates. Then improved method is proposed for optimization of high-speed induction motors modelled with 2D finite element analysis software.

- *Finite element method with non-obtuse triangulation.* A new refinement techniques called "quasi-red" and "quasi-yellow" are applied for the solution of a non-linear Poisson's equation describing magnetic flux density in 2-D. The equation is solved using the finite element method with non-obtuse triangulation. The main goal is to modify initial mesh in some non-standard way to have solely non-obtuse triangles. Usefulness of the method is demonstrated on the problem of optimal shape design of an electromagnet.
- *Experimental Design Method (EDM)/ optimization techniques.* Some results on optimization techniques and the Experimental Design Method (EDM) are presented and applied to a superconducting magnetic energy storage (SMES) device. The optimization process is realized through two different approaches. In order to solve easily this problem, an optimization manager based on numerical electromagnetic simulations is employed.
- *Silicon disc-type micromotors optimization.* The complex field and circuit modeling of silicon disc-type micromotors are proposed. 3-D motor structure is proposed to study. The field analysis and torque calculation results have given the knowledge about the changes of the motor structure in order to increase unit motor torque. Comparative analysis leads to the choice of optimal design of the structure.
- *Optimization techniques in magnetic component design.* Benefits of optimization techniques applied to magnetic component design are discussed using a comparison between an industrial hand-based design process and a methodology based on constrained optimization techniques. The results show dramatic design time decrease, as well as design process improvements in terms of reliability and formalism, which has to be balanced against encountered difficulties.
- *Optimization techniques applied to electromagnet pole-shape design.* Two optimization techniques applied to electromagnet pole-shape design are proposed and discussed. The aim was to find a shape, such that a desired magnetic field distribution is achieved in the air-gap. The chosen optimization techniques were simulated annealing and taboo search both applied to the aforementioned problem. For approximate solving of partial differential equations the finite element method was used, while MATLAB with PDE toolbox is exploited.

The last group of papers deals with specific methods and applications. The use of numerical methods are successfully applied to the solution of specific probelms, like:

- *Compensation of the gravity by means of a magnetic force acting on a small diamagnetic volume.* Generation of uniform field of magnetic forces in a given domain is a problem with various applications. The main idea is the definition of a way of identifying ideal current distributions able to generate a quasi- uniform force field. Based on one particular expression of the complex magnetic potential, an inverse methodology, which is valid for structures with plane symmetry (2-D field), has been worked out. Currents have been modelled by purely surface (or thin shell) current distributions.

- *Evaluation of the electromagnetic field emitted by an industrial source*. The design of a magnetic field shielding apparatus requires simulation both the source and the shield system and evaluation as well the main shield parameters like geometry, materials, position with respect to sources. An identification process of industrial magnetic fields with a negligible harmonic content is performed using a genetic algorithm applied to a simplified source. The proposed source is a suitable combination of three-phase ones and it seems able to create a realistic electromagnetic field distribution both in space and in time.

PART 3

APPLICATIONS

Introductory remarks

A knowledge of electromagnetism is necessary for understanding the operation of electromagnetic devices, electromechanical converters and electronic systems in general, such as sensors, actuators, solid state devices, integrated circuits and micro electro mechanical systems (MEMS); also electromagnetic compatibility represents an increasingly important topic. In general the co-existence of electric, magnetic, thermal and mechanical effects characterizes the global behaviour of any electromagnetic device or system.

There is a quite high degree of attention to applied electromagnetism due to the increasing number of new topics.

We could notice subjects like mechatronics, superconductivity, microelectronics and biotechnologies among emerging fields in which electromagnetism plays an essential role.

The complex geometry of devices and the non-linearity of magnetic circuits of devices represent unfavourable factors for field calculation. A very small air-gap with different number of slots in primary and secondary parts of converter produces high field gradients. Therefore the fine discretisation of the field region and, sometimes, the coupling of different methodologies of analysis are necessary.

Thirteen papers are included in this chapter; they may be divided into the following subgroups: Optimization Techniques of Devices (2,7,11), Drives (1,6,8,9,10,13), Transformers and Special Problems (3,4,5,12).

The *first group of papers* is focused on applying field theory and advanced software to the analysis of:

- *Induction heating system.* The hybrid FEM-DBCI (Finite Element Method - Dirichlet Boundary Condition Iteration) is exploited to evaluate the objective function. In order to reduce the computing time for the whole procedure, Design of Experiments (DOE) techniques are used to screen significant geometric parameters. The optimization procedure is carried out by means of the Hooke-Jeeves algorithm. The results are compared with those obtained by stochastic optimization of the device.
- *Solenoid electromagnet force characteristics.* The optimization of the solenoid force is carried out by varying the position and the dimensions of the disc and seeking maximal initial force. Finite element method has been used for field analysis and force computation.
- *Static converters by means of analytical models and a gradient optimization algorithm.* The aim is to carry out a compromise between the computation speed and the result accuracy, which allows treating more criteria in the same optimization process than a numerical or simulation approach.

This method is applied to optimise the volume of a PFC (Power Factor Correction) flyback converter under EMC (electromagnetic compatibility), loss and thermal constraints. In this way, an analytical model is developed for each optimization aspect. The use of these models in an optimization approach with a gradient algorithm under constraints, specific tools are developed and integrated in an adequate optimization environment. The optimization results are then validated by experiment

The *second group of papers* is devoted to general and specific problems of drives. For the optimization of electric drives, both simulation and analytical methods have been developed:

- *Simulation and optimization tools of electric drive.* Simulation and optimization tools are proposed, while a structure description based on XML. Then, it proposes a process and tools to transform them to obtain calculation component. In this way, both dedicated simulation components and dedicated calculation components are obtained. This is applied to the sizing of electric drive using optimization algorithms.
- *Rare earth permanent magnet generator optimization.* The suitability of evolutionary algorithms for design optimization of electromagnetic devices is discussed. Development and application of such an algorithm for such generator is then proved. Test results for a proof of concept generator are given for verification.
- *Permanent magnet DC motor optimization.* The performance of permanent-magnet DC motor can be improved by optimization of its magnetic circuit taking into account important effects of magnetic density distribution in motor air-gap and neutral zone. The optimization can be achieved by means of finite element method in field and torque calculation and statistical analysis using orthogonal arrays. The electromagnetic torque as an objective function was calculated directly from magnetic vector potentials of moving conductors as an average value through a slot pitch shift. A fast improvement in shape of some motor parts was achieved by a very limited number of field and torque calculations. The optimization results verified this approach in modeling of DC motors with permanent magnets.
- *Identification of turbine generator parameters.* Parameters of the mathematical model of the coupled - electromagnetic and temperature fields have been determined; using measured temperature data on the surface of the flange and fingers in the end zone. Different modes (rated, at the leading power factor and at the lagging power factor) are considered. Coupled field problem - electromagnetic and temperature field distribution in the end zone of a 200 MW turbogenerator is studied using FEM for forward problem. The secondary model of the temperature values at points with measuring control has been obtained using combination of response surface methodology and design of experiments.
- *Electrical machine design with new magnetic material characteristics.* The materials considered are amorphous alloy ribbons as well as Neodymium alloy permanent magnets involving very low eddy current losses. Such materials enable electric machine operation at higher frequencies compared with the standard iron laminations used in the traditional magnetic circuit construction and provide better efficiency.

- *Improvement in a six-phase flux-weakened permanent magnet synchronous motor with a third harmonic current injection.* It is shown that the flux-weakened permanent magnet synchronous motor with the third harmonic current injection is capable of producing more output torque per RMS current flowing through the windings than the same motor supplied with sinusoidal currents. The influence of the iron losses on the OPC and the constant-power speed range (CPSR) is included in the analysis by the posterior iron loss calculation. The improvement of the machine OPC with a third harmonic current injection depends on the level of flux weakening and it is higher at a high level of flux-weakening.

The *next group of papers* deals with specific methods and applications. Beyond the common use of numerical methods, analytical methods also are successfully applied to the solution of specific probelms, like:

- *Au and Al junction* in a current carrying contact for application to integrated circuits. The conductivity of the compound is identified by means of a field-inversion technique. The simulation of the growth process of the Au-Al compound relies on the transient field analysis of the current flow through the junction.
- *Multi-conductor transmission line model of a helical winding.* An analysis of the very fast electromagnetic transients will include the lumped voltage between the edges of the open electrostatic screen, previously found to be responsible for experimentally found anomalous distortion of the travelling waves. Contrary to the former model with distributed voltage and axial symmetry, the new model requires parameters beyond those of classical theories of wave propagation in transformer windings. The additional parameters are found by matching the simulated voltage transients with the experimental results.
- *Methodologies of preliminary design of a three-phase transformer.* Preliminary design requires low time cost. Although analytical models answer this issue, they brought some loss of quality, as they are subject to many hypotheses. Analytical model can be mixed with numerical ones to improve the modelling quality. The analytical model was used, with an automatic symbolic differentiation. Then FEM evaluation of the leakage reactance is introduced to the analytical model.
- *Non-destructive examination method for high temperature superconductors (HTSC's).* The scanning of the magnetic field distribution, which is trapped by a superconductor, is done. The identification of the positions, dimensions and orientations of sub - domains within the HTSC has been reduced for the determination of the critical current density (J_c). The defined inverse field problem has been solved by use of numerical algorithm based on FEM.

BIOGRAPHIES OF THE EDITORS

Prof. Marek Rudnicki was born on July 23, 1951 in Siedlce. In the course of his academic and research career he has been working for numerous institutions, i.e. High School of Agriculture, Polish Academy of Sciences, University of Warsaw, Technical University of Lodz, and lately for the Academy of Humanities and Economics in Lodz, where he cducts lectures on artificial intelligence, optimization, numerical methods, evolutionary computing as well as computer simulation and modeling. His main research interests include evolutionary computing, computational optimization and artificial intelligence. He has been cooperating with numerous research centers in Poland and abroad, i.e. University of Pavia (Italy), National Technical University of Athens (Greece), University of Jyväskylä (Finland), Helsinki University of Technology (Finland), Aalborg University (Denmark) and Technical University of Madrid (Spain). He has taught a number of PhD courses, e.g. International Summer School of Mathematics in Jyväskylä (Finland), International PhD School of Engineering in Aalborg (Denmark) and University of Pavia (Italy). He has a number of affiliations and memberships in scientific bodies, e.g. Polish Society of Applied Electromagnetism, Association of Polish Electrical Engineers, memberships of International Steering Committee of International Workshop on Inverse Problems and Optimal Design in Electrical Engineering, International Editorial Board of International Journal of Electromagnetics and Mechanics and International Organising Committee of Symposium on Micromachines and Servodrives. He is the author and co-author of the following publications: 1 book (**Neittaanmaki, P., Rudnicki, M., Savini, A.,** *Inverse Problems and Optimal Design in Electricity and Magnetism,* Clarendon Press, Oxford, 1996, 380 pages), over 50 articles in Polish and foreign journals and post-conference monographs (COMPEL, International Journal of Applied Electromagnetics in Materials, Nuclear Magnetic Resonance, Wiley & Sons, Walter de Gruyter and Plenum Press).

Prof. Sławomir Wiak was born on August 14, 1948. Since the beginning of his academic career he has been working at the Technical University of Lodz. He specializes in technical electrodynamics, computer aided design, computer simulation and modeling, intelligent microsystems modeling, software engineering, databases and expert systems. Not only does he cooperate with numerous research centers all over the world, e.g. University of Southampton (Great Britain), University of Maribor (Slovenia), University of Pavia (Italy), University of West Bohemia (Czech Republic), University of Coimbra (Portugal) and the Czech Academy of Sciences, but also with commercial companies, e.g. Vector Fields Ltd. (Great Britain), where he collaborates with a team of software engineers on computer aided design software. He is an active member of a number of associations and foundations, e.g. IEEE (Institution of Electrical and Electronic Engineers), ICS (International Compumag Society – one of the founders), ICS IEEE (International Computer Society), SEP (Association of Polish Electrical Engineers), Polish Society of Applied Electromagnetism, Polish Society of Theoretical and Applied Electrical Engineering and Śniadeccy Foundation. He is regularly asked to write reviews of articles to be published in the renowned scientific journals, e.g. IEEE Transactions on Magnetism and Elsevier Studies in Applied Electromagnetism in Materials. He is the author and co-author of 3 monographs and scientific dissertations, 9 special issues, 44 articles in foreign journals (e.g. Archiv für Elektrotechnik, IEE, Iternational Journal of Numerical Methods in Engineering, IEEE Transactions on Magnetism and COMPEL), 146 conference papers (presented in COMPUMAG, ISEF, CEFC, IGTE, MIS, ACEMP, ICEM, OIPE). Prof. Wiak is author of over 200 referee reports of papers published in international journals and contributed to International Conferences.

HOW TO FIND NEW AND FRESH IDEAS

Tapani Jokinen
Helsinki University of Technology, Laboratory of Electromechanics, Finland

Abstract: The paper deals with how people get ideas and how we can help the team members to co-operate in a group and see the positive sides of other people's ideas. Logical thinking is essentially ineffective in discovering new ideas, and sometimes it even prevents the birth of ideas. When the problem is left out of mind, it is all of a sudden worked out, although we have not given it a thought. Various methods have been developed for idea generation. The problem analysing and solving method, AIR-OPERA, is presented in the paper.

Key words: Problem solving, team work, idea generation

1. BIRTH OF IDEAS

Some people seem to be better equipped and talented than others in discovering new ideas. It would seem fairer if new ideas were born as the result of hard work and struggle. This, however, is not the case. **It is often typical that ideas suddenly pop up into the mind. Work of years may hamper the creation of ideas**, because thinking has stuck into a groove and old ideas seem to be the only possible ones. The world is full of hardworking scientists who advance logically and meticulously in their research, but new discoveries and ideas are beyond their reach.

The creative and the logical approaches are not mutually exclusive; on the contrary, they are complementary. Creative thinking is used when no logical solution is found to a problem. After the kernel of the problem is discovered, it can be logically grown into a final solution.

M. Rudnicki and S. Wiak (eds.), Optimization and Inverse Problems in Electomagnetism, 1–11.

Figure. 1. The birth of ideas is many times a sudden flush into to the mind.

New ideas often occur when **observations do not agree with an old theory** and it is necessary to review the old thoughts. The new information then usually leads to new ideas. On the other hand, an old theory can often explain even new knowledge, which then supports the old, possibly incorrect theory.

New knowledge is not necessary for new thoughts. **Old knowledge can be considered from a new viewpoint**. An example: Research on smallpox naturally attempted to find out why people caught the disease. But *Edward Jenner* started to examine why milkmaids apparently did not catch smallpox. Jenner noted that the relatively harmless cowpox protected milkmaids against the fatal smallpox. As a consequence of this observation, inoculation with a vaccine was discovered and smallpox was defeated.

Logical thinking is essentially ineffective in discovering new ideas, and sometimes it even prevents the birth of ideas. A case in point:

Guglielmo Marconi developed wireless equipment and managed longer transmission distances by increasing transmitter power and designing receivers that are more sensitive. Marconi started to plan wireless communication across the Atlantic. Experts considered this impossible, because according to their observations, radio waves propagated in a straight line as light does. Therefore, logically thinking, radio communication across the Atlantic was impossible. Marconi did not heed the argumentation but continued to try and finally succeeded. Neither the experts of that day nor Marconi himself knew that sufficiently long radio waves are reflected from the ionosphere, which has an electrically charged layer. This layer made radio communication across the Atlantic possible.

Chance is the beginning of very many ideas. Two examples of coincidences: Hertz discovered radio waves when he began to wonder about a small spark in a device not connected to any of the test devices he was using across the room. *Röntgen* discovered X-rays when he forgot to remove a fluorescent plate from a table where he was making tests with cathode-ray tubes. Hertz and Röntgen did not understand to search for radio waves or X-rays, as these were not known to exist.

The purpose of chance in creating new ideas can be thought as bringing forth something that otherwise would not be looked for. **Chance happenings can be multiplied by playing games**. The game should lack a purpose, be without plans or directions. The seeming fruitlessness is awkward and prevents many persons from playing. In play, ideas pop up spontaneously and will generate new ideas. The ideas are not in any logical order. If the player does not in any way try to direct the ideas and is sufficiently curious to follow up, ideas are generated in abundance. They may not necessarily seem very useful, but they may flash back and then be useful.

One way to promote the random interaction of ideas is the following old game. A group of people gather together to discuss a problem. They try to ignore the logical obstacles by telling each other whatever comes to mind. All thoughts are acceptable and nothing is too absurd. If everybody is enthusiastic, the probability increases of random thoughts and ideas being born, ideas which nobody would have otherwise thought of.

A good way to obtain new ideas is to expose oneself to random stimuli by visiting places like exhibitions or department stores, which are full of things one would never look for consciously. Strolling should be done in the manner of a scrap collector, i.e. by collecting everything, whether useful or not, that for some reason or other attracts one's attention. In the background is the thought about a problem, which needs a new solution. Such strolling often generates associations or new approaches to work out a problem.

Everybody probably knows the following situation. We have thought about a problem but could not solve it. **When the problem is left out of mind, it is all of a sudden worked out**, although we have not given it a thought. The problem has brewed in our subconscious, and new associations unfolding the solution flash into our mind. Familiar to many is the situation when we have decided something, but wish to sleep on it. The following morning we find completely new viewpoints. The new, improved decision is better than the one from the previous day.

1.1 Conscious mind and subconscious mind

Letting a problem brew and solve itself later can be explained (but not proved) with the aid of the following model of human thinking.

Human thinking can be considered to occur on three different levels. Knowledge for immediate use is in the **conscious awareness**, where thinking is analytical and uses logical reasoning. The other level of thinking is in the subconscious, which contains earlier experiences but is not immediately accessible. Between the subconscious and the conscious awareness is a **preconscious**, which transfers knowledge from the subconscious to the conscious awareness. The conscious mind usually interprets the information from the subconscious as illogical and insensible. A **filter** is formed between the subconscious and the conscious awareness. Filtering increases with age and prevents knowledge from the subconscious entering the conscious awareness. When the problem is left to brew, the preconscious and subconscious still work, and new possibilities may flash into mind when the filter is caught unprepared.

1.2 Dreams and problem solving

The human brain is never at rest, not even while sleeping. There is always different information processing going on, which during sleep differ very much from daytime thinking. This we can note from our dreams. They are wild processes, often bouncing back and forth in an illogical manner.

A diary-like record kept of dreams reveals that dreams often deal with the information and material experienced during the day. Experiences, worries and problems are transferred from our conscious thinking to our dreams. During sleep, we also try to work out problems left over from the day, but in a completely different way than while awake. Dreams do not pay heed to rules or inhibitions, which chain us while we are awake. The most peculiar matters are interconnected, and things may turn upside down.

It was noted in a preceding section that a problem left to brew often solves itself during sleep. The following method can make the solving of problems while sleeping more effective.

1. Analyse the problem during the day and try to find alternative solutions. Write down the solutions.
2. When going to sleep, think intensively for a short while once more about your problem. Create in your mind some kind of "dream friend" to whom you tell your problems. Tell your friend how you want the problem to be solved what you aim at.
3. The solution may come to your mind during sleep and you wake up. You should write the solution down immediately, because when you fall asleep again you may easily forget it.
4. Even though you may not remember any special dreams when you wake up, the solution may come to you by itself, or when you begin to reflect over the problem, you find new solutions that never occurred to you the day before.

Figure. 2. Solve your problem with your "dream friend".

1.3 Methods and main rules of idea generation

Various methods have been developed for idea generation. The different search methods for ideas show some similarities:

- **Do not be satisfied with the first practical idea**. Try to find as many ideas as possible. Then the probability increases that there are also good ideas.
- **Make a difference between the search for and the evaluation of ideas**. This holds true especially for intuitive methods. Intuitive thinking requires a lively state of mind, and it takes time to achieve this. Criticism will obliterate the spirited state and condemn ideas, which at first appear unrealistic but could have become good solutions if further developed.
- **Try consciously to avoid habitual solutions**, especially when the problem is difficult. Then a good solution is usually found only after a completely new approach is adopted.

2. AIR-OPERA (TWO & TWO METHOD)

Teamwork has for a long time been a slogan for effective work. Combine the expertise, enthusiasm and energy of different people and you will get synergy, a whole that is more than the sum of its parts. Very often, this works only in theory. Instead of 1+1 equalling more than two, you often get

1+1 achieving even less than one. Different people with different ideas, values and needs destroy each other's thinking. Instead of improving each other's ideas, they spontaneously criticise new ideas and defend their own views. The more new different people you add to the team, the less they can achieve as a team, because of their differences.

The problem analysing and solving method, AIR-OPERA, helps the participants to co-operate in a team and see the positive sides of other people's ideas. It changes the participants' points of view from their personal ones to that of the team. AIR-OPERA is also called the Two & Two method. The Finnish management consulting company Innotiimi developed the method during the 1980s. The method is discussed below in some detail.

2.1 Three stages

The process contains three stages (Fig. 3): 1 **A**nalysis of the task, 2 **I**deas, and 3 **R**esults. The first letters of the three stages A-I-R form the first part of the name of the method. All three stages have a similar general structure O-P-E-R-A: **O**wn thoughts, **P**air's thoughts, **E**xplanations, **R**atings, and **A**lignment.

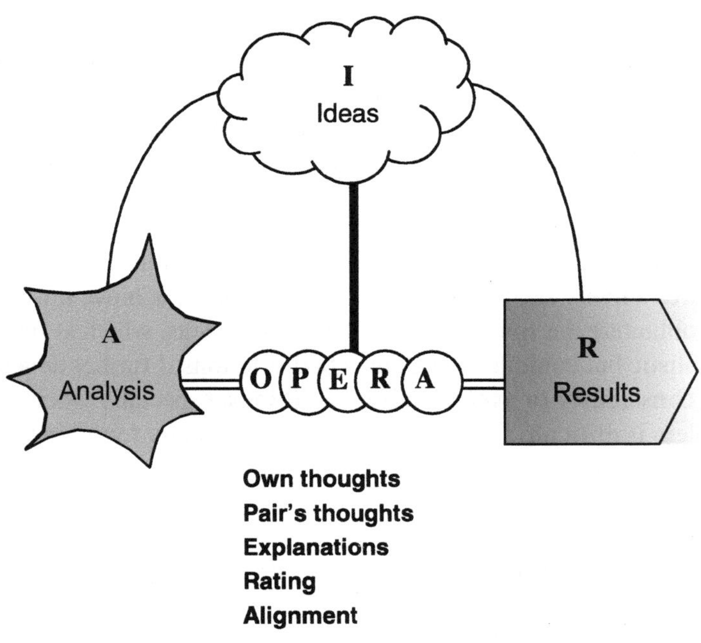

Figure 3. AIR-OPERA contains three stages (AIR) and every stage has a similar structure (OPERA).

2.2 Analysis of the task

In the analysis stage, the team is asked to find the most important problems or goals that it can think of. First, the leader of the team asks everyone to work alone and write down his or her **Own** thoughts.

In the second phase, the work is continued in **Pairs**. The task is to clarify through discussion three or four most important problems or goals the pair has found. Every item chosen should be written down on a separate sheet of paper. The papers are then taken to the board (Fig. 4). It is convenient to have a matrix board having six columns and six rows. A piece of a stick-on material has been glued in every matrix element so that it is easy to attach the papers to the board. Each pair has its own column.

Figure 4. AIR-OPERA works in pairs, and ideas are fixed on the matrix board.

In the third phase, the pairs **Explain** their thoughts. Every pair is given a couple of minutes to present their thoughts. No criticism or any kind of comments is allowed.

The fourth phase is **Rating**. Every pair chooses three or four best proposals on the board and marks them for instance with a plus (+). Only one of the marks is allowed to put on the pair's own column. After marking every pair tells which papers they have marked and why.

In the last, **Alignment**, phase the leader groups the papers according to the directions of the team so that the proposals with the most pluses are

placed on the top and similar ones underneath them (Fig. 5). The sugges-
tions that do not have any marks on them can be removed from the board or moved to the

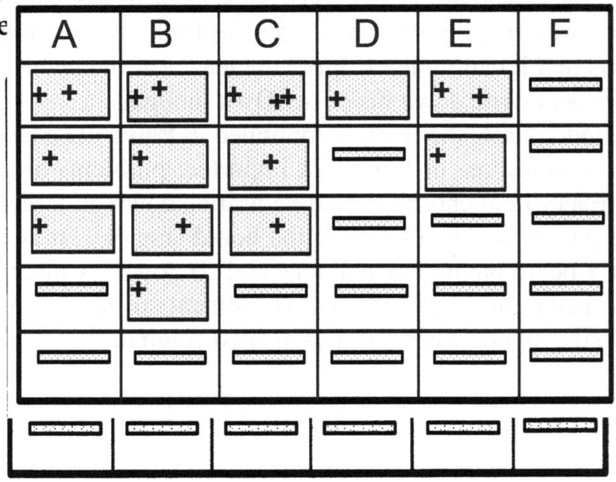

Figure 5. The best papers are marked and organised into groups.

2.3 Ideas

The starting point of the idea-generating stage can be either all the problems or goals that came up in the analysis, or it can be limited to one or two of the items. The choice depends on the extent of the issue and the time that is available. As in the analysis stage, also the idea-generating stage consists of five parts (O-P-E-R-A).

The idea-generating stage begins with the individual work. The leader asks everyone to think of and write down several of their **Own** suggestions and ideas on how to improve things that have come up (problems, weaknesses) or how to reach the goals.

Then working in **Pairs** follows. Every pair should put 3-5 of their best suggestions and ideas on the matrix board. Every idea is written on a separate sheet of paper and again each pair gets their own column.

In the next, **Explanation** phase, every pair presents their ideas.

After this, the **Rating** process begins. Each pair puts five pluses on their favourite ideas; one of these may be their own idea. After marking, the pairs argue their points. Then the leader asks the pairs to mark the two top ideas with hearts (triangles, whatever... Fig. 6 (a)). One of these may again be their own. Again, the pairs argue their points.

In the **Alignment** phase the ideas are organised on the board in a similar way to that of the analysis stage (Fig. 6 (b)). The leader puts the unmarked papers on the side and lifts the top ideas to the first row. The team guides the leader in the alignment process by telling him which ideas belong together in the same group. Finally, it is good to clarify the grouping by giving names to every group. The names must be short and they must represent essential features of the group.

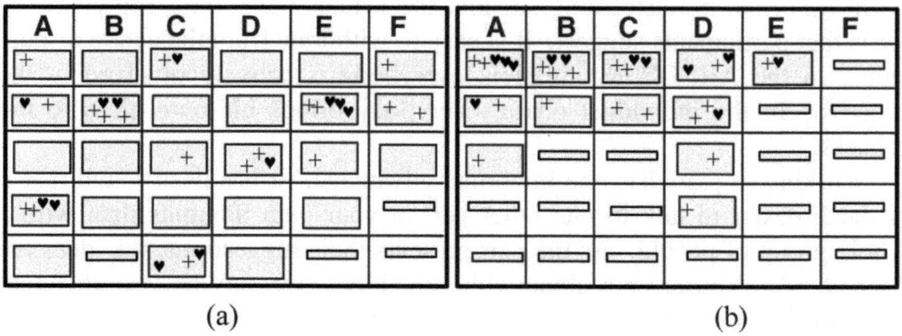

Figure 6. Collected ideas are marked on the matrix board (a) and organised into groups (b).

2.4 Results

In the result stage, the action plan is made. The OPERA process can be used also at this stage. First, the participants are asked to write down their own suggestions, how to advance and what action must be taken. Then the process continues in pairs. The leader asks the pairs to write down on the papers their best suggestions and to attach them to the board. After the explanations, the pairs give their pluses to the suggestions they consider the best ones.

In the last, alignment, stage, the suggestions are arranged in similar groups and the action plan is done. The leader writes on a flip chart the steps that will be taken, the names of the persons to carry out each step, and the timetable (Table 1). The group must make all the decisions, and the leader must be as neutral as possible.

Table 1. Agreement on further actions in AIR-OPERA.

ACTIVITY	CLIENT	SCHEDULE
1. Write memo	George	Feb. 27th
2. Present decision	Paul	Mar. 1st
3. XXXXX XX XXXX	John Doe	Month, Day
4. XXX XX XXXXXX	Jane Doe	Month, Day
5. ---		

2.5 Strengths and weaknesses of the AIR-OPERA method

In a normal brainstorming session, the leader writes on a flip chart the ideas that participants suggest. It is easy to withdraw from the active discussion. In the AIR-OPERA session, everybody is actively involved in the process. In the individual phase, everybody brings his own ideas to the process.

Working in pairs has one important advantage. It has been found that it is much easier to talk to one other person your own thoughts than with a whole group of people. As the pair presents their suggestions, the ideas are "our ideas". You are not alone with your ideas that may be very different from the usual solutions. This encourages you to tell your wildest thoughts. Working in pairs has also another advantage: namely the first evaluation is done during working in pairs and only the best ideas are put on the board.

The evaluation is done with a positive criticism. Only the best ideas are marked and discussed. No criticism is given to the less favourable ideas. Also the cross-evaluation, i.e. giving marks to the ideas of the others, has proved to be successful.

The matrix board, onto which the ideas are attached, makes it possible for everybody to see all the ideas continuously. Our brains are working effectively as we see all the alternatives at the same time.

The AIR-OPERA method has proved to be successful in building strong commitment among the participants. The method is relatively easy to learn. It always gives results.

The method has three disadvantages. Firstly, the whole process takes time, three to four hours. That is a general disadvantage of brainstorming methods. Secondly, the method requires a trained and experienced leader for guaranteed success. The third disadvantage is that the method is not the most creative problem solving method. If we need very unusual solutions for a problem, a better method is e.g. Synetics, developed by Gordon and Prince [2].

3. CONCLUSION

In the paper, the birth of ideas and the analysing and problem solving method, AIR-OPERA, are discussed and presented. The main rules of idea generation are following. Do not be satisfied with the first practical idea. Make difference between the search for and the evaluation of ideas. Try consciously to avoid habitual solutions. AIR-OPERA has proved to be successful in building strong commitment among the participants. It always gives results.

LITERATURE

[1] de Bono, E.: Serious Creativity, Harper Business, New York, 1992.
[2] Gordon, W.: Synetics. The Development of Creative Capacity, Harper & Row, New York, Evanston, London, 1961.
[3] Lumsdaine, E., Lumsdaine, M.: Creative Problem Solving. Thinking Skills for a Changing World, McGraw-Hill, New York, 1995.
[4] Nolan, V.: The Innovator's Handbook, Sphere Books, London, 1991.

MIMICKING NATURE

How nature inspires methodologies in computational electromagnetics

Antonio Savini
Department of Electrical Engineering, University of Pavia, I 27100 Pavia, Italy

e-mail: savini@unipv.it

Abstract: After a brief overview on the attempts to apply mathematical models to investigate nature and, reciprocally, to deduce mathematical tools from the study of nature, a survey of methodologies in computational electromagnetics inspired by nature is given. Present state and new trends are discussed

.Key words: nature-inspired optimization, non-deterministic algorithms.

1. HISTORICAL OVERVIEW

The attempt to apply mathematics in order to produce quantitative models of the natural world lies at the origin of the scientific revolution which began over four centuries ago.

Galileo, the founder of modern mechanics together used to say that " the book of nature is written in mathematical language" [1].

One century later, and more intensively in the 19th century, scientists were busy discovering the general mathematical principles underlying natural processes. Euler and Lagrange proposed the principle of minimal action in mechanics [2], while Clausius was the first to interpret thermodynamic phenomena in terms of increasing entropy.

The world of electricity remained hidden for a long time until Volta' experiments made the vital breakthrough with the invention of the battery.

M. Rudnicki and S. Wiak (eds.), Optimization and Inverse Problems in Electomagnetism, 13–19.

Maxwell, the great theoretician of this world, put forward the general laws governing it in his treatise [3].

In more recent times it has been pointed out how, in computational electromagnetics, it is possible to solve analysis problems by minimising a suitable functional and, similarly, how problems of synthesis can be reduced to the minimisation of adequate objective functions.

2. STUDYING NATURE

The application of mathematics to the modelling of natural systems is an endless process and much work still has to be done, in particular, to create mathematical models of biological systems.

Biology, which perhaps will be the science of the 21st century, needs mathematics.

The human brain, for instance, is still almost a mystery. It is guessed that it resembles an electric network as many as 10^{10} neurons with a huge amount of interconnections: each neuron is connected to about 10^4 others. The speed of information transfer is slow but the operation is massively parallel. Each neuron has the complexity of a modern microprocessor and learning basically consists of establishing new, or modifying previous, interconnections between neurons.

The scientific revolution had a turning point some fifty years ago, when the progress of electronics made it possible to manufacture computers that were able to perform increasingly complex computations.

The advancement in the power of computing facilities over the past twenty years can be interpreted, for instance, by means of the following figures.

In 1981 the first IBM personal computer used a 8-bit Intel 80888 processor working at 1.5 MHz, had a 16 kB of memory and was sold for 2495 US dollars.

Today a commercial personal computer equipped with a 32-bit Intel Pentium 4 processor working at 1.5 MHz, has a 128 Mb of memory and can be bought for half the price of the former one.

The advent of personal computing indeed can be considered the starting point of the era of information society, dominated by electronics and communications.

3. LEARNING FROM NATURE

The tremendous evolution of computing facilities has been accompanied by a no less amazing evolution of mathematical models.

Starting from the pioneering works of Alan Turing on Intelligent machines [4] and going through the work of N. Wiener on Cybernetics [5], in the past forty years a new discipline has grown up, called artificial intelligence. It addresses the mathematics of reasoning and learning with the ultimate goal of creating something, a program or a machine, which is "intelligent", that is to some extent capable of thinking and taking decisions, of solving a problem without explicitly programming it, in other words, capable of "intelligent" behaviour.

Artificial intelligence methodologies are introducing a revolution in very many fields including defence, medicine, economics, semiology, entertainement, besides science and engineering.

Within this frame, on the basis of the observation that natural and in particular living processes are "intelligent", intensive attempts have been made in order to extract, from nature, laws governing best reasoning and optimal behaviour.

In this process mathematics and biology, in a sense, exchange their reciprocal role in a sort of feedback operation: not just mathematics for biology, but also biology for mathematics; not only learning nature but also learning from nature. Mathematics needs biology indeed.

4. METHODOLOGIES INSPIRED BY NATURE IN COMPUTATIONAL ELECTROMAGNETICS

In the field of artificial intelligence one of the first attempts made in the early 1960s was that of creating expert systems, i.e. artificial experts in a specific field, collecting enough expertise in a knowledge base and then searching it by a programme called inference engine. To this purpose suitable programming languages were invented like LISP and Prolog, and successful applications were made in particular for diagnostics. The knowledge-based approach was the subject of criticism in the mid 1980s because ot its procedural difficulty as well as its domain specificity.

Based on a similar idea of training a programme to set up a mapping between inputs and ouptputs whose correlation is unknown a priori [6], artificial neural networks had a resurgence of interest in the 1980s thanks to Kohonen and Hopfield [7]. Actually the idea that a network of processors with controlled interconnections, i.e. a sort of rudimentary abstraction of biological neurons with their connections, could implement any logical function and therefore approximate any mapping function goes back to the pioneering work of Mc Cullock et al [8] and to that of Rosenblatt [9] of 1943 and 1957, respectively. During the last decade artificial neural networks of various types with multiple layers, both feedforward and recurrent, for supervised and unsupervised learning, have been developed, tested and applied; they have proven to be, in particular, an attractive tool for the solution of optimization problems.

As early as in 1965 Lofti Zadeh, the pioneer of soft computing, proposed fuzzy logic [10] in order to solve problems that imply uncertain, imprecise and insufficient information, as it often happens in practice. Today fuzzy logic is thoroughly applied to model and govern complicated control processes described by fuzzy variables and rules like, for instance image processing or even the anti-lock braking system for a car.

Fuzzy reasoning belongs to the class of approximate, probabilistic or stochastic reasoning, opposite to the deterministic one. As very often available data are insufficient or uncertain, then by attributing them probabilities as measures of degree of belief and defining the so-called information entropy, the probability theory based on the original work of Rev. Thomas Bayes (1763) can be applied. In this connection the maximum informational entropy principle formulated by E. Jaynes [11] can find a number of applications for self-organising systems.

The principle of natural evolution inspires a large family of search metodologies that today are very popular and are well suited for optimisation problems. Evolution is a self-organising, adaptive and intelligent process consisting of random processes like reproduction and mutation followed by selection based on best fitting. Evolutionary computation methodologies mimic this process. They include many slightly different algorithms, the most important of which are listed here.

Evolutionary Programming [12], developed by Fogel in 1962, simulates the evolution of a population and emphasizies mutation.

Evolution Strategies [13] proposed by Rechenberg in 1965 emulate the evolution of individuals.

Genetic Algorithms are based on the seminal work [14] by Holland on Adaptation in Natural and Artificial Systems and were lateron developed by Goldberg [14]; they include cross-over and operate on a binary string of chromosomes.

Genetic Programming [15], which is an extension of Genetic Algoruthms, was introduced by Koza and is based on the principle of the survival of the fittest.

The list of evolutionbased methodologies is far from being complete. The underlying idea to all of them is that the solution to a problem can be found by letting a population of possible solutions evolve according to some rules that are supposed to mimic the behaviour of populations in nature.

Recently, the observation of a population of ants establishing the shortest paths from their colony to the feeding sources has inspired the so-called Ant Systems [16], another method for "natural optimisation" : artificial ants exploring a region lay some information on the paths ; this information is used by the next ants moving along the same paths in order to decide where to go.

Another type of random search methodollogy is Simulated Annealing, developed by Kirkpatrick et al. [17]. It reproduces the evolution of a physical system towards the thermodynamic equilibrium: tremperature decreases through different energy states until the final configuration corresponding to the minimum energy is reached.

Again a natural process inspires an algorithm that, through a procedure of self adaptation in an intelligent way, leads to a natural optimal result.

All these families of lowest order heuristic random or probabilistic search methodologies, whose mathematical foundations are still under investigation, have been successfully applied to solve a number of problems in engineering science, in particular problems of optimisation and design.

5. PRESENT STATE

In the field of applied electromagnetics over the years the topic of automated aptimal design has gained a role of strategic importance in research communities [18].

To date the state of the art in this field is characterised by an extensive application of stochastic optimisation methodologies, mainly those based on the principle of natural evolution. For the latter, the natural law of survival of the fittest in a given environment is the model to find the best design configuration fulfilling given constraints.

In nature, individuals of a population in a given domain reproduce and cross among themselves in order to adapt to the natural habitat, so that the most favourable characteristics for fitting survive in the new generation.

Lkikewise, in computation a population of various design configurations, after reproducing, combine the most favourable characteristics resulting in a new population of design configurations which better fulfils the fitness goal.

Such stochastic optimisation methods offer remarkable advantages over well known gradient-based methods that use only local information to improve the current solution; this applies particularly for complex problems with numerous local minima, when it is required to reach the global minimum. They are robust, reliable and suitable for nonconvex and discontinous functions, also with discrete type variables. Another remarkable adavantage is that they exhibit a inherent parallelism. The only drawbacks are the huge computational effort and the slow convergence. As concerns convergence, it has to be pointed out that for such heuristic methods convergence is proven just in numerical terms and not on theoretical basis.

Recently hybrid methodologies have been proposed for global optimisation, combining the best quality of gradient-based methods, i.e. exploitation of local information to reach an accurate local solution, with that of stochastic methods, i.e. broad exploration leading to a rough and reliable solution.

6. LOOKING AT THE FUTURE

Looking in front of us at the frontiers of artificial intelligence, what we can see is a great deal of research work with the ultimate goal, which is still ambitious, of getting "intelligent" computers, capable of solving problems without explicitly programming them. This is the current challenge of artificial intelligence. As predicted already as early as in 1962 by Marvin Minsky in his pioneering paper [19], "we are on the threshold of an era that will be strongly influenced and quite probably dominated by intelligent problem-solving machines".

Thanks to the great advances in molecular biology and biotechnologies of the last two decades, inspiration to get this goal will probably be taken increasingly from nature and biology in particular.

On the one hand, from the point of view of software, new biological processes like that of molecules of the immune systems combating the invading pathogenic micro-organisms are going to stimulate new search methodologies like artificial immune systems.

On the other hand, quite a new scenario has been opened by the seminal paper of L. Adleman [20]. The idea of using the DNA of biomolecules, which occur abundantly in nature, to build biocomputers at nanoscales, exploiting the inherent massive parallelism of natural processes, is attracting the attention of researchers in advanced laboratories. What is more, evolutionary computation seems the natural way for programming and implementing biocomputers.

This way, the circle of inspiration from nature is closed. Indeed, the book of nature is inherently written in mathematical terms.

REFERENCES

1. G. Galilei: Opere (in Italian), A. Garbasso and G. Abetti (eds), Barbera, Firenze, vol. vi, p. 232, 1966
2. L. Lagrange: Mechanique Analytique, Paris, 1788
3. J.C. Maxwell : A Treatise on Electricity and Magnetism, Reprint, Dover, New York, 1954
4. D.C. Ince (ed): Mechanical Intelligence:Collected Works of A.M. Turing, North Holland, Amsterdam, 1992
5. N. Wiener: Cybernetics, MIT Press, Cambridge MA, 1961
6. T. Kohonen : Correlation Matrix Memories, IEEE Trans. On Computers, vol 21, pp. 353-359, 1972.
7. J.J. Hopfield: Neural Networks and Physical Systems with Emergent Collective Computational Abilities, Proc. Nat. Acad. of Sciences, vol. 79, pp. 2554-2558, 1982.
8. W. Mc Culloch et al: A Logical Calculus of the Ideas Immanent in Nervous Activity, Bull. of Mathem. Biophysics, vol.5, pp. 115-133, 1943.
9. F. Rosenblatt: Thre Perceptron: A Probabilistic Model for Information Storage and Organization in the Brain, Psychol. Rev., vol. 65, pp 386-408, 1958.
10. L. Zadeh; Fuzzy sets, Inf. Control, vol.8, pp. 338-353, 1965.
11. E.T. Jaynes; On the rationale of Maximum-Entropy Methods, Proc.IEEE, vol. 70, pp. 939-952, 1982.
12. L.J. Fogel: Autonomous Automata, Ind. Res. vol. 4, pp. 14-19, 1962.
13. I. Rechenberg: Evolutionsstrategie: Optimierung Technischer Systeme nach Prinzipien der Biologischer Evolution, Fromman Verlag, 1973.
14. D.E. Goldberg: Genetic Algorithms for Search, Optimization and Machine Learning, Addison Wesley, Reading MA, 1989.
15. J.R. Koza: Genetic Programming: On the Programming of Computers by Means of Natural Selection, MIT Press, Cambridge MA, 1992.
16. M. Dorigo et al: Ant System: Optimization by a Colony of Cooperating Agents, IEEE Trans. on Systems, Man and Cyber., part B, vol. 26, pp. 29-41, 1996.
17. S. Kirkpatrick et al: Optimization by Simulated Annealing, Science, vol. 220, pp: 671-680, 1983.
18. P. Neittaanmaki et al: Inverse Problems and Optimal Design in Electricity and Magnetism, Clarendon Press, Oxford UK, 1996.
19. M. Minsky: Steps Towards Artificial Intelligence, Proc. IEEE, pp.8-30, 1961
20. L. Adleman: Molecular computation of solutions of combinatorial problems, Science, vol. 226, pp. 1021-1024, 1994.

COMPARING STOCHASTIC METHODS ON SMES OPTIMIZATION

O. Hajji, S. Brisset, P. Brochet
L2EP

Abstract: In this paper, the optimization of a superconducting magnetic energy storage (SMES) device is performed by means of three stochastic methods: genetic algorithms, simulated annealing and tabu search which are then compared. The parameter sensitivity is studied and improvements are proposed.

Key words: genetic algorithm, simulated annealing, tabu search, SMES

1. INTRODUCTION

Optimization of electromagnetic devices can be made by different methods, either deterministic or stochastic. Most of the optimal design problems involve objective function with more than one optimum and with very little knowledge about its behavior. So, using deterministic methods may converge to one of the function local minima and then fail to solve the problem.

By introducing random movement, stochastic methods would explore different regions of the solutions space. They are rather simple to implement, stable in convergence and able to find the global optimum with quite good probability but the major drawbacks is the high number of function evaluation required. Genetic algorithm (GA), simulated annealing (SA) and tabu search (TS) are three classes of stochastic methods that are widely used in engineering. Here, they are applied to superconducting magnetic energy storage (SMES) optimization.

In the next section stochastic algorithms in use here are briefly described and the SMES optimization problem is presented. Finally, the results of

M. Rudnicki and S. Wiak (eds.), Optimization and Inverse Problems in Electomagnetism, 21–32.

optimization led by the three stochastic methods are compared and the main conclusions are given.

2. SIMULATED ANNEALING

SA is a procedure for optimization that has been firstly proposed to solve complicated combinatorial problems [1]. It has proved to be an effective and powerful global optimization algorithm in many combinatorial optimization problems such as the traveling salesman problem [1]. It is based on an analogy with thermodynamics where a system is slowly cooled in order to achieve its lowest energy state. The main stochastic rule used in SA is the Boltzmann probability [2].

The optimization process is still mainly controlled by the parameter T, i.e. the temperature. During SA optimization, new configurations are created with random moves in all directions. They are accepted or rejected according to Boltzmann probability distribution:

$$p \leq \exp\left(\frac{f(X_k)-f(X_{k+1})}{T}\right) \tag{1}$$

Where p is a random number between 0 and 1, X_k is the best point, X_{k+1} is the current point generated from X_k with a random move which length is lower or equal to the step size and f is the objective function to minimize.

After a certain amount of trials, the step length is reduced and the random moves restart. As the step length gets very short, a quasi equilibrium state is reached. Then, the control parameter T is reduced by multiplying its value by a factor slightly below one and the whole process is repeated with this new control parameter affecting the Boltzmann probability. The optimization stops when the control parameter has decreased sufficiently.

The fundamental structure of SA algorithm here in used is derived directly from the one applied to multimodal functions of continuous variables proposed in [2] but improvements are brought as detailed in the next section.

2.1 Improved simulated annealing

A new method to determine the step vector is proposed here, based on the proposition of Hu [3]. The steps lengths are taken from a list of realizable steps computed with formula (2).

All the step vectors are calculated in the beginning of algorithm. The procedure to calculate the different step vectors is:

Let n be the number of design variables and Ns the number of step size reductions.

1. First, consider that the objective function is defined as a function of n continuous variables x_i, $i \in [1,\dots,n]$ and each $x_i \in [x_i^L, x_i^U]$ where x_i^L and x_i^U are respectively the lower and upper boundaries of the design space.
2. Now, for each design variable *xi*, the component of each step vector is:

$$h_{i,j} = \frac{(x_i^U - x_i^L)}{c^j} \tag{2}$$

where the factor c is greater than 1 and determines the search space.

Numerical results on test functions show that the convergence speed of SA using this list of realizable steps is faster than the one used in [2]. The advantage of this algorithm according to standard SA is its low sensibility. At the beginning, the search starts with large steps in order to cover all the search space. Then the decreasing of the vector step and of the control parameter performed the search in a smaller space.

The SA algorithm is as follows:

1. Initialization: set a starting point X_0, a starting temperature T_0, the number of step size adjustments N_s, the number of temperature reduction N_t, and let the starting point be the best point.
2. Build the matrix of step vectors with (2).
3. Generate a random move in the current direction using the current step size.
4. If the objective function of the new point is lower than the best one then store the new point and its objective function into the best ones.
5. Compute the Boltzmann probability with (1) and the current temperature. If the test is true then store the new point and its objective function into the best ones.
6. Change the current direction of moves.
7. Return to 3. n times.
8. Change the current step size using (2).
9. Return to 3. N_s times.
10. Reduce the current temperature.
11. Return to 3. N_t times or stop if stop criterion applies.

SA algorithm is stopped if:

1. The maximum number of temperature reductions N_t is reached,
2. Or, the objective function no longer varies significantly.

2.2 SA with auto-tuning steps

In standard SA, while the temperature is kept constant, the step vector is periodically adjusted in order to fit well the objective function variation. After thermal equilibrium, the temperature is reduced. The step vector is changed a certain amount of times set up by the user depending on the optimization problem. Therefore, a more efficient criterion is searched to determine if the thermal equilibrium is reached.

The criterion proposed here depends on the step vector. In order to increase the speed of convergence and decrease the number of evaluations, some improvements are introduced. In optimization process, the numbers of accepted and rejected configurations must be preferably balanced in order to explore all the search space. Therefore, the step vector $h_{i,j}$ is changed as proposed in [2]:

$$h_{i,j} = h_{i,j}\left(1 + c\,\frac{\frac{n}{N} - 0.6}{0.6}\right) \quad \text{if } n > 0.6\,N$$

$$h_{i,j} = \frac{h_{i,j}}{1 + c\,\frac{0.4 - \frac{n}{N}}{0.4}} \quad \text{if } n < 0.4\,N \tag{3}$$

$$h_{i,j} = h_{i,j} \quad \text{else}$$

Where N is the total number of points in the neighborhood, n is the number of points accepted, and c is a constant. Equations (3) ensure that the step increases if the number of accepted points represents more than 60 % of the neighborhood and is reduced if the number of accepted points is lower than 40 %.

But this way to change the step is efficient only if the center of the neighborhood is very close to the global optimum. If not, the number of accepted point is low and (3) leads to a lower step and an even lower number of accepted points. Finally, the step becomes too small. The following criterion can be used to detect such problem:

$$\frac{h_{i,j}}{(x_i^U - x_i^L)} \leq \varepsilon \tag{4}$$

The problem can then be solved by resetting the step to the initial value and restarting from the best point previously found. The temperature is also decreased so that the probability to accept a point according to Boltzmann probability (1) is lower and the probability to find a point with a better objective function increases.

2.3 Parameters adjustment

The algorithm, like most stochastic algorithms, contains several user controlled parameters that require adjustment. The quality of results found is strongly depending on the adequacy of the adjustments made. Parameter values has been determined by means of sequential tests. Sensitivity analyses were performed for all main parameters. No claim is made that the resulting choice of parameter values is the best one and should be the same for all problems. However, they just work well on test problems.

The proposed algorithm contains three influent parameters:

- N_s : number of step size adjustments
- ε : epsilon is a prescribed small positive number
- g : the tolerated number of consecutive cycles with no improvement of the objective function

In order to find the best value of parameters, a search process is performed. For each parameter, an interval of possible values is defined and the best value is searched inside it. Table 1 gives the best parameters and their interval of variation.

The parameter ε was set equal to 0.008, a value which has already produced good compromise between computing time and solution quality. The value of N_s was set equal to 10 after some testing. Selecting a too small value of N_s , increases the risk of not exploring all the search space. To end with, g was set to 5 after some experimentations.

Table 1. intervals of algorithm parameters

parameter	best	min	max
ε	0.008	0.006	0.1
N_s	10	5	15
g	5	2	6

3. TABU SEARCH

TS is a metaheuristic method recently developed by Glover [4], [5] specifically for combinatorial optimization problem. It guides the search for the optimal solution making use of memory systems, which exploit the history of the search. To prevent cycling and to avoid the risk of trapping in local optimum, TS introduces the concept of an attribute, based memory. New solutions are searched in the neighborhood that is defined as the set of the points reachable with a suitable sequence of local perturbations, starting from the current solution.

Among all the visited solutions the best one is chosen. This strategy can lead to cycling on previously visited solutions. To prevent this effect, the algorithm set as taboo moves that has led to improvement of the objective function for a certain number of iterations, depending on the Tabu list size. The performance of a TS algorithm depends on the size of neighborhood of a solution, and on the number of iterations for which a move is kept as taboo. The neighborhood $N(X,h)$ of a point X is defined as:

$$N(X,h)=\{X' \ \ such \ \ as \|X'-X\| \le h\} \tag{5}$$

Where h is the step size and X' is any point of the space.

There are two fundamental structures of TS algorithms implemented here: Hu's TS (HuTS) and Universal TS (UTS). Based on the idea of Glover, Hu has implemented a Tabu search for optimization with continuous variables. An universal Tabu search is developed based on the study of the available TS and improvements on Hu TS. They are derived respectively from [3] and [6].

3.1 Hu Tabu search

The Hu TS is as follows:

1. Set the size of tabu list T, the size of the neighborhood N_n, and build the matrix of step vectors with (2).
2. Initialization: set a starting point X_0, set T as empty, and let the best point be the starting point: $X_{best}=X_0$, $f(X_{best})=f(X_0)$.
3. Generate a new random feasible move in the neighborhoods $N(X_{best},h_i)\ i=1,N_s$ of the current best point using non-taboo steps. Let X_j be the best point in all neighborhoods, $f(X_j)$ its objective function, and h_j the step used to generate X_j.
4. If $f(X_j)\le f(X_{best})$ then $X_{best}= X_j$, $f(X_{best})=f(X_j)$, and add the step h_j to the Tabu list T.
5. Update all step lengths using (2), and return to 3.

6. If all step lengths are in the Tabu list T and the termination criterion is not verified then set T as empty, and return to 3.

HuTS algorithm is stopped if:

1. The objective function no longer varies significantly, or
2. A large number of points are generated in the neighborhood and none has a lower objective function.

3.2 Universal Tabu Search

The UTS algorithm is as follows:

1. Set the size of the neighborhood N_n , the maximum number of exploration cycles N_e , the maximum number of cycles N_t , the number of step sizes N_s , and build the matrix of step vectors with (2).
2. Initialization: set a starting point X_0, set the number of cycles nc to zero, and let the starting point be the best point: $X_{best}=X_0$, $f(X_{best})=f(X_0)$.
3. Increment the number of cycles: $nc=nc+1$, and set the number of exploration cycles ne to zero.
4. Generate N_n x N_s new points in neighborhoods of the current point using N_s step lengths. Let X_n be the best point in all neighborhoods and $f(X_n)$ its objective function.
5. If $f(X_n)\leq f(X_{best})$ then $X_{best}= X_n$, $f(X_{best})=f(X_n)$, the best point becomes the current point, and return to 3.
6. The current point becomes X_n , increment the number of exploration cycles: $ne = ne +1$, and if $ne \leq N_e$ then return to 4.
7. The best point found previously becomes the current point.
8. If the termination criterion is not verified then return to 3.

UTS algorithm is stopped if:

1. The objective function no longer varies significantly, or
2. The maximum number of cycles is reached: $nc = N_t$.

4. GENETIC ALGORITHM

GA is an iterative procedure which maintains constant the size of population of candidate solutions [7], [8]. During each iteration, called a generation, the individuals of the current population are evaluated, and on the basis of those evaluations, a new population of candidate solutions is generated. By a randomized selection procedure a new population is selected that ensures that the expected number of times an individual is chosen is approximately proportional to its relative performance. In order to search other points in the search space, the two genetic operators' crossover and mutation are applied. To maintain the focus of the processes on the best solution found so far, an elitist approach can be built into the selection process.

4.1 New stop criterion

As it is well known, the main disadvantages of the standard GA (SGA) are the convergence accuracy and the long time taken. So that, using the number of generation as a stop criterion is an inadequate procedure. The optimization process can continue uselessly arising then a waste time. There is a strong necessity to propose a new stop criterion that detects as early as possible the global optimum with a given accuracy.

The proposed solution is to build an approximation of the objective function with special properties. If the approximation is accurate enough then it can be used to locate very quickly the optimum in the domain of feasibility.

The optimization process is performed in two steps. In the first step, the standard GA process is performed until the error between the objective function f and its approximation $\hat{f}$ is sufficiently low. In the second step, the approximation of the objective function $\hat{f}$ is used to locate the global optimum point in the confidence domain of the approximation function using recursive canonical analysis [9].

The performance of the proposed method depends on the selection of individuals kept to build the Response Surface (RS) and also on the formula chosen to compute its accuracy.

4.2 Response surface methodology

A second order approximation function of all design variables including interaction between two variables is built using the experimental design method and the response surface methodology [9]:

$$\hat{f} = a_0 + \sum a_i \cdot \overline{x_i} + \sum a_{ij} \cdot \overline{x_i} \cdot \overline{x_j} + \sum a_{ii} \cdot \overline{x_i}^2 \tag{6}$$

Where $\overline{x_i}$ is the centered reduced value of the design variable x_i, a_0 is the estimation of the objective function at the center of design space, a_i are the design variable effects, a_{ij} interactions between design variables and a_{ii} parabolic effects. The centered reduced values of the design variables are as follows:

$$\overline{x_i} = \frac{2}{x_i^U - x_i^L}\left(x_i - \frac{x_i^U + x_i^L}{2}\right) \qquad -1 \leq \overline{x_i} \leq +1 \tag{7}$$

Where x_i^U and x_i^L are the upper and lower boundaries of the design space. A matrix X is built; every line describes the centered reduced values of the design variables for each individual of the population that participate in building while every column is related to effect or interaction [9]. The method of the least squares is used to calculate all effects:

$$\hat{a} = \left(X^t X\right)^{-1} X^t f \tag{8}$$

Where f is the column vector of responses obtained from every individual that participate in the building of RS and $\hat{a}$ is the column vector of effects.

In order to compute the RS, individuals close to the best one in the current generation are chosen. A given proportion of the best individuals are selected whatever their position is. The RS built has a confidence domain defined by the maximum and minimum values of the design variables for selected individuals. The equation used to compute the accuracy of RS is the maximum relative error of the objective function for all selected individuals.

5. SMES OPTIMIZATION

To compare algorithms in solving engineering optimization problems, they are applied to the Superconducting Magnetic Energy Storage (SMES) benchmark described in [10].

5.1 Optimization problem

SMES consist of two solenoids: one inner main solenoid and one outer shielding solenoid. This second one is placed to reduce the stray field and its dimension have to be optimized to do so. The dimensions of the inner solenoid are fixed. The stray field is evaluated along 22 equidistant points as shown in Fig. 1.

Furthermore, the stored energy should be as closed as possible to 180 MJ. A correct design of the system should then couple the right value of energy to be stored with a minimal stray field and must fulfill the quench condition:

$$|J| \leq (-6.4|B| + 54.0) A/mm^2 \tag{9}$$

The quench condition is setting up a relation between the current densities J and the maximum magnetic induction $|B|$ within the coils. As the current density in coils is fixed to $22.5\, A/mm^2$, (9) can be changed to $|B| \leq 4.92\, T$.

The following objective function (OF) is proposed [10]:

$$OF = \frac{B_{stray}^2}{B_{norm}^2} + \frac{|Energy - E_{ref}|}{E_{ref}} \tag{10}$$

$$B_{stray}^2 = \frac{1}{22}\sum_{i=1}^{22} B_{stray,i}^2 \text{ , } E_{ref} = 180\, MJ \text{ , } B_{norm} = 3 \cdot 10^{-3}\, T$$

OF is computed with the finite element software Opera2D from Vector Fields.

5.2 Results

The proposed algorithms are employed to optimize the superconducting magnetic energy storage problem to demonstrate its efficiency in electromagnetic engineering problems. Results obtained by algorithms are presented in table 2. The time is expressed as the number of evaluations of the objective function (OF).

The three stochastic methods in their original forms (SGA, SA, HuTS) lead to a large number of evaluation of the objective function and a weak precision on the location of the global optimum. After improvements, these stochastic algorithms (GA+RS, ISA, UTS) are significantly faster and more accurate. They converge to similar solutions, taking similar times.

Table 2. Solutions of SMES optimization problem

Method	R1	H2/2	D2	OF	Energy	Bstray²	Time
Unit	m	m	m	-	10^6 J	10^{-6} T^2	-
SGA	3.040	0.240	0.386	0.0901	179.19	7.668	2400
GA+RS	3.084	0.238	0.395	0.0862	180.02	7.775	945
SA	3.078	0.237	0.390	0.0870	179.36	7.590	5025
ISA	3.089	0.246	0.381	0.0864	179.95	7.790	859
HuTS	3.080	0.246	0.381	0.0864	179.79	7.670	3821
UTS	3.077	0.245	0.385	0.0868	179.75	7.700	901

The convergence trajectories of the optimization solution for different stochastic methods are depicted in Fig. 2.. The fields contours under optimized geometry is given in Fig. 3. It can be seen that the Quench condition, i.e. $|B| \leq 4.92\, T$, is fulfilled.

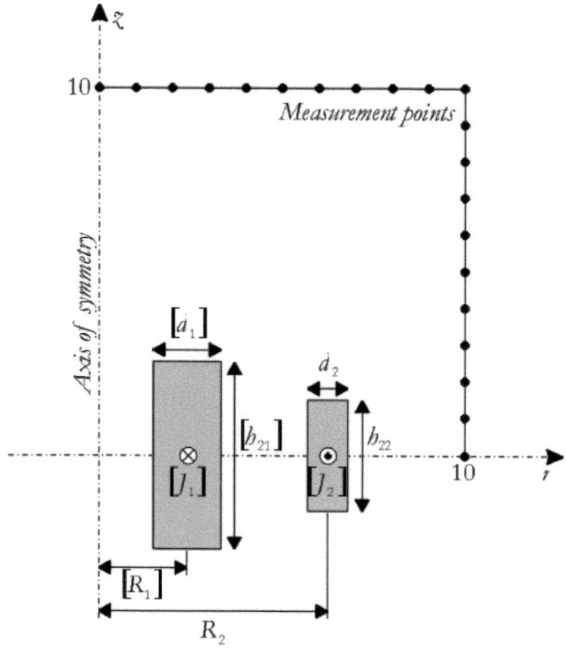

Figure 1. Design parameters R2, h22 and d2 of the Superconducting Magnetic Energy Storage Optimization Benchmark

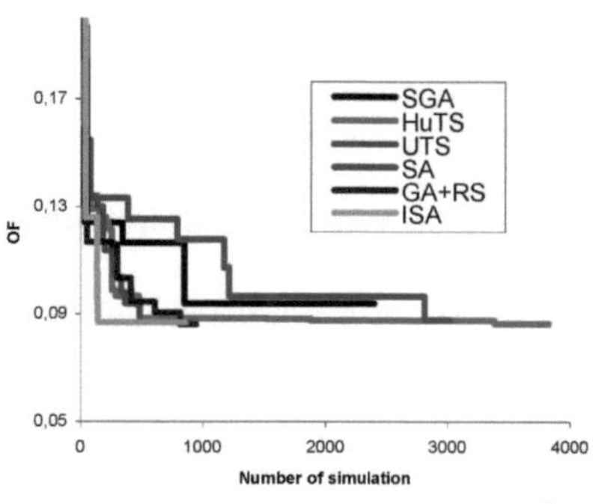

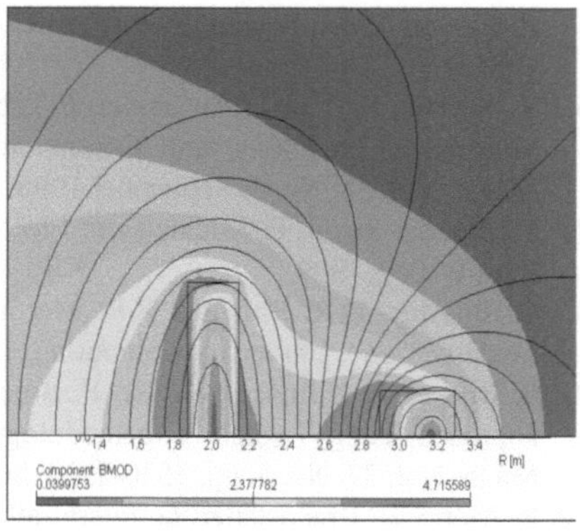

Figure 2. OF versus number of simulation *Figure 3.* Fields plot of the optimal configuration

6. CONCLUSION

Efficient uses of stochastic optimization algorithm in electrical engineering have been described. Three different optimization strategies (GA, SA, TS) has been compared on the superconducting magnetic energy storage (SMES) benchmark.

Stochastic methods rely on random. This, on one hand, allows them to explore the optimization domain better than the deterministic ones, but on the other hand, a large number of objective function evaluations is required.

To decrease it, a new stop rule for GA, based on the response surface methodology, is proposed. A table of step vectors is applied to accelerate the convergence of SA. For the optimization of the SMES, the improved versions of stochastic algorithm have led to better solutions using significantly less time.

REFERENCES

1. R.E. Randelman and G.S. Grest, "N-City Traveling Salesman Problem: Optimization by simulated Annealing", Journal of statistical Physics, Vol. 45, N° 5/6, 1986.
2. A. Corana et al, "Minimizing multimodal Functions of Continuous Variables with the Simulated Annealing Algorithm", ACM Trans, Math. Soft, Vol. 13, Sept. 1987, pp. 262-280
3. N.Hu, "Tabu search method with random moves for globally optimal design", Int, J: num, Meth Engineering, 35, pp. 1055-1070, 1992.
4. F. Glover, "Tabu Search -Part I", ORSA Journal on Computing, Vol.1, No. 3, pp. 190-206, 1989.
5. F. Glover, "Tabu Search -Part II", ORSA Journal on Computing, Vol.2, No. 3, pp. 4-32, 1989.
6. Y. Shiyou, L. Yan , "An universal Tabu Search Algorithm for Global Optimisation of Multimodal Functions with Continuous Variables in Electromagnetics", IEEE Trans. Magn., Vol. 34, No. 5, September 1998.
7. Y. Rahmat-Samii, E. Michielssen, Electromagnetic Optimization by Genetic Algorithms, John Willey, 1999.
8. D. Beasley, D.R. Bull and R.R. Martin, "An overview of genetic algorithms, Part 1 : fundamentals", University Computing, 15, pp. 58-69, 1993.
9. S. Brisset, F. Gillon, S. Vivier, P. Brochet, "Optimization with experimental design: an approach using Taguchi's methodology and finite element simulations", IEEE Trans Magn, Vol. 37, No. 5, pp. 3530-3533, September 2001.
10. P. Alotto, B. Brandstätter, G. Fuerntratt, Ch. Magele, G. Molinari, M. Nervi, M. Repetto, K. R. Richter, "Some Results on a SMES Device Optimization Benchmark Problem", International Journal of Applied Electromagnetics and Mechanics 9 (1998), pp. 315-324, IOS Press.

USING QUASI RANDOM SEQUENCES IN GENETIC ALGORITHMS

Heikki Maaranen, Kaisa Miettinen, and Marko M. Mäkelä
University of Jyväskylä, Department of Mathematical Information Technology, FIN-40014 University of Jyväskylä, Finland

Abstract: The selection of initial points in a population-based heuristic optimization method is important since it affects the search for several iterations and often has an influence on the final solution. If no a priori information about the optimization problem is available, the initial population is often selected randomly using pseudo random numbers. Many times, however, it is more important that the points are as evenly distributed as possible than that they imitate random points. Therefore, we have studied the use of quasi random sequences in the initialization of a genetic algorithm. Sample points in a quasi random sequence are designed to have very good distribution properties. The modified genetic algorithms using quasi random sequences in the initial population have been tested by solving a large number of continuous benchmark problems from the literature. The numerical results of three genetic algorithm implementations using different quasi random sequences have been compared to those of a traditional implementation of using pseudo random numbers. The results are promising.

Key words: random numbers, global continuous optimization, genetic algorithms.

1. INTRODUCTION

We consider optimization problems, where a continuous real-valued objective function is minimized over a feasible region in $\mathbf{R}^n$. The feasible region is defined as upper and lower bounds for each variable. Many practical optimization problems are nonconvex, and applying traditional local optimization methods may cause the search to stop when a local minimum is found. The local minimum, however, may be far from a global

M. Rudnicki and S. Wiak (eds.), Optimization and Inverse Problems in Electomagnetism, 33–44.

minimum. Furthermore, by only studying the neighborhood of the solution there is no way, in a general case, to validate, whether the current function value is a global minimum, see, for example, Törn and Žilinskas, 1989.

The non-trivial task of finding a global minimum of a continuous function can be in some specific cases solved using deterministic optimization methods. The deterministic methods, however, make strong assumptions on the objective function (see, for example, Androulakis et al., 1995) and, therefore, are not applicable in a more general case. This has led to the development of stochastic methods, which require less restrictions for the objective function.

One subgroup of stochastic methods is called metaheuristics. Usually, metaheuristics converge to the global optimum with probability one, meaning the methods will find a global minimum with a probability that approaches one as the algorithms runs on. However, as noted in Törn et al., 1999, it is normally impossible to estimate the probability of finding a global optimum in a finite number of iterations. Therefore, in many practical cases, metaheuristics cannot guarantee to find a global optimum. Instead, they usually find a good solution relatively fast.

Metaheuristics can be divided into single-solution-based and population-based methods. The single solution based methods consider one solution at a time and iteratively search for the new solution from the neighborhood of the current solution. Some popular single-solution-based metaheuristics for global continuous optimization are simulated annealing, see, for example, Decker and Aarts, 1991, Özdamar and Demirhan, 2000, and tabu search, see Battiti and Tecchiolli, 1996, Chelouah and Siarry, 2000, Simulated annealing and tabu search were originally developed for problems of discrete variables and the continuous variable variants differ in some aspects from the original versions.

In Michalewicz, 1994, the population-based methods are called evolution-based systems or evolution programs. The population-based methods consider several solutions at the same time and the population is evolved by changing some attributes of a single solution or combining two or more solutions according to prescribed rules. Some popular population-based methods are clustering methods, see Törn and Žilinskas, 1989, genetic algorithms, see, for example, Goldberg, 1989, Michalewicz, 1994, and scatter search, see Glover, 1977, Glover, 1999, Glover et al., 2000. There exist also population-based variants of simulated annealing, see, for example, Ali et al., 2002 and the references therein.

The distribution of the population at different stages of the search is essential for population-based methods. Often a population with a lot of diversity is preferred in the beginning and a more condensed population at the end of the search is desirable. In this paper, we concentrate on examining the effect that the initial population of a genetic algorithm has on the final objective function value and the consumed CPU time. Furthermore, we restrict to study the case when there is no a priori information available about the values and the distribution of the local minima or the size of their regions of attraction (for definition, see for example, Törn and Žilinskas, 1989).

Usually, genetic algorithms use pseudo random numbers both in initializing the population and in genetic operations (including selection, crossover and mutation). In genetic operations, the use of pseudo random numbers is natural, since the operations usually assume randomness. The use of pseudo random numbers in the initialization, on the other hand, needs closer examination.

The goal of pseudo random numbers is to try to imitate independent random numbers. However, the initial points in the population do not need to be random. Nevertheless, independent points have some useful properties: their distribution over the feasible region is uniform and also their projection to a lower dimensional space is uniform (in contrast to rectangular lattice points, see, for example, Sobol', 1979). Hence, the use of pseudo random numbers in the initialization can be justified because of their distribution properties, not because of their virtual independence (randomness). On the other hand, if some other point set has a better distribution, then it should be used in the initialization instead of a set constructed by pseudo random numbers.

One way to generate a sequence of points that has very uniform distribution is to use quasi random sequences. The sample points in a quasi random sequence are designed to maximally avoid each other, see Press and Teukolsky, 1989, and, therefore, to cover the search space thoroughly. The quasi random sequences have been successfully used in numerical integration and random search methods, see, for example, Niederreiter, 1992. Motivated by these encouraging results we here apply quasi random numbers for the initial population of a genetic algorithm and we study the effects that the different initial populations have on the search process.

The rest of the paper is organized as follows. In Section 2, we consider different aspects of pseudo and quasi random numbers from the viewpoint of optimization. Then, in Section 3, we briefly describe the genetic algorithm used in terms of genetic operations and introduce the optimization parameter values used. We also discuss how and where the quasi random numbers are used in our implementation. In Section 4, we introduce the test problems and some numerical and graphical results. In Section 5, we discuss some general issues related to quasi random number generation. Finally, we draw conclusions from the results and point out some future directions of our research.

2. QUASI AND PSEUDO RANDOM NUMBERS

A random number (or a random point or a random variable) is a mathematical abstraction, which cannot be generated using an algorithm. The traditional, algorithmically generated random numbers are more accurately called as pseudo random numbers. They are deterministic, but they try to imitate an independent sequence of random numbers. Common pseudo random number generators include, among others, linear congruential, quadratic congruential, inversive congruential, parallel linear congruential, additive congruential, lagged Fibonacci, and feedback shift register generators (see, for example, Gentle, 1998, Niederreiter, 1992, Wikramaratna, 1989). In addition, there are numerous modifications and combinations of the basic generators.

In cases where randomness is not required, a quasi random number generator may be used to generate a sequence of points in a desired dimension so that the points are as evenly distributed as possible over the feasible region. Common quasi random number generators include Hammersley, Faure, Halton, Sobol, Niederreiter, and SQRT generators. Good general references for quasi random number generation are, for example, in Gentle, 1998, Niederreiter, 1992.

Figures 1 and 2 illustrate samples of 200, 500, 1000, and 2000 points of pseudo and quasi random sequence, respectively. The pseudo random sequence is generated using MatLab's routine (in MatLab 5 and above) which combines a lagged Fibonacci generator with a shift register random integer generator. The quasi random sequence is generated using Sobol' generator (see, for example, Gentle, 1998).

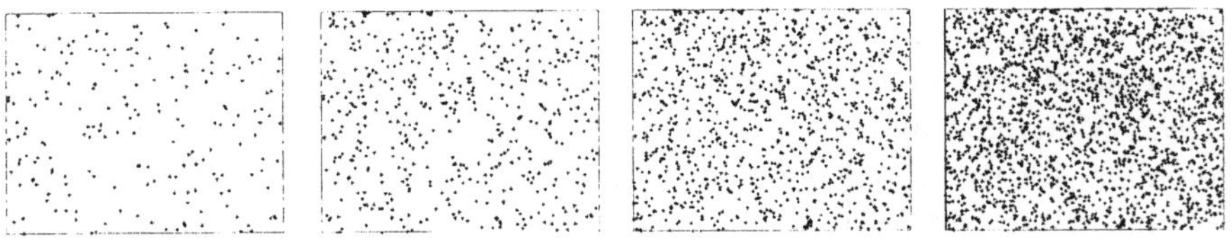

Figure 1. Two-dimensional pseudo random sequence of 200, 500, 1000, and 2000 points.

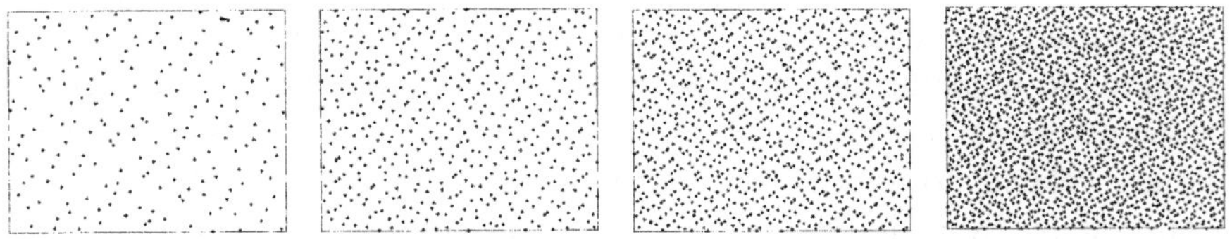

Figure 2. Two-dimensional quasi random sequence of 200, 500, 1000, and 2000 points.

Figures 1 and 2 are good examples of how the different generators are supposed to work. The different goals become obvious. There is no clear pattern to be detected in the distribution of the points in a pseudo random sequence. Thus, the goal to imitate independent random points is well attained. On the other hand, the points in the pseudo random sequence tend to clump and form clusters leaving other areas relatively unexplored. This is characteristic of an independent sample. The points in the quasi random sequence are more evenly distributed and a certain pattern can be detected, especially when the sample size is large. Hence, replacing a pseudo random sequence with a quasi random sequence is well-grounded in cases where the points do not need to be independent but good uniform distribution is required.

Quasi random sequences are commonly used in numerical integration (see, for example, Bratley and Fox, 1988, Niederreiter, 1992, Press and Teukolsky, 1989, Sobol', 1998, Tuffin, 1996). The error analysis guarantees that for a function with bounded variation the integration error is small if point sets with a low star discrepancy (or a small extreme discrepancy) are used, see Niederreiter, 1992. Different discrepancies are defined, for example, in Gentle, 1998, Kuipers and Niederreiter, 1974, Niederreiter, 1992. Quasi random sequences have low discrepancy values and, therefore, good error bounds can be obtained. In practice, this means that good estimates for integrals can be obtained with smaller sample sizes than when using pseudo random sequences.

In global optimization, quasi random sequences have been used in some random search methods (see, for example, Niederreiter, 1983, Sobol' and Bakin, 1994). Random search methods are most suited when optimizing functions of low degree of regularity, see Niederreiter, 1992, that is, for example, nondifferentiable or discontinuous functions. As a distinction from random search methods that apply pseudo random numbers, the methods applying quasi random numbers are called as quasi random search methods or quasi-Monte Carlo methods for optimization. Equivalently to numerical integration, better error bounds can be obtained using quasi random search methods than when using pseudo random search methods, see Niederreiter, 1992. In optimization context, this means obtaining a good approximation of the optimal objective function value. In general, however, both pseudo and quasi random search methods have a slow convergence, see Niederreiter, 1992, and require a very large number of function evaluations, see Niederreiter, 1983.

In Niederreiter, 1983, some refinements are introduced to speed up the basic quasi random search. Even with the refinement strategies the random search is still best suited for objective functions with a low degree of regularity. If the objective function is more regular, other methods are usually applied. Popular methods, not requiring gradient information, are metaheuristics, among others genetic algorithms. In the next section, we describe a novel way to apply quasi random sequences in genetic algorithms.

3. GENETIC ALGORITHM IMPLEMENTATION

In our implementation, we use quasi random sequence when selecting the initial population of a genetic algorithm. The selection of the initial population can be considered as one (initial) generation out of hundreds or thousands of generations. Moreover, selecting the initial population takes even less time than a regular generation in the genetic algorithm. Nevertheless, the initial population has a special role since all the populations in the iterative search process depend, to some extent, on the preceding population and, eventually, on the initial population.

The genetic algorithm used in this paper is described in Miettinen et al., 2000. We apply tournament selection, heuristic crossover, see, for example, Michalewicz, 1994, and Michalewicz's nonuniform mutation, see, for example, Deb, 2001. The optimization parameters used are given in Table 1. The parameter *Population size* is the number of individuals in a population, *Tournament size* defines the number of randomly selected individuals out of which the one with the best objective function value is chosen as a parent, and *Elitism size* is the number of the best individuals directly carried over to the next generation. The parameter *Crossover rate* regulates the probability on which the selected parents are crossbred and *Mutation rate* the probability on which an individual is mutated.

The parameters *Max generations*, *Steps*, and *Tolerance* are attached to the two-fold stopping criterion: the algorithm is stopped after the maximum number of generations (*Max generations*) is reached or if there has been no change larger than *Tolerance* in the best objective function value for a predefined number (*Steps*) of generations. The maximum number of generations used is set to be quite large, and hence, the latter part of the two-fold stopping criterion becomes often decisive in the tests to be reported.

Table 1. Optimization parameters for the genetic algorithm

Description	*Value*
Population size	201
Elitism size	21
Tournament size	3
Crossover rate	0.8
Mutation rate	0.1
Max generations	2000
Steps	100
Tolerance	10^{-7}

Several different parameter values were tested and the values given in Table 1 were chosen since they provided relatively good results for a large number of test problems. Four different versions were implemented. The first version, serving as a benchmark, is the original genetic algorithm using pseudo random numbers also in the initialization. The three other versions use Niederreiter, see Niederreiter, 1992, and two different Sobol' generators, see Bratley and Fox, 1988, when selecting the initial population. The difference between the two Sobol' generators is that one uses gray coding and the other one does not. In addition, there are differences in numerical computations. For further information about implementing a Sobol' generator, see, for example, Bratley and Fox, 1988.

4. NUMERICAL RESULTS

A test suite of 32 optimization problems was selected for testing the genetic algorithm implementation using different initial populations. The test problems were taken from the literature with a view to having a broad selection of different problems. All the problems were solved a hundred times with each implementation and the average values are reported. The names of the test problem families (if existing) along with the problem dimensions, box constraints, and the references are given in Table 2.

The box constraints are given as a cross product of intervals defined by the lower and the upper bound of each variable. If the lower and upper bounds are the same for each variable, only one interval is given. Note that the feasible region may differ depending on the source of reference.

Table 2. Test problems

#	*Name*	*Dimensions*	*Box constraints*	*Ref*
1-3	Michalewicz	2, 5, 10	$[0,\pi]$	Matlab toolbox, 2002
4	Rastrigin	6	[-600,400]	Matlab toolbox, 2002
5-7	Schwefel	6, 10, 20	[-500,500]	Matlab toolbox, 2002
8	Branin rcos	2	$[-5, 10]\times[0,15]$	Matlab toolbox, 2002
9-12	Griewangk	2, 6, 10, 20	[-700,500]	Matlab toolbox, 2002
13	Ackley's Path	2	[-32.768,32.768]	Matlab toolbox, 2002
14	Easom	2	[-100,100]	Matlab toolbox, 2002
15	Levy	4	[-10,10]	Madsen, 2002
16-18	Levy	5, 6, 7	[-5,5]	Madsen, 2002
19	P8	3	[-10,10]	Decker and Aarts, 1991
20	P16	5	[-5,5]	Decker and Aarts, 1991
21	Hansen	2	[-10,10]	Madsen, 2002
22-24	Corona	4, 6, 10	[-900,1100]	Dykes and Rosen, 1994
25-27	Katsuura	4, 6, 10	[-1,1]	Dykes and Rosen, 1994
28-31	-	3, 4, 10, 20	[-20,20]	Trafalis and Kasap, 2002
32	Epistatic Michalewicz	2	$[0,\pi]$	Roli, 2002

The other Sobol' sequence starts with a zero vector. In many of the test problems, the optimal point was originally located in the center of the box defined by lower and upper bounds of each variable. This caused the implementation using the Sobol' sequence to find the optimal point immediately. To obtain more comparable and informative results we shook the box so that the optimal point was no longer in the center of the box.

In many of the test problems, the four implementations obtained similar results with respect to objective function values and CPU times. Figures 3 and 4 illustrate the problems, where the differences were noteworthy. In Figures 3 and 4, *Nieder, Sobol' 1* and *Sobol' 2* denote the Niederreiter and the two Sobol generators, respectively. The objective function values were considered the same if the difference was less than 0.04 and the respective tolerance interval for the CPU times was 0.06 seconds. The bars denoted as *Avg.* in the histograms are the average values over the problems illustrated in the histogram.

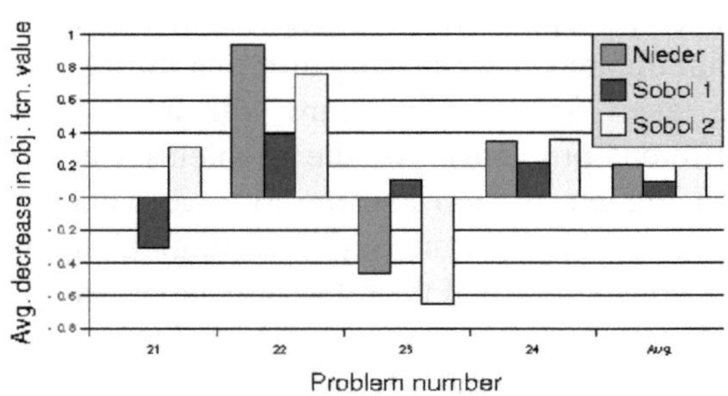

Figure 3. Decrease in objective function values

Figure 3 illustrates the difference in objective function value with respect to original genetic algorithm using pseudo random numbers. We can see that, on the average, the use of quasi random numbers improved the objective function value over the original genetic algorithm. Each of the quasi implementation obtained a better objective function value in three cases out of four.

Figure 4 illustrates the increase in CPU times. The Niederreiter implementation performs about the same as the benchmark. Also the CPU times for the two Sobol' implementations are about the same as the benchmark's except for the problems 26 (0.28, and 1.11 seconds) and 27 (0.31, and 0.35 seconds). For problems 26 and 27 there is a 2-8% and 0.7-0.8% increase with respect to the original CPU time, respectively.

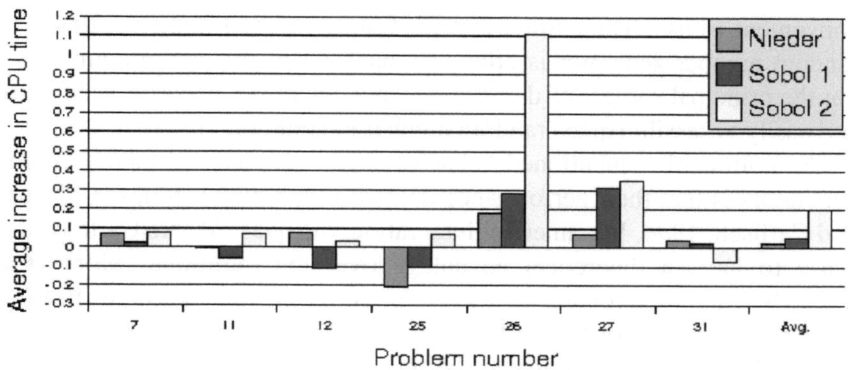

Figure 4. Average increase in CPU time

We also studied the effects that different initial populations have on the number of generations used. Since there is a strong dependence on CPU time and the number of generations, also the results were somewhat similar. The implementation using Sobol' 2 quasi random sequence needed on the average the most generations. However, this time Sobol' 1 performed better than Niederreiter. On the average, the implementation using Sobol' 1 quasi random sequence used the same amount of iterations as the benchmark.

The results indicate that different initial populations have an effect on the final objective function value and the consumed CPU time. The use of quasi random sequences on the average improved the robustness over the traditional genetic algorithm implementation using pseudo random numbers. The convergence improvement was most significant on the Corona function. The Corona function has approximately 10^{5n} local minima and is considered to be a very difficult problem for most optimization methods, see Dykes and Rosen, 1994.

While the convergence improved for the Corona function, the CPU time consumption slightly increased for the Katsuura function. However, many times when using genetic algorithms the major importance is put in convergence whereas speed is considered a secondary criterion of method evaluation, see, for example, Andre et al., 2001. Due to this, we consider the trade-off between reliability and computing time favorable for the implementations using quasi random sequences.

5. DISCUSSIONS

In general, the quasi random number generators have not been tested as elaborately as the pseudo random number generators. In addition, their absence from the standard mathematical software libraries makes them less convenient to use. The extra effort for testing or even implementing a new random number generator is often too much when the random numbers are not the principal subject of the study. Hence, there is a need for standardized and easily accessible quasi random number generators.

A commonly mentioned disadvantage of quasi random number generators is that their performance degrades with dimension, see Morokoff and Caflisch, 1994. In numerical integration, the quasi-Monte Carlo methods cease to have a theoretical advantage over the traditional Monte Carlo methods when the problem dimension goes over 12 (according to Bratley et al., 1992) or 40 (according to Mascagni and Karaivanova, 2000). According to Sobol', 1998, the former is correct only for integrands with all the variables equally important. Nevertheless, the use of quasi random sequences for large-dimensional problems needs further research both in numerical integration and in optimization.

Another disadvantage of quasi random number generators is that there exist no standard tests to measure their quality. Therefore, the results obtained in this paper should be validated by statistical methods guaranteeing the optimal distribution of the point set.

6. CONCLUSIONS

Quasi random sequences have been applied successfully in numerical integration and random search methods. However, they have not been used in more sophisticated methods designed for global optimization. In this paper, we extended the use of quasi random sequences to population-based metaheuristics. We studied the effects that different initial populations of a genetic algorithm have on the final objective function values and the CPU times.

We selected a test suite of 32 problems and applied three different quasi random sequences when selecting the initial population. The benchmark was a genetic algorithm using pseudo random numbers. The goal of using the quasi random sequences was to provide a set of initial solutions that is as uniformly distributed as possible. The results indicate that the distribution of the initial population has an effect on both the final objective function value and CPU time.

The use of quasi random numbers in the initialization made the implementation slightly more robust. For most of the problems the objective function values were about the same for all the implementations, but for the problems where the difference was noteworthy, the quasi implementations performed better on the average. On the other hand, the CPU times slightly increased for the quasi implementations. These results indicate that a more diverse initial population both increases the probability for the search to converge to the global optimum and slows down the speed of convergence.

In the near future, we will apply our quasi implementations for larger problems up to 50 dimensions. Our motivation is the following. The theoretical error analysis gives upper bounds for the errors. However, the bounds may be loose and, therefore, numerical analysis is needed to confirm the theoretical results. Moreover, the genetic algorithm implementation differs significantly from a random search method. We will also validate the results reported in this paper by applying other deterministic methods to generate points with good distribution properties.

Acknowledgements: This research was supported by Finnish Academy grants #8583, #65760, and #71822.

REFERENCES

Ali, M. M., Törn, A., and Viitanen, S. (2002). A direct search variant of the simulated annealing algorithm for optimization involving continuous variables. *Computers & Operations Research*, 29(1):87–102.

Andre, J., Siarry, P., and Dognon, T. (2001). An improvement of the standard genetic algorithm fighting premature convergence in continuous optimization. *Advances in Engineering Software*, 32(1):49–60.

Androulakis, I. P., Maranas, C. D., and Floudas, C. A. (1995). αBB : A global optimization method for general constrained nonconvex problems. *Journal of Global Optimization*, 7:335–363.

Battiti, R. and Tecchiolli, G. (1996). The continuous reactive tabu search: Blending combinatorial optimization and stochastic search for global optimization. *Annals of Operations Research*, 63:153–188.

Bratley, P. and Fox, B. L. (1988). Algorithm 659: Implementing Sobol's quasirandom sequence generator. *ACM Transactions on Mathematical Software*, 14(1):88–100.

Bratley, P., Fox, B. L., and Niederreiter, H. (1992). Implementation and tests of low-discrepancy sequences. *ACM Transactions on Modeling and Computer Simulation*, 2(3):195–213.

Chelouah, R. and Siarry, P. (2000). Tabu search applied to global optimization. *European Journal of Operational Research*, 123(2):256–270.

Deb, K. (2001). *Multi-Objective Optimization using Evolutionary Algorihtms*. John Wiley & Sons.

Decker, A. and Aarts, E. (1991). Global optimization and simulated annealing. *Mathematical Programming*, 50:367–393.

Dykes, S. and Rosen, B. (1994). Parallel very fast simulated reannealing by temperature block partitioning. In *Proceedings of the 1994 IEEE International Conference on Systems, Man, and Cybernetics*, volume 2, pages 1914–1919. IEEE Press.

Gentle, J. E. (1998). *Random number generation and Monte Carlo methods*. Springer-Verlag.

Glover, F (1977). Heuristics for integer programming using surrogate constraints. *Decision Sciences*, 8(1):156–166.

Glover, F. (1999). Scatter search and path relinking. In Corne, D., Dorigo, M., and Glover, F., editors, *New Ideas in Optimization*, pages 297–316. McGraw Hill.

Glover, F., Laguna, M., and Marti, R. (2000). Scatter search. Manuscript via private communication.

Goldberg, D. E. (1989). *Genetic Algorithms in Search, Optimization, and Machine Learning*. Addison-Wesley.

Kuipers, L. and Niederreiter, H (1974). *Uniform Distribution of Sequences*. John Wiley & Sons.

Madsen, K. (2002). Test problems for global optimization. http://www.imm.dtu.dk/~km/GlobOpt/testex/.

Mascagni, M. and Karaivanova, A. (2000). What are quasirandom numbers and are they good for anything besides integration? In *Proceedings of Advances in Reactor Physics and Mathematics and Computation into the Next Millenium (PHYSOR2000)*.

Matlab toolbox (2002). Genetic and evolutionary algorithm toolbox for use with matlab. http://www.geatbx.com/.

Michalewicz, Z. (1994). *Genetic algorithms + data structures = evolution program*. Springer-Verlag.

Miettinen, K., Mäkelä, M. M., and Toivanen, J. (2000). Comparison of four penalty function-based methods in handling constraints with genetic algorithms. Technical Report B17/2000, University of Jyväskylä, Department of Mathematical Information Technology.

Morokoff, W. J. and Caflisch, R. E. (1994). Quasi-random sequences and their discrepancies. *SIAM Journal on Scientific Computing*, 15(6):1251–1279.

Niederreiter, H. (1983). Quasi-Monte Carlo methods for global optimization. In Grossmann, W., Pflug, G., Vincze, I, and Wertz, W., editors, *Proceedings of the 4th Pannonian Symposium on Mathematical Statistics*, pages 251–267.

Niederreiter, H. (1992). *Random Number Generation and Quasi-Monte Carlo Methods*. SIAM.

Özdamar, L. and Demirhan, M. (2000). Experiments with new stochastic global optimization search techniques. *Computers & Operations Research*, 27:841–865.

Press, W. H. and Teukolsky, S. A. (1989). Quasi- (that is, sub-) random numbers. *Computers in Physics*, 3(6):76–79.

Roli, A (2002). Test problems in R^2. http://iridia.ulb.ac.be/aroli/ICEO/Functions/Functions.html.

Sobol', I. M. (1979). On the systematic search in a hypercube. *SIAM Journal on Numerical Analysis*, 16(5):790–793.

Sobol', I. M. (1998). On quasi-Monte Carlo integrations. *Mathematics and Computers in Simulation*, 47(2–5):103–112.

Sobol', I. M. and Bakin, S. G. (1994). On the crude multidimensional search. *Journal of Computational and Applied Mathematics*, 56(3):283–293.

Törn, A., Ali, M. M., and Viitanen, S. (1999). Stochastic global optimization: Problem classes and solution techniques. *Journal of Global Optimization*, 14(4):437–447.

Törn, A. and Žilinskas, A. (1989). *Global Optimization*. Springer-Verlag.

Trafalis, B. and Kasap, S. (2002). A novel metaheuristics approach for continuous global optimization. *Journal of Global Optimization*, 23(2):171–190.

Tuffin, B. (1996). On the use of low discrepancy sequences in Monte Carlo methods. Technical Report 1060, I.R.I.S.A., Rennes, France.

Wikramaratna, R. S. (1989). ACORN — a new method for generating sequences of uniformly distributed pseudo-random numbers. *Journal of Computational Physics*, 83(1):16–31.

SYMBOLIC TREATMENT ON EQUATION BASED MODELS TO PREPARE CODE GENERATION FOR OPTIMIZATION PROCESS

L. Allain (*), L. Gerbaud (*), C. Van Der Schaeghe (**)
(*)Laboratoire d'Electrotechnique de Grenoble, CNRS UMR 5529 INPG/UJF, ENSIEG BP 46, 38402 St Martin d'Hères cedex, Grenoble, France, E-mail : Laurent.Gerbaud@inpg.fr, Loig.Allain@leg.ensieg.inpg.fr

(**) Somfy Industrie, 50 Avenue du Nouveau Monde, BP 152, 74307 CLUSES Cédex France,
E-mail : christian.vanderschaeghe@somfy.com

Abstract: The paper deals with a symbolic treatment of analytical models, that are used to describe systems to be size using methods with optimization under constraints techniques. Such models are equations based. They are made of relations between parameters and system performances. In many cases, they have to be arranged by the designer, to get a well organised and oriented description for their valuing in the optimization process. Indeed, a complete physical description for a system is often made of equations, algorithms and functions. To optimise on many criteria, the computer code that values them has to be performed. The paper aims to provide an automatic help to organise the model elements, and to create a computer science code that carries out the model valuing. The description of these models is stored in a format that will allow easier code generation, for their valuing in different kind of tools, especially in optimization one : PRO@Design / EDEN [1]. These treatments are applied on electrical systems, mainly power electronics structures, electric drives, electromechanical systems, but also any similar system (in the modelling view point described in the paper).

Key words: Symbolic treatment, equation based optimization

M. Rudnicki and S. Wiak (eds.), Optimization and Inverse Problems in Electomagnetism, 45–52.

1. ELEMENTS FOR MODEL DESCRIPTION

1.1 Model mainly based on analytical equations

In an optimization process, the first task the designer has to face with is the description of the system he wants to size. In the specific approach proposed in the paper, these systems are described by equations. These equations are relations between parameters of the system and performances expected from the system. If a large part of a system can be described by equations, some cases may appear to be more complex. This is especially true when performances are deduced from analysis of a simulation over time (e.g. mean value of a signal, or Fourier analysis for a signal). In such a case, systems are described with equations and sometimes algorithm to represent some control aspects or hybrid systems[2].

In order to size the system, a lot a numerical valuing are necessary; this introduces some different elements in the system description. But, because of their natures, they have to be separated from the model description. This may be implicit solving, interpolation, extrapolation, series, etc...

The elements of the model are described and requirements are proposed for these elements.

1.2 Supplementary aspect for dynamic representation of Power Electronics circuits

As it is demonstrated in the above paragraph, an equation based representation of systems may be insufficient. Algorithm and function appear to be convenient elements to represent the whole complexity of a studied system, i.e.: associations of electromechanical devices with their power supply and control aspects. Power supply may be represented through hybrid models, i.e. that power electronics structures are represented by different electrical circuits, each of them associated with a specific state conduction of the switches. But, the model would not be complete if the algorithm describing the conditions under which a circuit state may be considered or another one. In this way, the models considered in this paper are fully described by equations and algorithm.

1.3 Model representation for capitalisation

Moreover, equations and algorithms use for their description variables, they have to be part of the model description.

In order to comply with standard description language for such data base, an XML representation is proposed. It is supposed to better suit with such description than classical text based representation [2, 3 and 4]. Moreover, models are described without any orientation i.e. that input, output or parameters elements are not yet specified.

As example, the permanent magnet synchronous machine on Figure 1.a is represented by its equations, complying with the proposed representation Figure 1.b. The model is taken from [5]

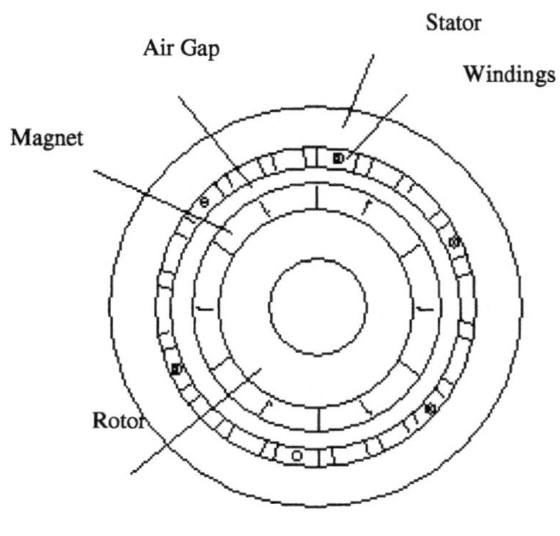

Fig. 1- a: Permanent Magnet Synchronous Machine Schematic description.

```
<MODEL name= “Permanent Magnet”>
<EQUATION value="Gem = pi/(2*lambda)*(1-
Kf)*sqrt(kr*beta*Ech*E)*pow(D,2)*(D+E)*Be" />
<EQUATION value="Ech=A*Jcu"/>
<EQUATION value="A=kr*E*Jcu" />
<EQUATION value="Kf = 1.5*p*beta*(e+E)/D" />
<EQUATION value="Be=(2*la*M)/(D*log((D+2*E)/(D-2*(la+e)))" />
<EQUATION value="C=pi*beta*Be*D/(4*p*Bfer)"/>
<EQUATION value="p=pi*D/deltaP" />
<EQUATION value="Vu=pi*D/lambda*(D+E-e-la)*(2*C+E+e+la)" />
<EQUATION value="Va=pi*beta*la*D/lambda*(D-2*e-la)" />
<EQUATION value="Pj=pi*rhoCu*D/lambda*(D+E)*Ech" />
<EQUATION value="lambda=D/L"/>
<EQUATION value="Fobj = cva*Va+cvu*Vu+cpj*Pj"/>
</MODEL>
```

Fig. 1- b: Permanent Magnet Synchronous Machine XML representation

1.4 Orienting the model

Once, the model is described, the designer has to decide what he expects from the model. I.e., the designer has to make a choice of input and output variables, parameters. Once this is defined, it is necessary to rebuilt the description of model, equations are manipulated to get a description suitable for a computer. Equations have to be oriented, systems have to be extracted and numerical method may be proposed to solve problem (interpolation, Newton-Raphson, ...). This is a repetitive and annoying task, hopefully, computer can automated a large part of such a symbolic treatment. The proposed treatment will be performed on models dedicated to the optimization, but all these operation can be performed in the same way on power-electronics model used for simulation.[6]

2. SYMBOLIC TREATMENT

Nowadays, the designer has to define himself the outputs and inputs of the model of the system to size. And then model has to be written to produce a description suitable for a numerical process based on computer resolution (figure 2). This work is painful, and often introduces human errors in the written equations and algorithms. Such errors can be avoided by automating these tasks.

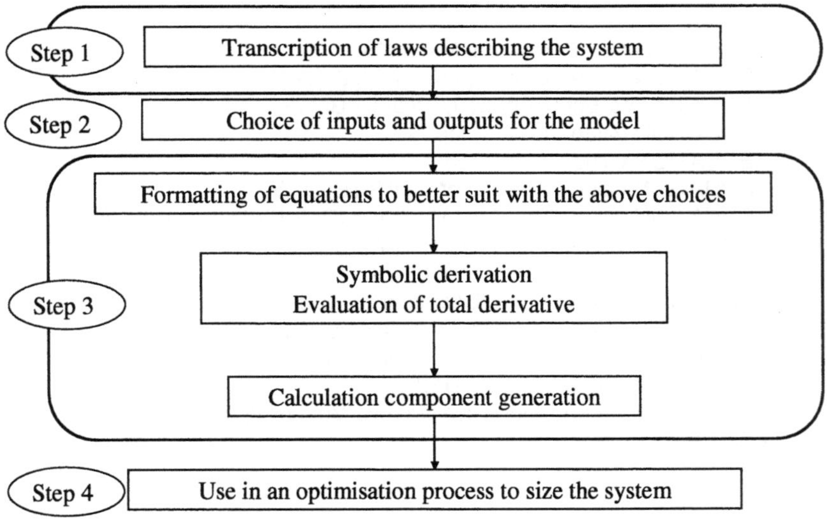

Figure 2 : *A way to get a sizing process*

First, the valuing order of the model part can be directly deduced from the equations and algorithms. Indeed, using Tarjan's algorithm [7], that allows to get strongly connected sub-graph from a graph, with the implementation proposed by I.S. Duff in [8] and that builds a block triangular form of a sparse matrix. To get a valuing order for a system, it is proposed to build an occurrence matrix for the system, in this description, its lines represent equations of the system and its columns are linked with dimensions (see figure 3).

Equation	beta	Bs	Bfer	C	cgj	cva	D	cvu	deltaP	e	E	Ech	Fobj	Jcu	Kf	Kr	lambda	L	la	M	p	Pj	Gamma	rhoCu	Va	Vu	A
Gam=pi/(2*lambda)*(1-Kf)*sqrt(kr*beta*Ech*E)*pow(D,2)*(D+E)*B	1	1	0	0	0	0	1	0	0	0	1	1	0	0	1	1	1	0	0	0	0	0	1	0	0	0	0
Ech=A*Jcu	0	0	0	0	0	0	0	0	0	0	0	1	0	1	0	0	0	0	0	0	0	0	0	0	0	0	1
A=kr*E*Jcu	0	0	0	0	0	0	0	0	0	0	1	0	0	1	0	1	0	0	0	0	0	0	0	0	0	0	1
Kf=1.5*p*beta*(e+E)/D	1	0	0	0	0	0	1	0	0	1	1	0	0	0	1	0	0	0	0	0	1	0	0	0	0	0	0
Bs=(2*la*M)/(D*log((D+2*E)/(D-2*(la+e))))	0	1	0	0	0	0	1	0	0	1	0	0	0	0	0	0	0	0	1	1	0	0	0	0	0	0	0
C=pi*beta*Bs*D/(4*p*Bfer)	1	1	1	1	0	0	1	0	0	0	0	0	0	0	0	0	0	0	0	0	1	0	0	0	0	0	0
p=pi*D/deltaP	0	0	0	0	0	0	1	0	1	0	0	0	0	0	0	0	0	0	0	0	1	0	0	0	0	0	0
Vu=pi*D*lambda*(D+E-e-la)*(2*C+E+e+la)	0	0	0	1	0	0	1	0	0	1	1	0	0	0	0	0	1	0	1	0	0	0	0	0	0	1	0
Va=pi*beta*la*D*lambda*(D-2*e-la)	1	0	0	0	0	0	1	0	0	1	0	0	0	0	0	0	1	0	1	0	0	0	0	0	1	0	0
Pj=pi*rhoCu*D*lambda*(D+E)*Ech	0	0	0	0	0	0	1	0	0	0	1	1	0	0	0	0	1	0	0	0	0	1	0	1	0	0	0
lambda=D/L	0	0	0	0	0	0	1	0	0	0	0	0	0	0	0	0	1	1	0	0	0	0	0	0	0	0	0
Fobj=cva*Va+cvu*Vu+cgj*Pj	0	0	0	0	1	1	0	1	0	0	0	0	1	0	0	0	0	0	0	0	0	1	0	0	1	1	0

Figure 3 : Global occurrence matrix for synchronous machine

In order to get the valuing order, it is useful to know which elements are inputs for a valuing step. Figure 4 shows a typical optimization process. It clearly appears that, on step A, input parameters are known i.e. that their values are yet defined.

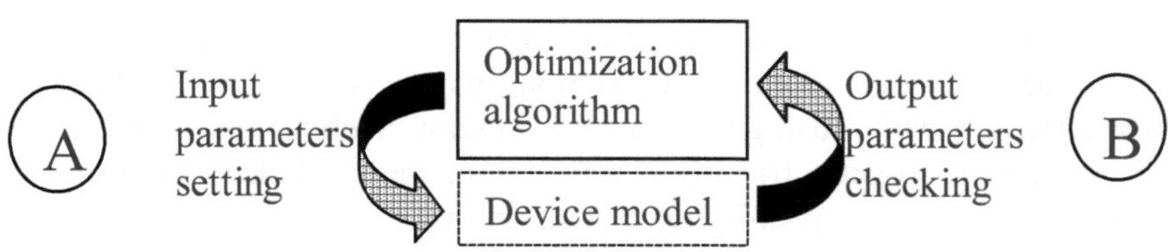

Figure 4 : Optimization process : Block solving

Let us assume that geometrical dimensions and physical constant are inputs for the model.

It results in a reduced occurrence matrix, which takes into account only unknown variables. Applying the Block Triangularization Algorithm on such a matrix produces the description on Figure 5. This one demonstrates that the occurrence matrix (which is a sparse one) can be transformed into a block triangular one.

Equation	lambda	deltaP	Kf	Be	Pj	A	Kr	Gem	Bfer	Vu	Va	Fobj
lambda=D/L	1	0	0	0	0	0	0	0	0	0	0	0
p=pi*D/deltaP	0	1	0	0	0	0	0	0	0	0	0	0
Kf = 1.5*p*beta*(e+E)/D	0	0	1	0	0	0	0	0	0	0	0	0
Be=(2*la*M)/(D*log((D+2*E)/(D-2*(la+e)))	0	0	0	1	0	0	0	0	0	0	0	0
Pj=pi*rhoCu*D/lambda*(D+E)*Ech	1	0	0	0	1	0	0	0	0	0	0	0
Ech=A*Jcu	0	0	0	0	0	1	0	0	0	0	0	0
A=kr*E*Jcu	0	0	0	0	0	1	1	0	0	0	0	0
Gem = pi/(2*lambda)*(1-Kf)*sqrt(kr*beta*Ech*E)*pow(D,2)*(D+E)*B	1	0	1	1	0	0	1	1	0	0	0	0
C=pi*beta*Be*D/(4*p*Bfer)	0	0	0	1	0	0	0	0	1	0	0	0
Vu=pi*D/lambda*(D+E-e-la)*(2*C+E+e+la)	1	0	0	0	0	0	0	0	0	1	0	0
Va=pi*beta*la*D/lambda*(D-2*e-la)	1	0	0	0	0	0	0	0	0	0	1	0
Fobj = cva*Va+cvu*Vu+cpj*Pj	0	0	0	0	1	0	0	0	0	1	1	1

Figure 5 Order occurrence matrix - BTF

Then, it can be extracted from such a matrix that one dimension is linked to an equation. That implies that the value of this variable will be extracted by solving an elementary problem of one equation. More complex case shows matrix where a list of variables is extracted from the numerical solution of a system. The last one could be a linear system. If not numerical method to solve the problem may be proposed to the user (e.g. linear or non linear solving of equation system).

When one variable could be associated with one equation, a symbolic solution is extracted if it is possible by using symbolic transformation. It results in a procedural description of the system. The term "procedural" has to opposite with the term "declarative". The first one is given to the description that embeds a kind of description for numerical processes.

Such a representation (Figure 6) is also available for model capitalization, because it is out of any solving environment (here optimization environment).

It results in a faster and more efficient computer science code. Indeed, the code produced from a description give a sequence of elementary operations instead of using big matrixes solving. Such an approach allows in the same time to get a friendly description of model. Indeed, user does not have to search for a specific representation that may comply with computer needs.

```
<PROCESS name="/Process">
    <ZONE >
<RELATION expression="lambda=D/L" order="0" resolution="explicit">
<PARAMETER name="lambda"/></RELATION>
<RELATION expression="p=pi*D/deltaP" order="1" resolution="explicit">
<PARAMETER name="deltaP"/></RELATION>
<RELATION expression="Kf = 1.5*p*beta*(e+E)/D" order="2" resolution="explicit">
<PARAMETER name="Kf"/></RELATION>
<RELATION expression="Be=(2*la*M)/(D*log((D+2*E)/(D-2*(la+e)))" order="3" resolution="explicit">
<PARAMETER name="Be"/></RELATION>
<RELATION expression="Pj=pi*rhoCu*D/lambda*(D+E)*Ech" order="4" resolution="explicit">
<PARAMETER name="Pj"/></RELATION>
<RELATION expression="Ech=A*Jcu" order="5" resolution="explicit">
<PARAMETER name="A"/></RELATION>
<RELATION expression="A=kr*E*Jcu" order="6" resolution="explicit">
<PARAMETER name="Kr"/></RELATION>
<RELATION expression="Gem = pi/(2*lambda)*(1-Kf)*sqrt(kr*beta*Ech*E)*pow(D,2)*(D+E)*Be" order="7"
resolution="explicit">
<PARAMETER name="Gem"/></RELATION>
<RELATION expression="C=pi*beta*Be*D/(4*p*Bfer)" order="8" resolution="explicit">
<PARAMETER name="Kr"/></RELATION>
<RELATION expression="Vu=pi*D/lambda*(D+E-e-la)*(2*C+E+e+la)" order="9" resolution="explicit">
<PARAMETER name="Vu"/></RELATION>
<RELATION expression="Va=pi*beta*la*D/lambda*(D-2*e-la)" order="10" resolution="explicit">
<PARAMETER name="Va"/></RELATION>
<RELATION expression="Fobj = cva*Va+cvu*Vu+cpj*Pj" order="11" resolution="explicit">
<PARAMETER name="Fobj"/></RELATION>
</ZONE>
</PROCESS>
```

Figure 6 : *Model representation describing a numerical process*

The file resulting from the symbolic analysis is then used in order to get a calculation component (here called COB) with its entire API that can be embedded in a calculation environment (e.g. Matlab/Simulink, MathCAD, Pro@Design) for optimization. This includes the calculation of the gradients thanks to the differentials (see figure 7).

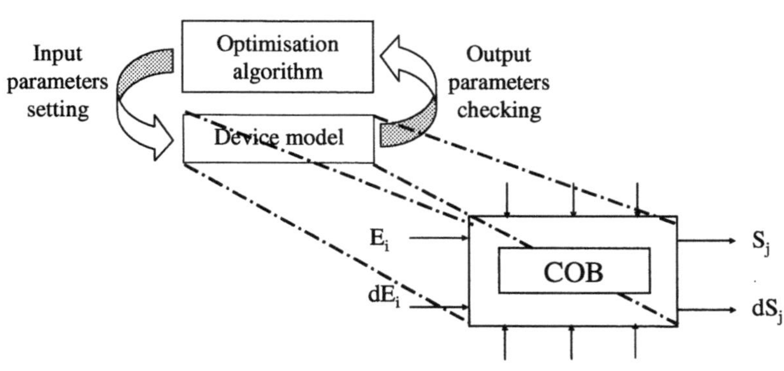

Figure 7 : *Use of a calculation component (COB)*

3. CONCLUSION

The paper describes the results of the model analysis and especially the way to order the model parts from their description. This description is implemented in an XML. Finally, a tool, called MAEL (for Module for Analysis of Equation and Logic) has been developed in our lab that embeds the suggested model treatments. A tools embedding code generation is being developed that allows producing computer science code dedicated to each targeted environment.

REFERENCES

[1] – E. Atienza, J. Bigeon, F. Wurtz and B. Belhabib, "Steps to an Electrical Design Environment", IEEE-IECON'99, San Jose, CA, USA, November 29th – December 3rd, 1999, pp° 815-820

[2] – D.A. van Beek 1, J.E. Rooda "Languages and applications in hybrid modelling and simulation: Positionning of Chi" Control Engineering Practice vol. 8, nr. 1, 2000, 81-91.

[3] – M. Anderson – "Object-Oriented Modelling and Simulation of Hybrid System", Ph D Thesis, ISRN LUTFD2/TFRT—1043—SE, December 1994

[4] – Luc Thevenon, "Représentation des Systèmes Hybrides Complexes par Flux de Données : Développement d'un Outil de Modélisation et de Simulation des Procédés Batch", Ph. D. Thesis, INP Grenoble, France Octobre 2000

[5] - A. D. Kone, B. Nogarede et M. Lajoie Mazenc, "Le dimensionnement des actionneurs électriques: un problème de programmation non linéaire", Journal de Physique III, 1993, page 285-301

[6] – L.Allain, L.Gerbaud, C. VanDerSchaeghe « Capitalisation and treatment of models for the optimization of electric drives », OIPE'02, Lodz, Poland, 2002

[7] – R.E. Tarjan, "Depth first search and linear graph algorithms", SIAM Journal of Computing, vol. 1, n°2, 1972.

[8] – I.S. Duff et J.K. Reid, "An implementation of Tarjan.s algorithm for the block triangularisation of a matrix ", ACM Transactions on Mathematical Software, vol 4, n°2, 1978, p.137-147.

MULTIOBJECTIVE SHAPE DESIGN IN ELECTROMECHANICS: AN APPROACH TOWARDS BENCHMARKING

Paolo Di Barba
Department of Electrical Engineering, University of Pavia, I-27100 Pavia, Italy

e-mail: di_barba@etabeta.unipv.it

Abstract: The shape synthesis of a linear actuator and a permanent-magnet motor is revisited in terms of multiobjective optimisation using Pareto criterion of optimality. Best-compromise solutions are approximated by means of sample-and-rank method. The two case studies are proposed as model problems for inverse electromechanics.

Key words: multiobjective optimization, inverse electromechanics, benchmarking.

1. INTRODUCTION

The definition of model problems is important for development and comparative application of methods and algorithms for field inversion. In the middle nineties a first group of benchmarks was proposed, focusing on inverse magnetostatics (Loney solenoid - University of Pavia, SMES device - TEAM Workshop): they were characterised by either a single objective or

M. Rudnicki and S. Wiak (eds.), Optimization and Inverse Problems in Electomagnetism, 53–61.

by the scalar formulation of a two-objective problem; consequently, the application of numerical procedures gave rise to a single solution, that was supposed to be the optimum. However, it is believed that multi-objective optimization is truly meaningful if it aims at the identification of the full set of best-compromise, or non-dominated, solutions in the objective space (Pareto optimal front, POF). In the case of two objectives, the POF is represented by a curve that can show non-trivial topological properties, like non-convexity and non-connection, or numerical features like non-uniform spacing. Recently, literature has offered issues both clarifying mathematical fundamentals of multi-objective optimisation [1] and developing methodologies for applications to many branches of engineering science [2].

In particular, a collection of multiobjective problems defined analytically is available; they should be regarded as the preliminary step when testing a multiobjective optimization method against topological properties of POF. Nevertheless, there is a need to go beyond abstract examples based on mathematical functions towards industrial problems characterized by finite-element models of devices; from this viewpoint, reference problems or benchmarks for multi-objective optimisation are still lacking in electrical engineering. In this context, electromechanics represents a challenging area for multiobjective design due to its complexity: in fact, direct models imply repeated non-linear magnetic analyses for simulating forces and torques, whereas inverse problems involve both field-dependent and energy-dependent criteria.

2. THEORETICAL ASPECTS OF BENCHMARKING

The definition of benchmarking is not obvious, because it can be interpreted in at least two senses:

-at a practical level, as a ground for comparing and validating different optimization algorithms;

-at a more abstract level, as a way to enhance the prior information that is given about the design problem.

The first view is the more traditional one; however, it suffers from the fact that solutions may be compared from very many different points of view, thus producing different results of comparisons. At a first glance, it might be stated that test problems are considered to be meaningful if they give rise to relevant topological properties of the POF, the approximation of which represents a challenge for any algorithm to converge.

The second view of benchmarking is less trivial to think about. In multiobjective optimisation, it could be exploited e.g. by measuring the distance of the current design from the POF based on a suitable metrics;

another possibility could be to identify the region of unfeasible design, i.e. the subset of the objective space in which no design is physically possible: this region is bounded by the POF that represents a theoretical barrier to optimization.

Both interpretations are valid and result in useful applications; therefore, revisiting traditional design problems in terms of multi-objective optimization could give new information. Moving from this background, two case studies dealing with electromechanics have been proposed as possible reference problems.

3. CASE STUDIES

Linear actuator

An electromagnet composed of an excitation winding included in a fixed magnetic core and a plunger, representing the movable core, is first considered. The force acting on the plunger appears when the winding is supplied by rectangular pulses of current with density equal to 2 10^5 Am^{-2}. The direct model is based on two-dimensional finite-element analysis of magnetostatic field [3], taking into account the non-linear behaviour of iron. The force is calculated by means of virtual work, using central finite differences with a step Δ=0.5 mm. Additional losses are evaluated in an indirect way, as the stray field in the winding. Though being a simple device, its design optimisation gives rise to a non-trivial inverse problem, characterised by three conflicting criteria. It can be cast as follows:

starting from a set of initial solutions randomly generated, find the shapes of device corresponding to best compromises among:

- *force when the air-gap is 6 mm wide, to be maximized;*
- *maximum radial component of induction in the winding, to be minimized;*
- *cost of materials, to be minimized, defined as* $c_1V_1+c_2V_2$ *where* $c_1=1$, $c_2=3$, V_1, V_2 *are per-unit costs and volumes of iron and copper, respectively.*

Five design variables control the shape evolution of the device; i.e.: length and cross-section of the magnetic circuit, radius of the winding, inclination angle of the air-gap profile. A set of bounds preserves the geometrical congruency of the model; in particular, the minimum value of winding radius is prescribed.

Permanent-magnet motor

The device considered is a six-pole permanent-magnet, having air-gap equal to 6 mm and rating a torque of 1 Nm at a speed of 3000 rpm [4]. The inverse problem can be cast as follows:

starting from a set of initial solutions randomly generated, find the shapes of device corresponding to best compromises between:

- *the cogging torque, to be minimised;*
- *the energy stored in the air-gap, to be maximised.*

For the sake of field analysis, the cross-section geometry has been rectified as shown in Fig.1. The shape of the tooth corners is varied, introducing a triangular profile at the ends; the triangle has length equal to 2.2 mm and height h_s varying between $h_s= -5$ mm and $h_s=5$ mm (when $h_s<0$ a claw-tooth geometry is originated); θ_m is the ratio of magnet gap to 1/8 of the pole pitch: h_s and θ_m have been assumed as design variables, having fixed all the other geometric variables. In particular, when $h_s<0$ a claw-tooth geometry is originated. A set of bounds preserves the geometrical congruency of the model.

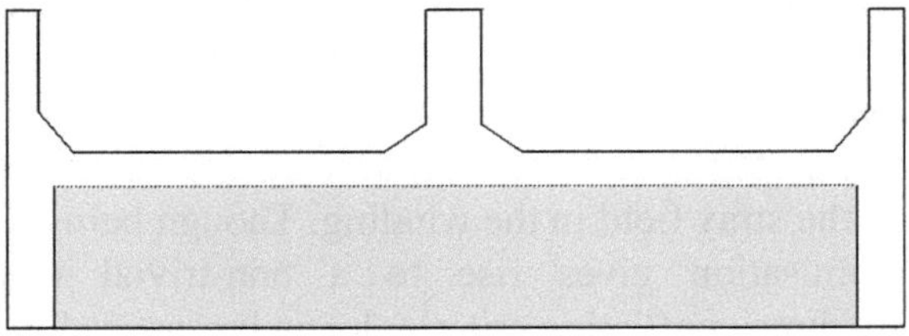

Figure 1. Permanent-magnet motor: (pole-pitch rectified geometry).

Iron is assumed to be ideal, while the magnet (remanence 400 mT, coercive field 282 kAm^{-1}, uniform radial magnetization) is modeled by means of an equivalent current sheet. The torque is obtained by means of the virtual work principle, considering a rotation step equal to 0.5 deg.

4. METHODOLOGICAL ASPECTS

An effective approach to the identification of non-dominated solutions is given by the sample-and-rank method, that might represent a cost-effective alternative to a true method of optimisation. To this end, a distribution of samples is generated randomly in the feasible region of the design space, based on a uniform probability density function. Samples are then mapped into the objective space via calculation of objectives, giving rise to a distribution that in general is a non-uniform one. The criterion of Pareto optimality [5] is then applied to the latter distribution; this way, non-dominated solutions are recursively extracted from the set of samples in the objective space; the corresponding points in the design space identify the device geometries.

The accuracy of the method depends on the density of samples discretizing the design space; the computational cost of this method can be estimated by

$$c_1 = np^{nv} \times nobjf \times c \tag{1}$$

where *np* is the number of points discretizing each design variable, *nv* is the number of variables, *nobjf* is the number of objective functions, *c* is the cost of a single field analysis; the cost depends exponentially on the number of variables; it should be compared with the typical cost of an evolutionary optimization method [5]

$$c_2 = npop \times niter \times nobjf \times c \tag{2}$$

with *npop* number of individuals and *niter* number of convergence iterations at a given tolerance. Considering the ratio of c_1 to c_2, the sample-and-rank method seems to be advantageously applicable for a fast approximation of best compromises in the case of two or three objective functions.

5. RESULTS

Linear actuator

Resorting to sample-and-rank method, the projections of the three-dimensional objective space on force-cost subspace and on force-stray field subspace have been represented in Fig.2 and Fig.3, respectively. In both cases, the local Pareto front is approximated by the south-eastern boundary; the true representation of best-compromise front should be a surface in the space force-cost-stray field.

Geometries corresponding to extreme non-dominated solutions in the aforementioned subspaces are reported in Fig. 4. Remarkable differences in objective functions and in device geometry can be noted.

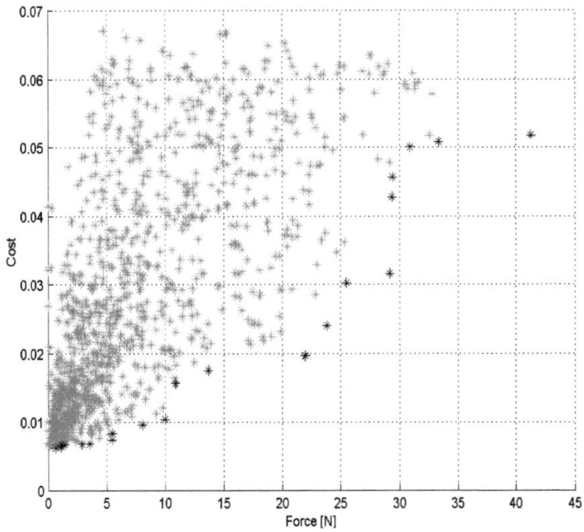

Figure 2. Sampling of force-cost subspace.

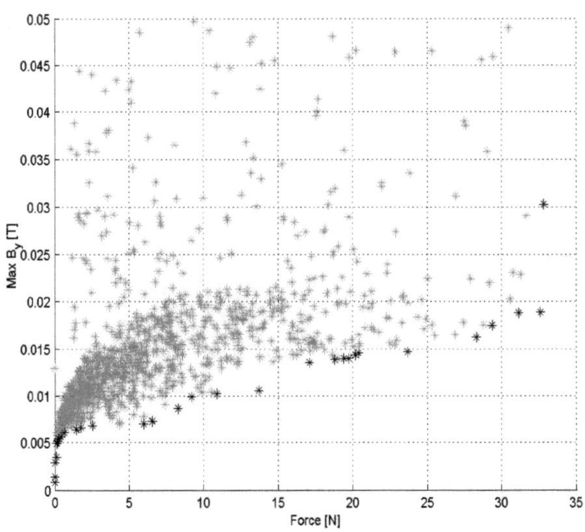

Figure 3. Sampling of force-stray field subspace.

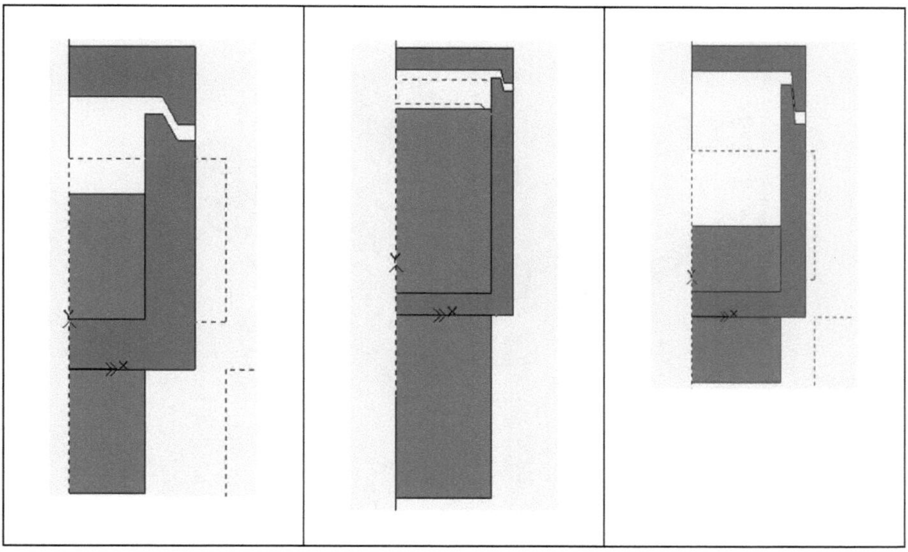

Figure 4. Extreme geometries: min force-min cost (left), max force-max cost (middle) coincident with max force-max stray, min force-min stray (right).

Permanent-magnet motor

After repeated field analyses, the error surface showing maximum cogging torque against design variables has been obtained as in Fig.5. In particular, it can be noted that a strong reduction of cogging effect is obtained when the periodicity of magnets is comparable with the periodicity of teeth; in that condition, it happens that positive and negative half-waves of torque tend to cancel each other within a polar pitch. The sample-and-rank approach was able to identify an approximation of Pareto front, represented by south-eastern boundary of the sampled region shown in Fig.6. Geometries characterizing the two extreme non-dominated solutions are summarized in Table I.

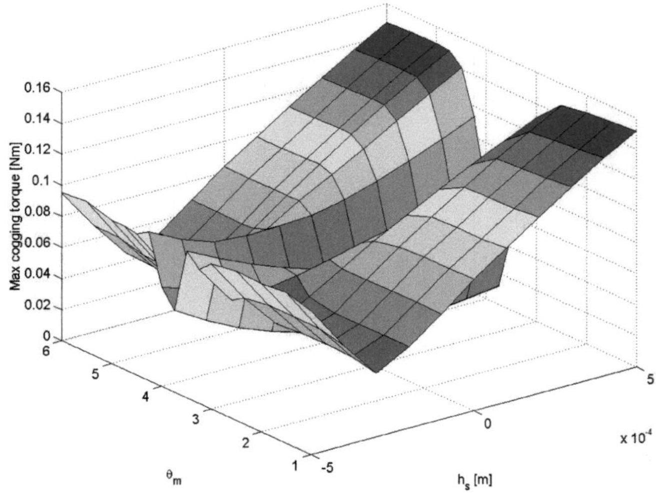

Figure 5. Torque and energy *vs* design variables.

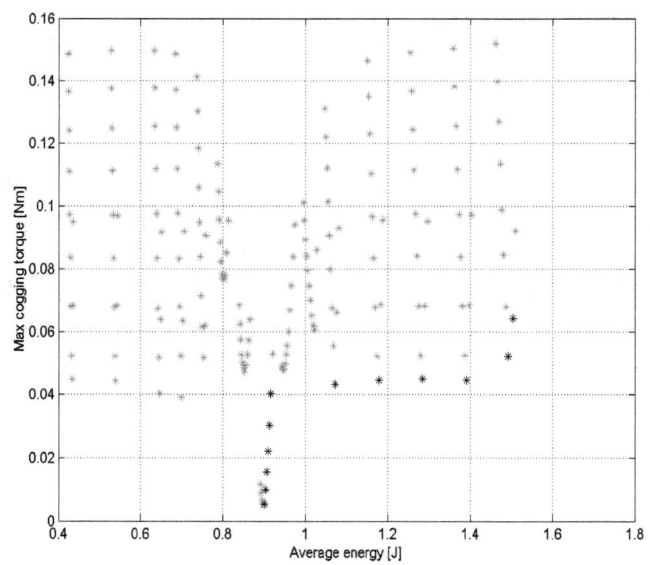

Figure 6. Sampling of energy-torque space
with non-dominated solutions.

Table 1 Extreme non-dominated solutions.

	h_s [mm]	θ_m	Energy [J]	Torque [Nm]
Max energy	-0.4	1	1.504255	0.064343329
Min torque	0.1	3.75	0.900288	0.005313580

The best-compromise front is represented by a curve; it appears to be non-convex, spaced in a rather uniform way, possibly non-connected. In fact, the front seems to be composed of two different branches: the former, poorly sensitive with respect to energy, is characterized by torque in the interval 0.005-0.04 (Nm) approximately; the latter, poorly sensitive with respect to torque, is characterized by energy variable in the interval 1.1-1.4 (J), approximately. Non-connection is a remarkable property; from the physical viewpoint, it should be originated by the oscillating behaviour of cogging torque versus geometry in the region $h_s<0$ (claw-tooth motor).

Always, variability in both objective space and design is strong.

6. CONCLUSION

The shape synthesis of a linear actuator and a permanent-magnet motor has been revisited in terms of multiobjective design. In both cases, the sample-and-rank method has been effective by applied to identify an approximation of best-compromise solutions.

REFERENCES

1. K. Miettinen: Non-linear Multiobjective Optimization. Kluwer Academic Publishers, 1999.
2. C.A. Coello Coello, D.A. van Veldhuizen and G.B. Lamont: Evolutionary Algorithms for Solving Multi-Objective Problems. Kluwer Academic Publishers, 2002.
3. /www.infolytica.com/ (visited in July 2002).
4. E. Costamagna, P. Di Barba and A. Savini: Numerical Schwarz-Christoffel Transformations for the Optimal Shape Design of a Permanent Magnet Motor. In: Electromagnetic Fields in Electrical Engineering - ISEF01. IOS Press, 2002, pp.39-43.
5. P. Di Barba: "A fast evolutionary method for identifying non-inferior solutions in multicriteria shape optimization of a shielded reactor", COMPEL, Vol.20, No.3, 2001, pp.762-776, 2001.

A NEW METHOD FOR OPTIMIZING THE DESIGNING OF GROUNDING GRIDS

Giuseppe Delvecchio, Michele Castellitti, Nello Medoro, Marcello Sylos Labini and Domenico Verde

Department of Electrotechnics and Electronics - Polytechnic of Bari - Via E. Orabona 4 - 70125 Bari (Italy) - E-mail: g.delvecchio@area-tecnica.uniba.it

Abstract: In the various designing steps of a grounding grid, the grids are gradually chosen by the designer according to his experience. A method for choosing, in a quite automatic way, the various grids and so for speeding up the designing process is given in this paper. The method is based on the determination of the maximum touch voltages, these being calculated by resorting to a genetic algorithm carried out by the Authors. Moreover, the building cost of the grounding grid is optimized thanks to the "Traveling Salesman" algorithm.

Key words: grounding grid, Maxwell's subareas method, Traveling Salesman, genetic algorithm

1. INTRODUCTION

The grounding grid designing made in the traditional way can be summarized according to the following points [1]:

1. On the basis of the value t_F of the time during which the fault persists, the designer must determine the maximum permissible touch voltage U_{Tp} fixed by Standards.
2. The designer chooses a grounding grid as starting grid.
3. The earthing voltage U_E of this grounding grid must be calculated. This voltage must be compared to the maximum permissible touch voltage U_{Tp}.
4. If $U_E < U_{Tp}$ the grounding grid which has been chosen turns out well.

M. Rudnicki and S. Wiak (eds.), Optimization and Inverse Problems in Electomagnetism, 63–70.

5. If $U_E > U_{Tp}$, there can be some dangerous points on the soil surface, that is points where the touch voltages $U_T(P)$ are greater than the maximum permissible touch voltage U_{Tp}. In this case the designer, according to his experience, must choose another grounding grid and he must do all the above-mentioned designing steps again, as from point 3 above until $U_E < U_{Tp}$.

The method carried out by the Authors automatically chooses the grounding grid at each designing step so as to speed up the whole designing process. Briefly, this method is based on both the search for the soil points in which the touch voltages are dangerous and on the choice, in each step of the design method, of an adequate grounding grid, this grid being obtained by suitably connecting to each other some previously defined points of the soil. The search for the dangerous soil points is made by an ad hoc genetic algorithm defined by the Authors in [2]; the connections of the soil points are carried out by a "Traveling Salesman" algorithm.

2. THE CRITERION FOR CHOOSING THE GROUNDING GRIDS

The optimization method proposed by the Authors is based on the following points:

1. it is necessary to determine, for each supposed grounding grid, both its earthing voltage U_E and the touch voltages $U_T(P)$ generated by the grounding grid on the soil surface points;
2. the soil surface points $P_{s\,max}$ where the touch voltages $U_T(P)$ are greater than the maximum permissible touch voltage U_{Tp} must be found;
3. the nodes of the grid under examination must be connected with the points $P_{g\,max}$, these points being put vertically under the points $P_{s\,max}$ at the same depth of the grounding grid (Fig. 1 gives an example of points $P_{s\,max}$ and $P_{g\,max}$ in the case of a grounding system consisting of three rectangular electrodes).
4. the connections mentioned above must be carried out by a "Traveling Salesman" algorithm.

As regards point 1 above, the well-known Maxwell's subareas method has been applied [3, 4, 5].

As for the points 2 and 3 above, they allow us to choose, in each designing step, the grounding grid relating to the next step. For, if the grounding grid of the *i*-th step doesn't turn out well, that is if the Maxwell's method gives $U^i_E > U_{Tp}$, it is necessary to find on the soil surface all the points $P^i_{s\,max}$ in which the touch voltages are greater than U_{Tp}. This can be done very quickly by resorting to the genetic algorithm method suggested in [2], this algorithm having been suitably modified in this paper, as we will see in the next section. Once all the points $P^i_{s\,max}$ relating to the whole soil surface are known, it is possible to determine the points $P^i_{g\,max}$ put vertically under the points $P^i_{s\,max}$ as well as it is possible to determine the new grounding grid of the $(i+1)$-th step by connecting the nodes of the i-th grid to the points $P^i_{g\,max}$. The Authors have found that the new grounding grid obtained according to points 2 and 3 above has an earthing voltage U_E lower than that of the previous grid, as well as the touch voltages are lower than the touch voltages generated by the previous grid. Therefore this criterion for choosing the grounding grids speeds the whole designing process up.

Finally, as regards point 4 above, the application of the "Traveling Salesman" algorithm allows us to make the connections between the various points, that is between the "old" nodes of the i-th grid and the new points $P^i_{g\,max}$, by the minimum path. In this way, we can reduce the length of the leaking conductors and, consequently, we can optimize the grounding grid cost, as we will see later on.

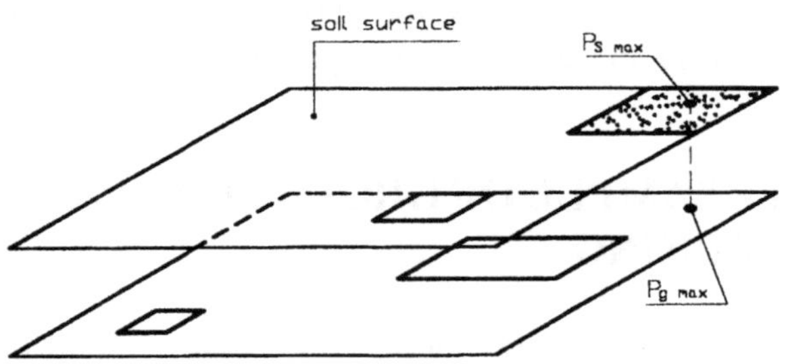

Figure 1. Test-area for determining the points $P_{s\,max}$ and $P_{g\,max}$

3. THE GENETIC ALGORITHM METHOD FOR DETERMINING THE POINTS $P_{S\,MAX}$

In [2] the Authors carried out a genetic algorithm method which allowed them to determine the maximum touch voltage $U_{T\,max}$ without calculating the touch voltages $U_T(P)$ in a very great number of points of the soil surface. Briefly, the optimization method consists of the following. A first sampling of the function $U_T(P)$ is done at random. This sampling concerns a small number of points P of the soil surface. These points represent the population to which we apply the genetic algorithm method. Each point P(x, y), in terms of genetic algorithms, is an individual having two chromosomes. A binary encoding is applied to the x and y co-ordinates of each point P. The binary numbers are considered to be string variables, so crossover and mutation are applied to them. The Authors have chosen the "one-point cross-over" technique and established that each individual thus generated has a 4% probability of undergoing a mutation. The rate at which the population improves is measured by calculating, in each *i*-th iteration, the difference ε_i between the maximum touch voltage obtained in the *i*-th iteration and the average value between all the touch voltages calculated in the same iteration. The algorithm is stopped when at least one of the two following conditions occurs: ε_i is smaller than a pre-arranged value ε; the number of iterations reaches a pre-arranged maximum value.

The genetic algorithm method we have now described is able to determine, in the soil surface under examination, only the point of maximum touch voltage. On the contrary, it is now necessary to determine all the points $P_{s\,max}$ of the soil surface where the touch voltages $U_T(P)$ are greater than the maximum permissible touch voltage U_{Tp}. This inconvenience has been solved by the Authors subdividing the whole soil surface into various areas (test-areas) and by applying the above genetic algorithm method to each test-area, to find, in each test-area, a probable point $P_{s\,max}$ (see Fig. 1).

4. AN EXAMPLE OF GROUNDING SYSTEM DESIGNING

We intend to design the grounding system of a group consisting of a transformer room and two buildings. The grounding system must leak a current $I_F = 50$ A; moreover, the time t_F during which the fault persists is known and is equal to 0.54 s. According to Standards, only one grounding system must be realized and, on the basis of the value of the time t_F, the value of the maximum permissible touch voltage is $U_{Tp} = 190$ V.

Let's consider the grounding system shown in Fig. 2 as the starting one, this being constituting of three rectangular electrodes connected by an insulated conductor. Each electrode is made up of horizontal cylindrical conductors having a section $S = 35$ mm^2 and is buried at a depth $h = 0.5$ m in a soil having resistivity $\rho = 300\ \Omega$m. Let's apply the Maxwell's subareas method to this system to calculate its earthing voltage U_E. We get $U_E = 268.5$ V. This value is greater than U_{Tp} and so the above starting grounding system is not good.

We can now take two ways:

- either we resort to the traditional designing method, that is the designer, according to his experience, chooses a grounding system, determines its earthing voltage U_E and gradually goes on choosing the systems until $U_E < U_{Tp}$;

- or we adopt the automatic designing method, that is to say, the calculation program finds the points $P_{g\,max}$ and, according to them, goes on choosing automatically another grounding system and so on until $U_E < U_{Tp}$.

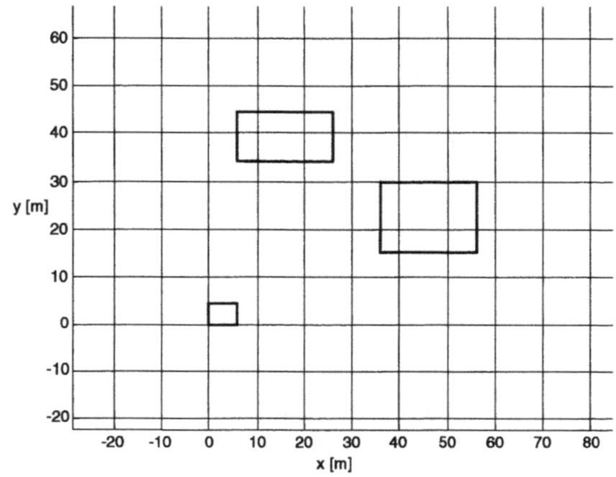

Figure 2. Starting grounding system (U_E=268.5V > U_{Tp}; € 953)

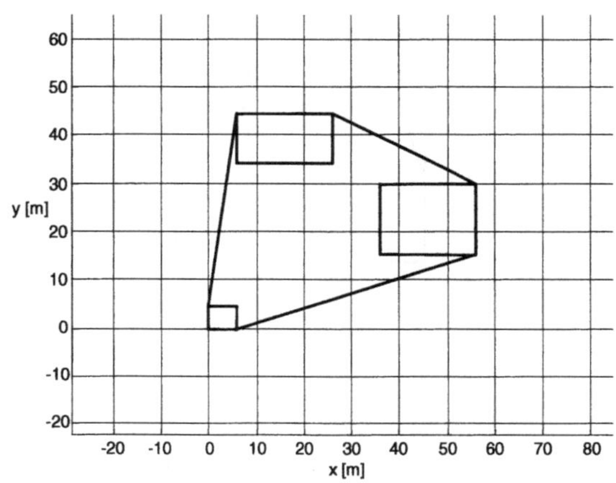

Figure 3. A tentative grounding system made by designer (U_E=199V > U_{Tp}; € 1747)

Figures 3 and 4 show some tentative grounding systems, chosen by the designer by the traditional designing method. Of course the number of these tentative systems can be high and depends on the designer experience. On the contrary, the calculation program carried out by the Authors performs the grounding system design in few steps: Figs. 5, 6 and 7 show the grounding systems chosen little by little by the calculation program, in only three steps.

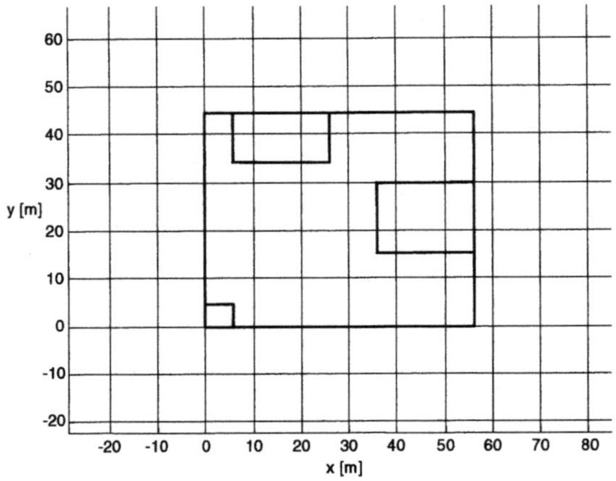

Figure 4. A tentative grounding system made by designer (U_E=176V < U_{Tp}: this system is good; € 2155)

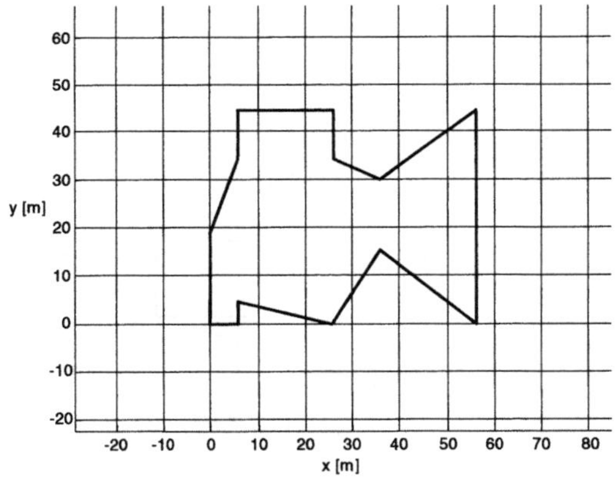

Figure 5. Grounding system: 1st step of the program (U_E = 198V > U_{Tp}; € 1456)

The grounding system in Fig. 7 is also optimized from the point of view of the total length of the leaking conductors, and so from the point of view of the building cost. For, the cost of the system in Fig. 4 is € 2155 while that of the system in Fig. 7 is € 1587.

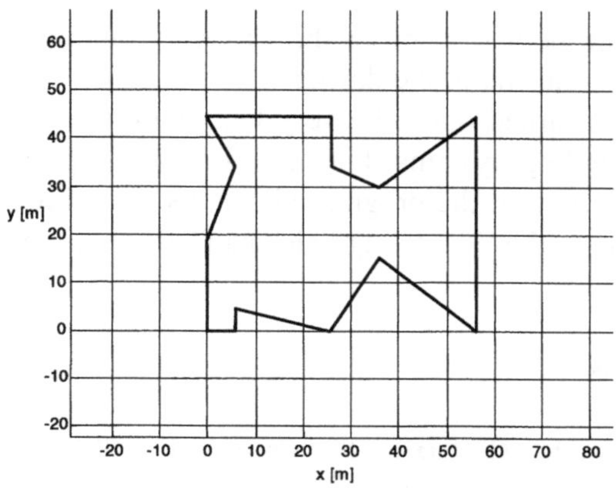

Figure 6. Grounding system: 2nd step of the program (U_E = 194V > U_{Tp}; € 1504)

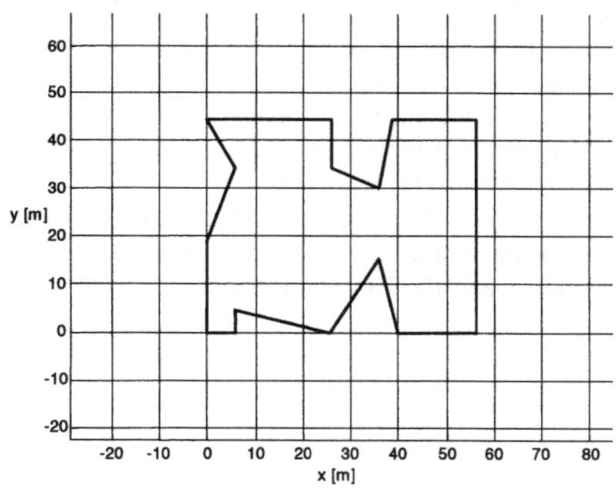

Figure 7. Grounding system: 3rd step of the program
(U_E = 187.8V < U_{Tp}: this system is good; € 1587)

5. CONCLUSIONS

In this paper the Authors have defined a completely automatic designing method of any grounding system. It asks the designer to plan only the starting grounding system, after that the program, according to the dangerous points existing on the soil surface, these points having been found thanks to a genetic algorithm method, chooses the various grounding systems and, in few steps, comes to the end of the design.

The method has turned out to be very fast, not only for the criterion for choosing the grounding systems, but also for the adoption of the genetic algorithm method. At last, the program allows us to optimize the grounding system cost, thanks to the "Traveling Salesman" algorithm carried out.

REFERENCES

1. A. Covitti, G. Delvecchio, C. Marzano, M. Sylos Labini: A Global Optimization Method for Designing Meshed Grounding Grids, Proceedings on Tenth biennal IEEE Conference on Electromagnetic Field Computation, CEFC 2002, Perugia (Italy), p 321, June 16-19 2002
2. G. Delvecchio, F. Neri, M. Sylos Labini: A genetic algorithm method for determining the maximum touch voltage generated by a grounding system, Proceedings on OIPE 2002, Lodz, Poland, September 12-14 2002
3. V. Amoruso, S. De Nisi, G. Negro, M. Sylos Labini: A complete computer program for the analysis and design of grounding grids, International Journal of Power and Energy Systems, Vol. 15, n. 3, pp.122-127, 1995
4. A. Covitti, S. De Nisi, F. Laddomada, M. Sylos Labini: A fuzzy-Maxwell combined method for simplifying the calculation of the current field, IEEE Transactions on Magnetics, Vol. 36, n. 4, pp. 708-711, July 2000.
5. M. Sylos Labini, A. Covitti, G. Delvecchio, C. Marzano: A Study for Optimizing the Number of Subareas in the Maxwell's Method, Proceedings on Tenth biennal IEEE Conference on Electromagnetic Field Computation, CEFC 2002, Perugia (Italy), p 320, June 16-19 2002

SOLVING ODE FOR OPTIMIZATION

Specific use of the matrix approach

Vincent Fischer, Laurent Gerbaud
Laboratoire d'Electrotechnique de Grenoble (INPG / UJF – CNRS . UMR 5529)
ENSIEG, BP 46, 38402 Saint Martin d'Hères CEDEX, France
e-mail : vincent.fischer@leg.ensieg.inpg.fr

Abstract: The paper deals with the sizing of some electrical devices, mainly electrical circuits, using optimization under constraints. More accurately, gradient optimization algorithms are considered. So, the derivatives of the outputs of the sizing model according to its inputs have to be valued. In electrical devices, some sizing criteria may require the solving of ordinary differential equation (ODE). E.g, this may concerns maximum values, r.m.s. values, values of some variables at a specific date, etc. Several methods may be used to solve such ODE : symbolic, numerical or a mix of them. In the specific case of systems modelled by linear state equations – e.g. electrical circuits- an approach based on matrix exponential is proposed to solve corresponding ODEs. After a presentation of the interests of the proposed approach compared to simulation in the field of gradient optimization, the paper presents the ODE solving and the chosen matrix exponential methods. Finally, the method is applied on some electrical circuits.

Key words: ODE, matrix exponential, optimization

1. THE PROBLEMATIC

This paper deals with the sizing of electrical devices (more accurately electrical circuits), in an optimization optic. Different approaches of optimization exist [1][2], which are based on the numerical simulation of the devices. These numerical approaches lead to important computational times, especially if the sizing model has many outputs based on the simulation, which depend on several inputs. The ODE solving based on matrix exponential is a good way to reduce the computing time.

M. Rudnicki and S. Wiak (eds.), Optimization and Inverse Problems in Electomagnetism, 71–78.

The optimization process of an electrical device is based on its sizing model, represented in Fig. 1.

(1)

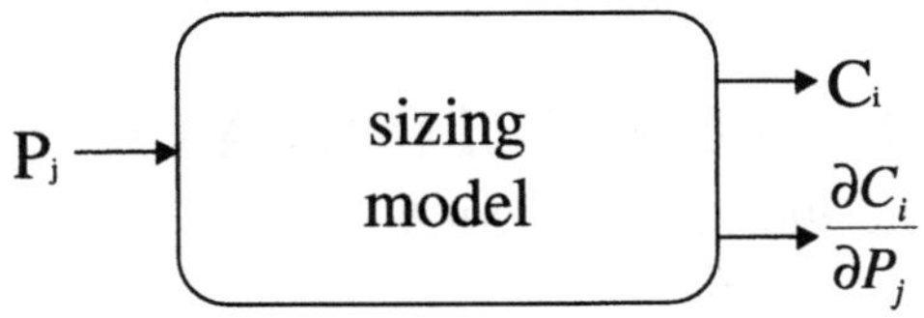

Fig. 1, the sizing model and its inputs and outputs

The output Ci of such a model are the values of the sizing criteria.. They are calculated from the inputs Pj of the sizing model, mainly the parameters of the device. These parameters are inductance, capacitance, frequency, etc. The sizing criteria may concern values of state variable at a specific date, extrema, r.m.s. or average values, etc. These criteria depend on the state variables X of the system to size, which satisfy a state equation :

$$\dot{X}(t) = A \cdot X(t) + B \cdot u(t)$$

where t represents time and u(t) the expression of the state input (e.g. : the sources of electrical circuits). A, B and u directly depend on the parameters of the circuits, $\lfloor P_j \rfloor_{j=[1..n]}$.

Furthermore, optimization gradient algorithms need the partial derivatives of the sizing criteria according to the sizing model inputs, i.e. the $\dfrac{\partial C_i}{\partial P_j}$.

These partial derivatives can be expressed with the partial derivatives of the states.

2. THE ODE SOLVING

As equation (1) is linear, its complete solution can be expressed as follows [3]:

$$X(t) = e^{At} \cdot X(0) + \int_0^t e^{(t-\tau)A} \cdot B \cdot u(\tau) \cdot d\tau \qquad (2)$$

In order to value expression (2), the exponential of the matrix A has to be defined. Several algorithms can be used to estimate it [4][5]. Some of these use the eigenvalues of A, but they are limited for our purposes. The selected algorithms do not use such methods. They are fast and they require few memory space.

2.1 The Taylor Series Development

The first selected algorithm is based on the Taylor Series Development of the exponential operator :

$$e^A = \sum_{n=0}^{+\infty} \frac{A^n}{n!} \tag{3}$$

Computing this infinite sum is evidently impossible. An upper bound must be chosen, depending on the desired calculus accuracy. To obtain the accuracy ε, the upper bound N must satisfy to the following criteria [4]:

$$0 < \left(\frac{\|A\|^{N+1}}{(N+1)!} \right) \left(\frac{1}{1 - \frac{\|A\|}{(N+2)}} \right) \leq \varepsilon \tag{4}$$

As long as the Frobenius norm of the matrix A is smaller than 1, this algorithm gives accurate results, but when the norm is greater than 1, the accuracy is lost. To compensate for that issue, scaling and squaring is used, based on the following relation :

$$e^A = \left(e^{\frac{A}{2^S}} \right)^{2^S} \tag{5}$$

S is chosen so that $\left\| \frac{A}{2^S} \right\| < 1$. Then the exponential of $\frac{A}{2^S}$ is computed, and finally the result is squared S times. This algorithm gives very accurate results for all kinds of matrixes.

2.2 The Padé Approximation

The (p,q) Padé approximation of the matrix exponential is :

$$R_{pq}(A) = \left[D_{pq}(A)\right]^{-1}\left[N_{pq}(A)\right] \tag{6}$$

where $N_{pq}(A) = \sum_{i=0}^{p} \frac{(p+q-i)!\,p!}{(p+q)!\,i!\,(p-i)!} A^i$

and $D_{pq}(A) = \sum_{j=0}^{q} \frac{(p+q-j)!\,q!}{(p+q)!\,j!\,(q-j)!} (-A)^j$.

This algorithm is very sensitive to the values taken for p and q. The diagonal approximants (e.g. $p=q$) give the more accurate results [4]. For this algorithm, the norm of the matrix A has to be small, so scaling and squaring is also used like for the Taylor Series.

2.3 The Dunford Decomposition

This method is based on the Dunford decomposition of the matrix exponential :

$$e^{A.t} = \sum_{i=1}^{k} \frac{e^{\lambda_i .t}}{P_A'(\lambda_i)} \cdot \prod_{j \neq i} (A - \lambda_j \cdot I) \tag{7}$$

with $P_A(\lambda) = \prod_{i=1}^{k} (\lambda - \lambda_i)$, where the λ_i are the eignevalues of A.

Knowing the eigenvalues of the system, an expression of the matrix exponential can be obtained :

$$e^{At} = Q(A) = \sum_{i=0}^{k-1} Q_i(t) \cdot A^i \tag{8}$$

Using this value of the matrix exponential in expression (2) gives :

$$X(t) = \sum_{i=0}^{k-1} Q_i(t).A^i.X(0) + \sum_{i=0}^{k-1} A^i.\int_0^t Q_i(t-\eta).B.u(\eta).d\eta \tag{9}$$

Without the knowledge of the symbolic eigenvalues, their partial derivatives can be obtained, as the characteristic polynomial depends on λ and on the parameters of the model, (the P_j). For every eigenvalue λ of A :

$$P_A(\lambda) = 0 \Rightarrow dP_A(\lambda) = 0 \tag{10}$$

Knowing that every P_j is independent of every other parameter $P_{i, i \neq j}$, we deduce for every input parameter of the device that :

$$\frac{\partial \lambda}{\partial P_i} = -\frac{\dfrac{\partial P_A}{\partial P_i}}{\dfrac{\partial P_A}{\partial \lambda}} \tag{11}$$

With this approach, obtaining the eigenvalues and their partial derivatives will be implicit and purely numeric. However, the ODE solving is explicit and symbolic, the eigenvalues of the state matrix being some elements of the input parameters. By composition of all the sub-expressions, $X(t)$ is differentiable according to every input parameter of the device.

2.4 Expression of the integral part

In order to value expression (2), the matrix exponential has to be estimated, and the integral part has to be symbolically expressed. As the sources encountered in electrical engineering are constant, polynomial or sinusoidal, the integral part will be expressed for only these kinds of sources. The complete solution of the state system is the sum of the solutions found for each single source.

The expression of (2) for a constant source $u(t)=U$ is :

$$X(t)=e^{At}X_0+\left[A^{-1}e^{At}-A^{-1}\right]BU \tag{12}$$

The expression of (2) for a polynomial source $u(t)=t^n$ is :

$$X(t)=e^{At}X_0+A^{-(n+1)}\left[n!e^{At}-\sum_{k=0}^{n}\frac{(At)^k n!}{k!}\right] \tag{13}$$

The expression of (2) for a sinusoidal source $u(t)=sin(\omega t+\varphi)$ is :

$$X(t)=e^{At}X_0+\left[A^2-\omega^2 I\right]^{-1}\begin{bmatrix} e^{At}\left[\omega cos(\varphi)+Asin(\varphi)\right] \\ -\left[\omega cos(\omega t+\varphi)+Asin(\omega t+\varphi)\right]\end{bmatrix}B \tag{14}$$

To sum the different sources, the matrix B is made of a column per source : $B=[B_i]_{i=[1..N_S]}$ where N_S is the number of sources.

The expression (2) becomes :

$$X(t)=e^{At}\cdot X(0)+\sum_{i=1}^{N_S}\int_0^t e^{A(t-\tau)}\cdot B_i\cdot u_i(\tau)d\tau \tag{15}$$

The integral part can be expressed with the knowledge of the type of every single source $u_i(t)$. With these expressions, the entire solution can be expressed symbolically. In the expression of the entire solution, the only numerical part is the valuation of e^{At}.

3. OBTAINING THE STATES DERIVATIVES

Obtaining the states partial derivatives symbolically is not as easy as it seems at first sight. The matrix product is not commutative, which implies that, in the Taylor Series optic, the formal derivative of the exponential operator must be expressed as a sum :

$$\frac{\partial\left(e^A\right)}{\partial x}=\frac{\partial\left(\sum_{n=0}^{+\infty}\frac{A^n}{n!}\right)}{\partial x}=\sum_{n=0}^{+\infty}\frac{\partial\left(\frac{A^n}{n!}\right)}{\partial x}=\sum_{n=0}^{+\infty}\frac{\frac{\partial\left(A^n\right)}{\partial x}}{n!} \tag{16}$$

The expression of $\frac{\partial\left(A^n\right)}{\partial x}$ can also be expressed as a sum :

$$\frac{\partial\left(A^n\right)}{\partial x}=\frac{\partial A}{\partial x}A^{n-1}+A\frac{\partial\left(A^{n-1}\right)}{\partial x}=\sum_{k=0}^{n-1}A^k\left(\frac{\partial A}{\partial x}\right)A^{n-k-1} \tag{17}$$

The symbolic expression of $\frac{\partial\left(e^A\right)}{\partial x}$ is :

$$\frac{\partial\left(e^A\right)}{\partial x}=\sum_{n=0}^{+\infty}\sum_{k=0}^{n-1}\frac{A^k\left(\frac{\partial A}{\partial x}\right)A^{n-k-1}}{n!} \tag{18}$$

The evaluation of this expression is easily done with a computer, but is accurate only when the norm of the matrixes A and $\frac{\partial A}{\partial x}$ are small. The problem comes from the fact that the scaling and squaring method cannot be used to compute the derivative, because both matrixes A and $\frac{\partial A}{\partial x}$ should be scaled in order to obtain a good accuracy. Scaling both matrixes is not possible, due to the fact that no relationship can be expressed between $\frac{\partial\left(e^{\frac{A}{m}}\right)}{\partial x}$ and $\frac{\partial\left(\frac{A}{m}\right)}{\partial x}$.

The derivative of the matrix exponential operator must be computed with another different method.

3.1 Numerical approach

The derivative of the matrix exponential operator can be obtained with the finite differences method, based on the following relationship :

$$\frac{\partial\left(e^A\right)}{\partial x}=\lim_{h\to 0}\left(e^{A+h\frac{\partial A}{\partial x}}-e^A\right)\frac{1}{h} \tag{19}$$

Obtaining $\frac{\partial A}{\partial x}$ is not difficult, either with a symbolic derivation tool, or with a finite difference approach as well. The issue of this method is the choice of the differentiation step h. If the step is too large, the derivative has no more sense, as h must tend to 0. If it is too small, numerical problems appear because the precision is not sufficient for the computation of $e^{A+h\frac{\partial A}{\partial x}}-e^A$.

3.2 Symbolic approach

Another possible approach for the computation of the derivatives of the matrix exponential operator is the symbolic recombination of the matrixes involved in the computation. This approach is actually studied. Once these derivatives are known, the states partial derivatives are easily computed.

4. APPLICATION EXAMPLES

The results obtained with a static converter plugged on a RSIL (fig. 2) are compared with numerical simulations, in terms of accuracy, computation time and memory occupation (fig. 3).

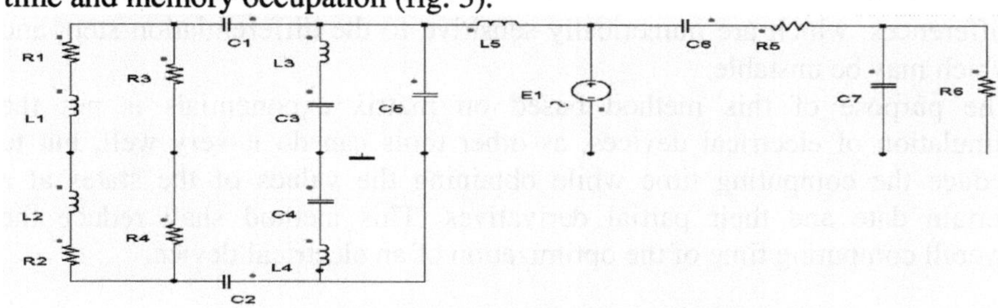

Fig. 2, the static converter with the RSIL

The results are :

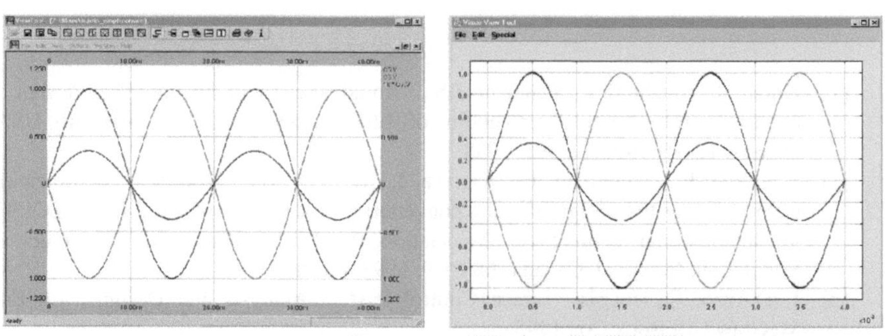

Fig. 3, results with Simplorer® (left) and with the matrix exponential (right)

The comparison is done for several frequencies of the voltage source, to get a good representation of the harmonic spectrum of the operating mode.

If the whole simulation takes more computation time with the matrix exponential approach, the time gain lies in the fact that each point of the simulation can be obtained without the knowledge of the others, whereas the numerical simulation needs the knowledge of the past to obtain the next point. As a consequence, memory occupation is far lesser with the matrix exponential method than with the numerical simulation. The accuracy is equivalent with both methods. For the Dunford approach, results are discussed in paper [8].

5. CONCLUSION

In this paper we have presented a method to solve the ordinary differential equations involved in the models of electrical devices (and more specifically electrical circuits). This method obtains directly the values of the states at a certain date and their partial derivatives for optimization purposes, to the contrary of numerical methods which must start from the origin and integrate the ODE step by step until they reach the desired date. To obtain the partial derivatives of the states with the finite differences, another simulation must be executed.

The presented method requires less computing time and less memory occupation. The states are obtained with a very good accuracy, as well as the partial derivatives, to the contrary of the derivatives obtained with finite differences, which are numerically sensitive to the differentiation step, and which may be unstable.

The purpose of this method based on matrix exponentials is not the simulation of electrical devices, as other tools can do it very well, but to reduce the computing time while obtaining the values of the states at a certain date and their partial derivatives. This method shall reduce the overall computing time of the optimization of an electrical device.

REFERENCES

[1] Kragh H., Blaabjerg F., Pedersen J.K., "An advanced tool for optimised design of power electronic circuits", proceeding of IEEE-IAS'98, Saint Louis, Missouri, USA, October 12-15, 1998, pp 991 – 998

[2] Viarouge P., Tourkhani F., Kamwa I., Le-Huy H., "Nonlinear optimization techniques for the design of static converters", proceedings of the IMACS-TC1'93, 4th international conference, Association for Mathematics and Computers in Simulation, Montreal, Canada, July 7-9, 1993, pp 543 – 547

[3] J. D'Azzo, C. Houpis, "Linear Control System Analysis and Design", 4th Ed., McGraw Hill Book Co., 1995

[4] C. Moler, C. Van Loan, "Nineteen dubious ways to compute the exponential of a matrix", SIAM Review 1978, Vol. 20 No.4, pp 801 – 836

[5] W. Harris Jr, J. Fillmore, D. Smith, "Matrix Exponentials – Another Approach", SIAM Review 2001, Vol. 43 No 4, pp 694 – 706

[6] I. Najfeld, T.F. Havel, "Derivatives of the matrix exponential and their computation" Advances in Applied mathematics, Vol. 16 No 3, September 1995, pp. 321 – 375 (doi:10.1006/aama.1995.1017)

[7] Dunford, Schwartz, "Linear operators, part I, General theory", John Wiley and Son, ISBN 0.470.22605.6

[8] L. Gerbaud, M. Barkatou, G. Villard, A. Bolopion, "Résolution d'équation différentielles en vue de l'étude de circuits d'électronique de puissance", RIGE Vol 4, N° 1-2/2001, pp 105-120, ISBN 2-7462-0233-6

USE OF DISCRETIZATION AND SOLUTION HISTORY IN STOCHASTIC OPTIMIZATION

J. Lähteenmäki
Helsinki University of Technology, Laboratory of Electromechanics, P.O. Box 3000, FIN-02015 HUT, Finland

Abstract: A practical improvement for stochastic algorithms used with numerical models is proposed. Combined use of solution space discretization and solution history is implemented. Discretization gives the possibility to use a solution history where all the solutions evaluated are stored. Use of the solution history guarantees that a single solution candidate is calculated only once. This approach is useful for the stochastic algorithms that typically evaluate many solution candidates. The approach is most useful in the cases where a stochastic algorithm decreases the search space during the optimization process. The improvement proposed was tested in optimization of high-speed induction motors modeled with 2D finite element analysis software. With a genetic optimization algorithm, the average time saving for seven separate optimization runs was 39%.

Key words: optimization, search space discretization, solution history

1. OPTIMIZATION OF ELECTRIC MOTORS MODELED WITH FEM

Electric motors are often modeled with finite element models (FEM). Properly used, these 2D or 3D models give the most accurate results especially in more demanding modeling cases. Theses cases could include difficult or novel topologies, non-linearities and different transient phenomena. In the research relating to this work, a 2D time-stepping FEM model [1] has been used to calculate the characteristics of high-speed electric induction motors.

M. Rudnicki and S. Wiak (eds.), Optimization and Inverse Problems in Electomagnetism, 79–84.

Heavy models like FEM take a considerable amount of time and efficient computers to solve. In optimization, the calculation of an objective function means that this kind of model has to be solved. Thus, each function evaluation can take several or several tens of minutes. In order to have a feasible calculation time for an optimization run, the number of function evaluations should be kept as low as possible.

The FEM models are not very informative as far as the needs of deterministic optimization algorithms are considered. They do not have the analytical presentation and first and second order derivatives of the objective function, which are often used by the optimization algorithms. Finite differences can be used instead of derivatives, but that may lead to case and variable specific tuning of algorithm specific parameters. Also, a numerical model can induce noise into the objective function value and the use of finite differences becomes more difficult [2].

2. STOCHASTIC ALGORITHMS IN OPTIMIZATION OF ELECTRIC MOTORS

Using a stochastic algorithm like a genetic algorithm (GA) is then justified for the following reasons:

- The GA does not need the derivatives of an objective function. The calculation of several solutions in order to determine the finite differences is avoided. Also, the problems associated with the finite differences are avoided.
- The objective function may have several local optima. As the algorithms based on a calculation of gradients or a Hessian matrix often converge to a local minimum, the global solution may not be found. The GA searches for the solution from the whole solution space determined by the limits for the variables and other constraints. Thus, it is more likely that the global optimum is found.
- When designing new kinds of electric motors like high-speed motors, the optimal designs may be quite different from the conventional designs. The GA is not sensitive to the initial point of the design, which is often the case with the deterministic algorithms. The GA is more like an intelligent stochastic algorithm, combining the 'freedom of prejudice' of a stochastic algorithm with the ability of an deterministic algorithm to 'concentrate on the relevant designs'.

Different stochastic algorithms are studied in optimization of electric motors. GAs have been studied and used in [2–5]. A fuzzy algorithm was used in [6] and a simulated annealing method was compared with other algorithms in [7].

Solving heavy models like FEM can take most of the time used in optimization. In this work, the FEM model uses roughly 95 % of the computation time in optimization. Hence, the lower number of designs calculated the better. Unfortunately, this is contradictory to the nature of the stochastic optimization algorithms. In order to decrease the optimization times and make more use of the time spent, a practical improvement is suggested. This improvement is applicable to any algorithm but it is most efficient when used with a stochastic optimization algorithm running a heavy model.

3. COMBINATION OF DISCRETE SEARCH SPACE AND SOLUTION HISTORY

The idea of the discretization of the search space is to avoid evaluation of designs that are practically the same. For example, an induction motor with an air gap of 2.0 mm is no different from a motor with an air gap of 2.001 mm. The discretization should be selected so that the smallest difference between two designs is noticeable and realizable. In the above example, the machine produced would probably have the air gap somewhere near 2.0 ± 0.05 mm. Higher tolerances would be possible but not necessarily cost-effective.

For optimization algorithms using heavy models, the discretization is not a problem since the gradient and Hessian information of the objective function is unavailable anyway. Finite difference approximations could still be used. A proper discretization could actually improve the use of finite differences because of the possible noise inherent in the models. However, the relation between discretization or relative perturbations and the accuracy of finite differences is a complicated issue [2] and avoided by using stochastic algorithms.

Another way to decrease the number of function evaluations is to make sure that the same design is not evaluated more than once. The stochastic algorithms have no deterministic path to follow and the probability of calculating the same point several times in a discretized solution space exists. This phenomenon can be significant when using algorithms that concentrate their search on a portion of the solution space only. For example, the simulated annealing algorithm and GA used in this work does so to some extent.

In order to avoid repeated evaluation of the same design, a search or solution history can be used. The history data includes the free variables and the corresponding objective function value, but it can also include more information on the designs calculated. Every design evaluated is stored in the memory or in a file. The new designs to be evaluated are compared with the designs in the history data. If the design is already evaluated, the value of the objective function is returned and the time consuming calculation with a heavy model is avoided. Even with a large history data, the time spent in the comparison of designs is a fraction of the time used in heavy modeling. Fig. 1 shows the flow chart of the optimization process with a discretized solution history (DSH) method.

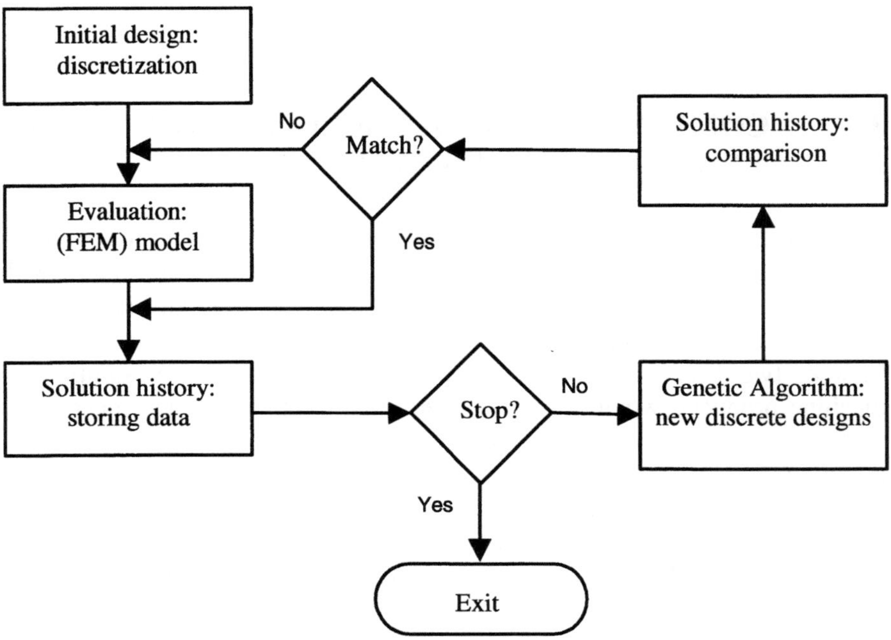

Figure 1. Flow chart of optimization with discretized solution history (DSH) method.

4. TEST RESULTS FOR THE DSH METHOD

To demonstrate the effect of using DSH, optimization run data for seven separate optimization cases are presented in Table 1. The results in the table show that the optimization time is reduced by 29 – 47 %, depending on the individual run. Summing up all the evaluated and calculated designs respectively, the average time saving is 39 %. The calculated designs cover the solution space more efficiently because the designs have noticeable differences between them. Thus, not only is time saved but the results contain more information for post-processing.

Table 1. Effect of DSH in decreasing the computation times, using GA with FEM model.

Optimization case	Free variables	Generations in genetic algorithm	Designs evaluated	Designs calculated with FEM	Optimization time [h]	Time saved [%]
1	2	8	323	181	46	44
2	6	14	593	412	101	31
3	8	44	1970	1049	153	47
4	7	33	1488	928	203	38
5	3	11	474	334	119	30
6	2	8	315	173	34	45
7	4	13	550	389	175	29

The time used in optimization relative to the amount of designs calculated is different from case to case. One reason for this is that different cases have different finite element meshes and mesh sizes to start with. The size of the mesh also depends on the values of free variables in some cases. Another reason is that in some cases the solution space of the free variables was not subjected to non-linear constraints. These constrains would have neglected the possibility of impossible combinations for values of the variables. Only linear constraints could be used with the present GA software. An improper value combination is detected by the FEM mesh generator and a penalty function value is returned for that design. Thus the design is considered evaluated but no time is consumed since no actual FEM calculations are made.

5. OTHER USES FOR THE DSH

Another use for the DSH is the opportunity to include more information on the calculated designs. In the case of an induction motor, different loss components, power factor, supply voltage and current, slip, torque and temperature rise data could be stored. Then, the solution history can be used for post-processing of the optimization results. Different objective functions can be valued simply by taking the data needed for the evaluation from the history data. Thus, the stored data can be used not only for a sensitivity analysis as a function of designs, but also as a function of the objective function, or indeed both. In addition, statistical analysis can be used to find relationships between designs and their characteristics in much the same way as with measured product data [8].

The DSH can be also used as a database for 'surface fitting' where the behaviour of the solved points is approximated by functions that are faster to solve. An example of this is the use of linear and cubic spline functions as a replacement for a FEM model as illustrated in [9]. One potential use for the solution data would be the teaching of a neural network system.

6. CONCLUSION

The use of design variable discretization together with solution history were proposed as a new improvement for the optimization algorithms used with numerical modeling. The combined use of discretization and solution history can significantly decrease the computation time and makes more informative post processing of the results possible. This improvement is applicable to any algorithm but it is most efficient when used with stochastic optimization algorithms like genetic algorithms, fuzzy logic or simulated annealing. There, the reduction of calculation time is likely to be most significant. The use of solution history in surface fitting is also utilized best with stochastic algorithms, since the fitting requires large amounts of data in order to give a good estimation of the real behaviour of the objective function.

REFERENCES

1. Arkkio A. 1990. Finite element analysis of cage induction motors fed by static frequency converters. IEEE Trans. on Magnetics, MAG-26. pp. 551–554
2. Palko S. 1996. Structural Optimization of an Induction Motor using a Genetic Algorithm and a Finite Element Method. Acta Polytechnica Scandinavica, Electrical Engineering Series No. 84. Helsinki, Finland. 99 p..
3. Bianchi N. and Bolognani. S. 1998. Design Optimisation of Electric Motors by Genetic Algorithms. IEE Proceedings of Electric Power Applications. Vol. 145. No. 5. September. pp. 475 – 483.
4. Lampola P. 2000. Directly Driven, Low-Speed Permanent-Magnet Generators for Wind Power Applications. Acta Polytechnica Scandinavica, Electrical Engineering Series No. 101. Espoo, Finland. 62 p (+106p.)
5. Lähteenmäki J. 1997. Optimisation of High-Speed Motors Using a Genetic Algorithm. IEE Conference on Electrical Machines and Drives EMD '97. Proceedings, August 12 – 16, Cambridge, UK. pp. 26 – 30
6. Bianchi N., Bolognani S. and Zigliotto M. 1996. Fuzzy Knowledge Based Design of PM Synchonous Motors. International Conference on Electrical Machines ICEM '96. Conference Proceedings. Vigo, Spain. Vol. 2. pp. 365 – 369
7. Toma C. F. and Lähteenmäki J. K. 2000. Comparison of optimization algorithms used with finite element analysis for designing high-speed induction motors. International Conference on Electrical Machines ICEM 2000, Proceedings. Vol 2. August 28 – 30. Espoo, Finland. pp. 642 – 644
8. Kytömäki R., Arkkio A., Niemenmaa A. and Suontausta A. 1997. Validity of Conventional and Modern Methods of No-Load Loss Calculation in Asynchronous Machines – Statistic approach. IEE Conference on Electrical Machines and Drives EMD '97. Proceedings, August 12 – 16, Cambridge, UK. pp. 185 – 189
9. Bianchi N. and Bolognani S. 1999. Design Optimisation of Electrical Motors by an Adaptive Model Based Optimisation Technique. Ninth International Conference on Electrical Machines and Drives. Conference Publication No. 468, IEE. pp. 291 – 296

A GENETIC ALGORITHM METHOD FOR DETERMINING THE MAXIMUM TOUCH VOLTAGE GENERATED BY A GROUNDING SYSTEM

Marcello Sylos Labini, Giuseppe Delvecchio and Ferrante Neri
Department of Electrotechnics and Electronics - Polytechnic of Bari - Via E. Orabona 4 - 70125 Bari (Italy) - E-mail: sylos@poliba.it

Abstract: The Authors suggest a new method for quickly determining the maximum touch voltage generated by a grounding system leaking a known current. The touch voltages in the points of the soil surface are calculated by the Maxwell's subareas method and the search for the maximum touch voltage is carried out by an ad hoc genetic program which is based both on the "one-point crossover" technique and on a mutation having a pre-arranged probability rate. The method has been successfully tested on some grounding systems.

Key words: grounding grid, Maxwell's subareas method, genetic algorithm

1. INTRODUCTION

As is well-known, when a grounding system leaks a fault current I_F, there are touch voltages $U_T(P)$ in all the points P of the soil surface area involved by the grounding system. These voltages can be dangerous. It is therefore very important to determine, by measurements and/or calculation, the touch voltages in the various points of the soil surface and, consequently, the maximum touch voltage $U_{T\,max}$.

A good method for calculating the touch voltages $U_T(P)$ is the Maxwell's subareas method [1, 2]. Figs. 1 and 2 show two examples of the voltages $U_T(P)$ trend, this trend having been calculated by this method in well-defined areas of the soil surface; the voltages are generated by a square grounding

M. Rudnicki and S. Wiak (eds.), Optimization and Inverse Problems in Electomagnetism, 85–92.

electrode whose sides are 50 m long and a rectangular meshed grounding grid whose size is 60 m × 40 m, respectively. With reference to Fig. 1, the $U_T(P)$ function shows a maximum in the middle of the figure and then, as we move from the center to the sides of the grounding electrode, $U_T(P)$ gradually decreases. Moreover, as we move away from these sides, $U_T(P)$ increases again and, close to the vertices of the soil surface area under examination, it takes new maxima which are higher than the previous one.

The manual search for $U_{T\,max}$ is hard, and so it is better to resort to the optimization methods. As regards these methods, it is well-known that the *hill-climb* methods, which are based on the gradient of the function under examination, are efficient if this function has only one maximum, but if the function has several maxima, like that which gives us the touch voltages, the hill-climb methods could find one of the local maxima instead of the absolute maximum. For example, the search for $U_{T\,max}$ in the soil surface area of Fig. 1, could lead us to find the maximum at the center of this area instead of finding the true maxima existing in the vertices of the same area. To get rid of the above-mentioned difficulty, the Authors have recently defined a stochastic-deterministic optimization method to automatically determine the true maximum touch voltage in the area involved by the grounding system [3]. This method turns out to be good, but rather long. On the contrary, the method suggested in this paper, since it is based on the genetic algorithms, is much faster than the previous one.

2. THE GENETIC ALGORITHM METHOD FOR DETERMINING $U_{T\,MAX}$

Briefly, the well-known Maxwell's subareas method consists in: subdividing the grounding grid into a number N of subareas; calculating the earthing voltage U_E taken by the grounding grid and calculating the subcurrents leaked by each subarea. These subcurrents allow us to determine, in each point P of the soil surface, the voltage U(P) as the sum of the voltages generated in P by each subarea and by its electrical image. Consequently, it is possible to calculate the touch voltage in P by the formula:

$$U_T(P) = U_E - U(P) \tag{1}$$

According to the above, a lot of calculation time is required for determining the trend of the touch voltages $U_T(P)$ relating to all the points of the soil surface area under examination. On the contrary, the genetic algorithm method carried out by the Authors allows to determine the maximum touch voltage $U_{T\,max}$ without calculating the touch voltages $U_T(P)$ in a very great number of points, as previously done in Figures 1 and 2.

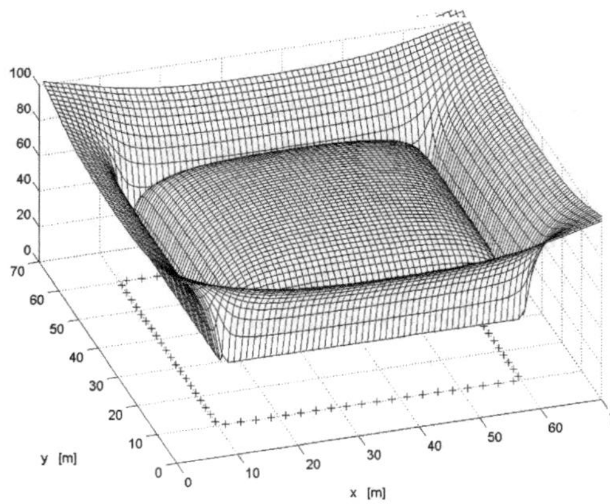

Figure 1. Touch voltage trend for a square grounding electrode

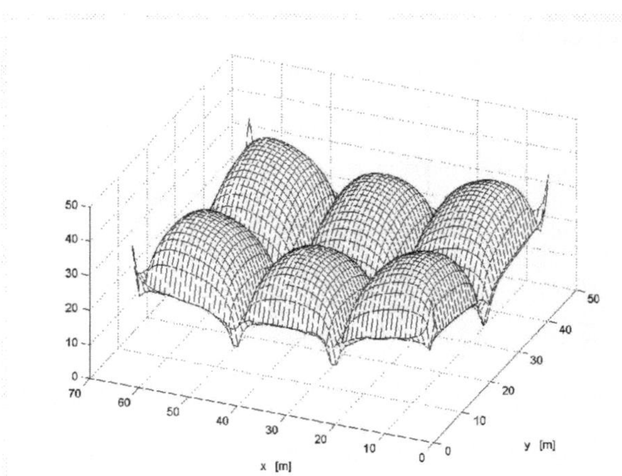

Figure 2. Touch voltage trend for a meshed grounding electrode

The optimization method consists of the following [4, 5]. A first sampling of the function $U_T(P)$ is done at random. This sampling concerns a small number N_p of points P of the soil surface. These points represent the population to which we apply the genetic algorithm method. Each point P(x, y), in terms of genetic algorithms, is an individual having x and y as chromosomes. The x and y co-ordinates of each point P are expressed in binary numbers. The binary numbers are considered to be string variables, so they undergo cross-over and mutation. The Authors have chosen the "one-point crossover" technique and established that each individual thus generated has a 4% probability of undergoing a mutation. As regards cross-over, the probability that cross-over will occur is equal to 100%. In other terms, in each iteration, N cross-over always occur, where N is a number chosen by us a priori. Therefore, the choice of the pairs of chromosomes which undergo cross-over occurs at random N times each iteration. This way of choosing the chromosomes implies the possibility that a chromosome will generate several times during a same iteration.

The number of individuals of the population that, at the end of each iteration, must survive is fixed a priori. For example, if it is arranged that T individuals must survive, the T individuals having a higher fitness are saved in a new memory location. In the case under examination the fitness relating to each individual or point P corresponds to the value taken by the touch voltage function $U_T(P)$ in the point P.

The rate at which the population improves is measured by calculating, in each *i*-th iteration, the difference ε_i between the maximum touch voltage obtained in the *i*-th iteration and the average value between all the touch voltages calculated in the same iteration. As an example, Fig. 3 gives the curve of the maximum touch voltage and that of the average value for the above square electrode.

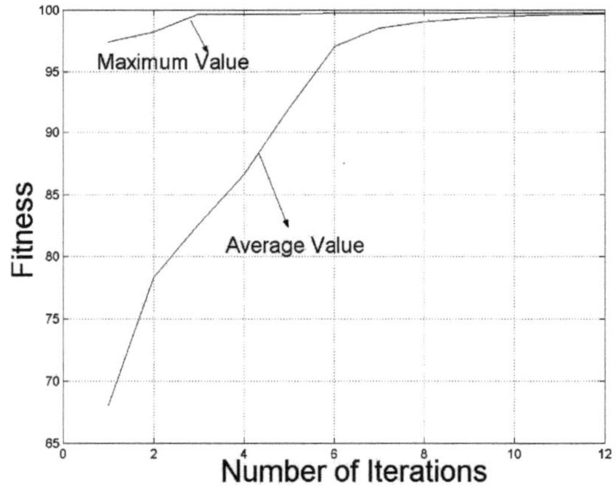

Figure 3. Maximum and average touch voltage curves vs. number of iterations

The algorithm is stopped when at least one of the two following conditions occurs: the value ε_i is smaller than a pre-arranged value ε (where ε represents an index of the accuracy we want to have in the result); the number of iterations N_{iter} reaches a pre-arranged value $N_{iter\,max}$. By this criterion it is possible to determine the processing instant in which it is superfluous to go on having the program run since the population has reached such an evolution level that further iterations would involve an insignificant advantage in terms of accuracy.

3. NUMERICAL RESULTS

We have studied a 50 m × 50 m square grounding electrode and a 60 m × 40 m rectangular meshed grounding grid having 6 meshes. Both the grids leak $I_F = 100$ A, consist of cylindrical conductors having a section $S = 50\ mm^2$ and are buried at a depth $h = 0.5$ m in a homogeneous soil having resistivity $\rho = 100\ \Omega m$. As regards the soil surface areas under study where we want to find the maximum touch voltage, we have considered the following areas:

- in respect of the square electrode, a square area whose sides are 70 m × 70 m (see Fig. 1);
- in respect of the meshed grid, a rectangular area whose sides are 61 m × 41 m (see Fig. 2).

Table 1. A square grid in a 70 m × 70 m area

x [m]	y [m]	$U_{T\,max}$ [V]	N_{iter}	time [s]
0	**70**	**99.7498**	**9**	**14.94**
0	**0**	**99.7498**	**12**	**11.25**
0	0	99.7498	11	13.89
70	**70**	**99.7498**	**10**	**15.16**
0	70	99.7498	11	13.51
70	**0**	**99.7498**	**10**	**14.12**

Table 2. A meshed grid in a 61 m × 41 m area

x [m]	y [m]	$U_{T\,max}$ [V]	N_{iter}	time [s]
8.02	**8.31**	**42.6592**	**18**	**12.41**
8.03	8.27	42.6591	13	8.84
8.02	**32.70**	**42.6592**	**24**	**16.75**
53.00	**8.33**	**42.6592**	**13**	**8.84**
52.98	**32.69**	**42.6592**	**15**	**12.46**
8.02	8.32	42.6592	16	13.07

As for the number of initial points to take into account (that is, the number of individuals constituting the population) $N_p = 100$ points has been chosen. As regards the number of cross-over to carry out in each iteration, $N = 100$ cross-over has been chosen. Moreover, a 4% probability of mutation occurring has been set. The process is stopped when the difference ε_i is smaller than $\varepsilon = 0.1$ V or when $N_{iter} = N_{iter\,max} = 30$.

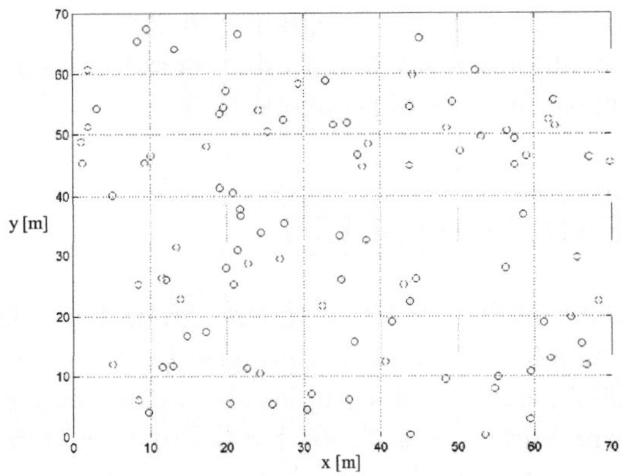

Figure 4. Initial population

The results got by an Intel CELERON processor having a clock frequency equal to 1000 MHz, relating to the two cases, are given in the Tables 1 and 2, respectively.

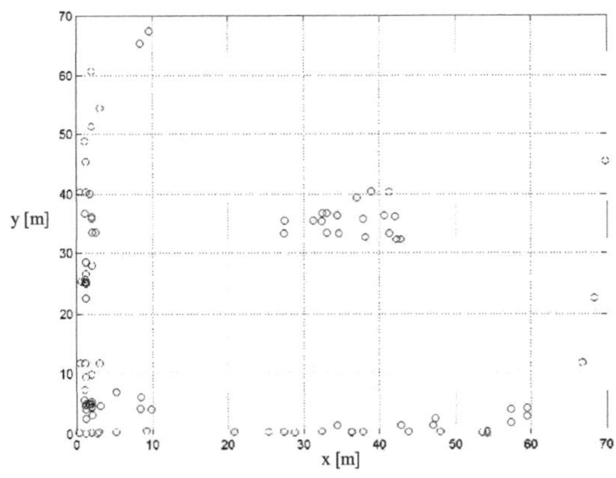

Figure 5. Population after three iterations

As regards the square grounding electrode of Fig. 1, Table 1 gives us the results got by having the optimization program run several times. As you can see, the maximum value of the touch voltage has always been found in the four vertices of the area under examination. This result is entirely in agreement with what is given in literature and the $U_T(P)$ trend shown in Fig. 1, this trend having been found by calculating $U_T(P)$ in a very great number of points.

Figures 4 to 7 give some phases of the search for $U_{T\,max} = U_T(0, 0)$.

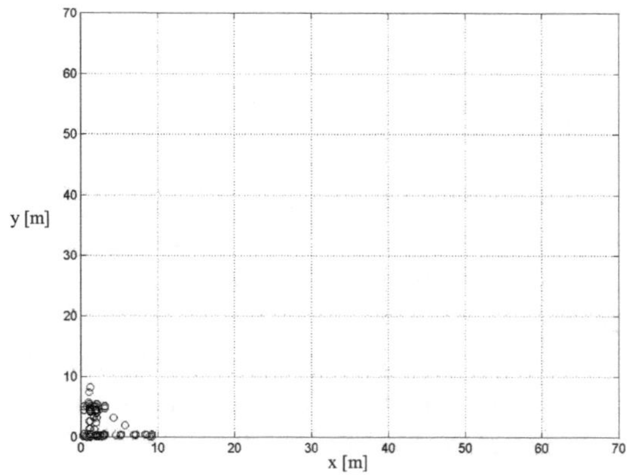

Figure 6. Population after eight iterations

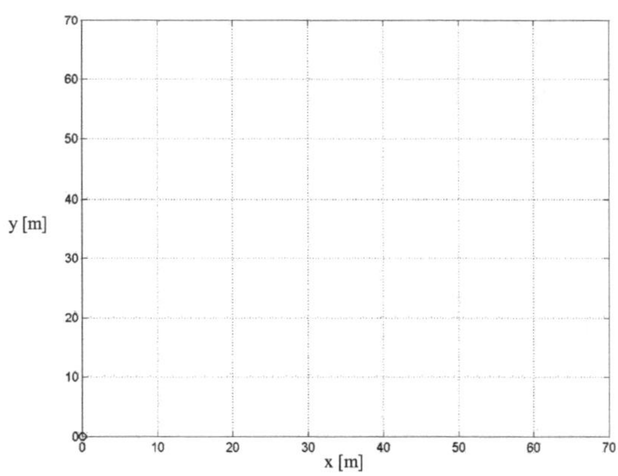

Figure 7. Final population: $U_{T\,max} = U_T(0, 0)$

As you can notice, at the beginning of the search process, there is a gathering of likely maximum points both round the center of the soil surface area and round the edges of this area, while at the end the optimization program "focuses" only on the search for the maximum round the point P(0, 0) and rejects all the probable local maximum points. In Fig. 7 all the individuals (or points) are in P(0, 0).

As regards the meshed grid, Fig. 2 shows us how $U_{T\,max}$ is found in the corner-meshes. This is confirmed by the results of the optimization program given in Table 2. For, the maxima, which are practically all identical ($U_T = 42.65V$), are in the four corner-meshes.

4. CONCLUSIONS

The Authors have dealt with the problem relating to the determination of the maximum touch voltage which can be found in the soil surface points put near the leaking grounding system. This problem is heavy and expensive if measurements must be taken on the spot, but it is also heavy if it must be solved analytically. As regards the analytical solution, the Authors have defined an ad hoc genetic program. This program has turned out to be reliable and much faster compared to the previous stochastic-deterministic optimization method defined by the same Authors. For, as regards reliability, all the simulations carried out on each grounding system have given good results. As for speed, the genetic optimization method leads to the final result through an extremely low number of iterations, and so through low calculation times.

The Authors think that the calculation method proposed is very useful, since it allows to previously calculate the maximum touch voltage, and so allows to avoid the measurements on the spot.

REFERENCES

1. A. Covitti, S. De Nisi, F. Laddomada, M. Sylos Labini: A fuzzy-Maxwell combined method for simplifying the calculation of the current field, IEEE Transactions on Magnetics, Vol. 36, n. 4, pp. 708-711, July 2000
2. M. Sylos Labini, A. Covitti, G. Delvecchio, C. Marzano: A Study for Optimizing the Number of Subareas in the Maxwell's Method, Proceedings on Tenth biennal IEEE Conference on Electromagnetic Field Computation, CEFC 2002, Perugia (Italy), p 320, June 16-19 2002
3. A. Covitti, G. Delvecchio, M. Sylos Labini, D. Verde: A global optimization method for determining the maximum touch voltage generated by grounding systems, Studies in Applied Electromagnetics and Mechanics,Vol 22, IOS Press, pp 373-378, 2002
4. M.Mitchell: Introduction to Genetic Algorithms, Boston (USA), MIT press 1996
5. M. Rudnicki: Evolutionary and genetic tools in optimization, Jyvaskyla 10th Summer School, Lecture Notes, August 2000

OPTIMIZATION TECHNIQUES BENEFITS FOR MAGNETIC COMPONENT DESIGN

David Magot, Frederic Wurtz, Bruno Cogitore, Jean-Pierre Keradec

Laboratoire d'Electrotechnique de Grenoble, INPG/UJF CNRS, ENSIEG BP 46
38402 St Martin d'Hères cedex, France

Microspire S.A, 16 Parc d'activités du Beau Vallon, 57970 Illange, France

Abstract: Benefits of optimization techniques applied to magnetic component design are discussed here, using a comparison between an industrial hand-based design process and a methodology based on constrained optimization techniques. The results show dramatic design time decrease, as well as design process improvements in terms of reliability and formalism, which has to be balanced against encountered difficulties.

Key words: Design automation, constrained optimization, magnetic component

1. INTRODUCTION

A new design methodology has been developed, based on the idea that merging in a friendly way a device model, a specification sheet and an optimization algorithm would increase design process efficiency. This has led to several successive pieces of software developed at our lab [1-2], based on constrained optimization algorithms. The purpose of this paper is therefore to show and analyze the benefits provided by such a methodology, applied to magnetic component industrial design. The results presented here are coming from a comparison between this

M. Rudnicki and S. Wiak (eds.), Optimization and Inverse Problems in Electomagnetism, 93–100.

methodology, based on constrained optimization, and a hand-based design process. Both have been applied to the design of the same test device, in order to show practically what can be achieved for industrial magnetic components. It is a current measurement transformer (Fig. 1), which is believed to be representative enough of the type of device we are currently dealing with, because of the problems its design involves. Among them, the device complexity, strong tolerance issues coming from a specified measurement precision, as well as some parameters evaluations like iron losses can be found.

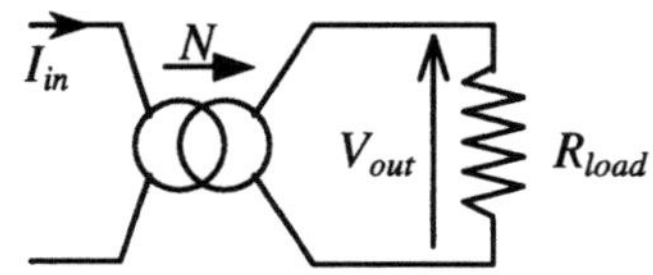

Figure 1. Current measurement transformer principle

2. DESIGN PROCESS

2.1 Sizing step

Talking about magnetic component design, what is directly concerned by the use of constrained optimization is the way the sizing step is done (Fig. 2). Hence, all the other steps, which are related to technological choices not yet supported by the methodology involved, will not be discussed here. The same model complexity has been managed for both design processes, to produce our results. In this respect, the full model required to handle the chosen test device, as currently used by Microspire staff, is made of equations mainly, as well as some algorithms. It involves about 30 input parameters to be set and 10 output parameters to be checked. However, only 10 input parameters are to be investigated, the remaining ones being fixed during the sizing step.

2.2 Hand-based sizing

The hand-based sizing is mainly a try-and-error type process (Fig. 3). As currently performed by Microspire staff, it uses transformer equivalent circuits and heavily relies on the designer experience [4]. Although the designer makes use of a piece of software to ease the parameters computation, this process is fully hand-based. Indeed, all choices are to be made by him at each step: to select an output parameter as a target, then an input parameter to modify as well as the nature of this modification, and eventually to check the result brought by this change. Difficulties are often rising, as an output parameter can depend on up to nine input parameters in the model involved. Moreover, strong interdependencies between parameters are often observed. This can thus easily become a brainstorming activity. Hence, designer experience helps to boost the convergence of the sizing process toward a solution.

Moreover, this process can hardly be well formalized. On the one hand, its starting point can dramatically change with the specification sheet provided. On the other hand, at a given iteration, a couple of input/output parameters can successively interchange themselves as a matter of value checking and refinement. As an example, the gain G of a current measurement transformer can roughly be given as (Fig. 1):

$$G = \frac{V_{out}}{I_{in}} = \frac{R_{load}}{N} \quad (1)$$

This gain being fixed by the specification sheet, the device has to be sized using R_{load} and N. Assuming a previous iteration has given a value for N, the designer deduces the value of R_{load} from equation (1). He has then to check this value, to be sure it is compatible with the specifications, and that it will not prevent the device to work properly. If this value is not correct in this respect, the designer would then fix a new value for R_{load} based on the last one, hence fixing a new value for N using equation (1), and so on.

Finally, three different models are successively processed to produce a suitable solution (Fig. 3), each being used as a starting point for the computation of its next, more complex child. This is required to deal with implicit parameters and parameter interdependency. At each stage, several iterations are required to get a satisfactory result.

2.3 Constrained optimization based sizing

The optimization-based sizing is an iterative process, which involves three steps (Fig. 4). The software developed at our lab offers facilities to help designers to perform all these, plus the model creation in a given format to allow its use in several software environments [2-3]. The model corresponding to the device to size is first loaded in an optimization module (Fig. 5). At this stage, the problem setting takes place. Here, the input parameters can either be fixed or constrained by interval, as well as the output parameters. Those last ones can also be let free, or defined to be minimized. Those configuration possibilities have shown to be sufficient for most problems. Then, an optimization algorithm is loaded and configured, before being launched. Since the model we had to deal with contained non-linear equations, non-linear constrained optimization algorithms had to be used [5-6]. The results can then be analyzed using a friendly graphic user interface. This enables the designer to diagnose some common issues, like incompatible constraints on different parameters, preventing a solution to be found for example.

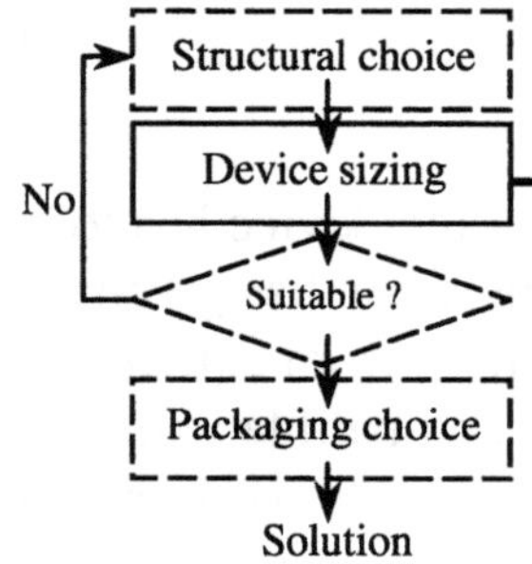

Figure 2. Magnetic component design

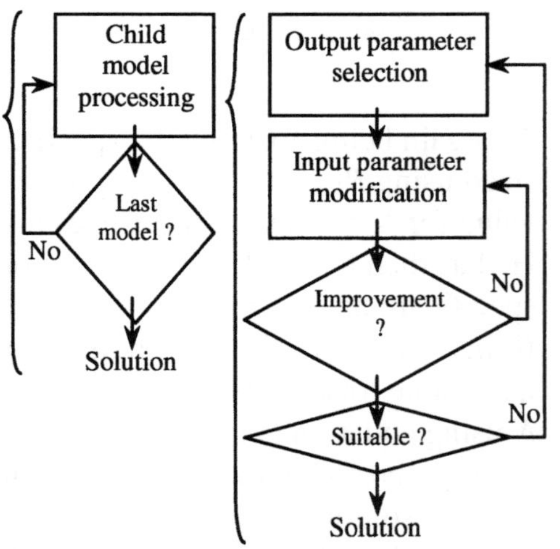

Figure 3. Hand-based sizing process

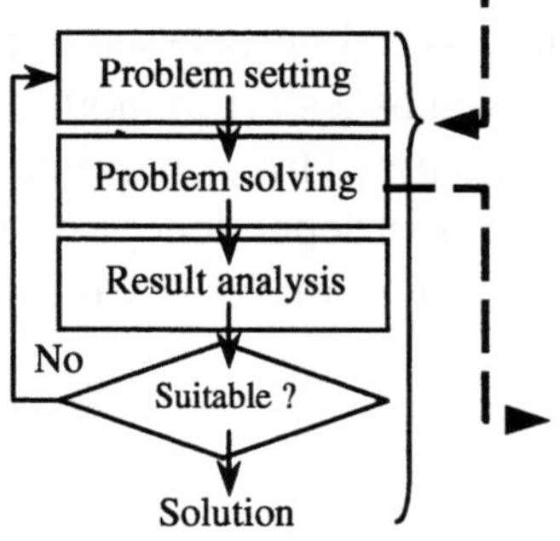

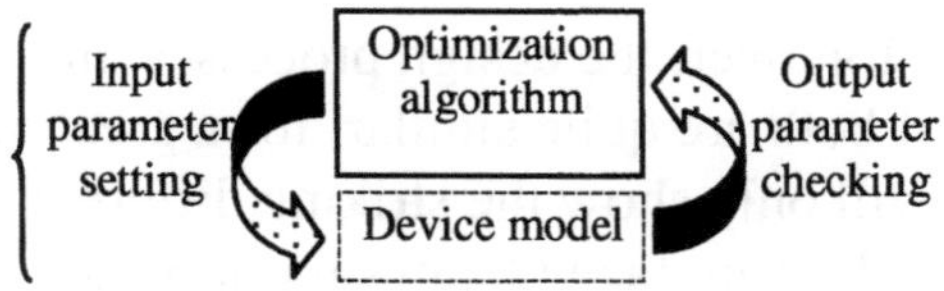

Figure 4. Optimization-based sizing process

Figure 5. Problem solving

3. RESULTS COMPARISON

3.1 Specifications

The current measurement transformer specifications, used as a test device to produce the results presented below, have been taken from an industrial specification sheet provided to Microspire by a customer. Its characteristics are:

- Input: 13A max, sinusoidal, 360-440Hz
- Output: 4.5V max
- Measurement precision: ±5%
- Core: ring shape
- Outer diameter: 30mm max
- Primary coil: 1 turn, external wire
- Secondary coil: 70% max of allowable winding surface (To save space for a third coil, used to test the transformer)
- Temperature: -40°C min, +75°C max
- Volume to be minimized

Microspire staff has done the hand-based design. The constrained optimization based sizing step has been performed on a single standard PC (Pentium 1GHz, 512Mo RAM).

3.2 Results

Although it appeared difficult to find precisely the time cost for the hand-based design, a rough estimate can be given as a minimum of five hours, which seems to be usual for this kind of device. Fortunately, it has been easier to estimate the time cost for the other methodology, which took about thirty minutes to be performed, hence taking more than ten times less.

Moreover, the design processes did not lead to the same solution, even if both are quite similar, for a great number of parameters. Hence, we will only show the striking differences between them. All values given below are taken from suppliers data sheets, except the wire length and copper volume that have been computed. Table 1 shows that the optimization-based design process allowed for a 24% copper saving. The core outer diameter could also be decreased, but its height has increased as well, keeping the core section at a similar value in both cases, giving almost the same core volume.

Table 1. Solution comparison for both design processes

Design process	Core				Secondary winding		
	Outer diameter (mm)	Inner diameter (mm)	Height (mm)	Volume (mm^3)	Wire diameter (mm)	Wire length (mm)	Copper volume (mm^3)
Hand-based	20	10	7	1465	0.125	27589	338
Optimization-based	18.8	11	8	1460	0.112	26148	258
Saving	6 %		-14 %	0.3 %			24 %

4. SIZING PROCESS COMPARISON

4.1 Benefits

Thanks to the use of a constrained optimization algorithm, the problem solving step automation is the key factor for the time saving, which could also be observed on other examples. It is also the cause for the solution improvement, which depends on the investigated example. It can even help to reject some common ideas in a given design domain, that were leading previously to non-optimal solutions. However, the results given by this algorithm are directly influenced by the quality of the problem setting task. This one involves to fix or to set bounds on some parameters, before the problem-solving step (Fig. 4). Those values can come from the specification sheet, from physical hypothesis required for the model validity, from technological limitations, or from the designer experience. As checking involving the same types of values has to be done all along the hand-based

sizing process, after each parameter modification, there is no time loss that can balance the gain described above.
Looking at the description of both methodologies, it is quite clear that the new sizing process is less confused than the other is. Indeed, its principle is simpler, as all the involved steps were taking place all along the hand-based process, and had to be done directly by the designer, as constraints and optimality checking. This should also increase the sizing process reliability. As well, it should be noticed that the designer did not make the optimality checking in a mathematical sense, using derivatives. The hand-based sizing process stopping criterion was therefore quite subjective, depending on the objective the designer had in mind mainly.
Another benefit comes from the complexity management offered by the optimization algorithm. Only one complex model with well defined input-output parameters is now required, compared to three different ones for the hand-based sizing process. It also allows for a clarification of the model being used. As an example, let's consider the R_{load} parameter (Fig. 1), which was both used as input and output parameter in the hand-based sizing process, as explained in section 2.2. As output parameter, equation (1), being involved for its evaluation, can be scientifically justified. However, used as input parameter to compute N, its value had to be defined using another equation, coming from empirical deductions, hence being less reliable. It had to be used because only some of the parameters were known to the designer at this step. However, using the other sizing process, this last equation is no more used, as R_{load} is clearly an input parameter of the model used by the optimization algorithm, which would then be too complex to be processed by hand.

4.2 Issues

Some difficulties have also to be faced with, partly due to the used optimization algorithms. Indeed, they only performed local optimization, which has not appeared as a critical issue here. They also only handled continuous parameters. Unfortunately, some parameters are discrete, such as wire diameters, when they have to be chosen among tabulated values given by suppliers. This has resulted in the need for a discrete parameter hand-based post-processing, which aims at selecting the best combination among the possible values. No optimization algorithms able to deal efficiently with those discrete parameters have been found so far, for our optimization problems. Researches are currently undertaken to address this issue. Additional difficulties may also arise during the problem setting step from the starting point computation, required by the optimization algorithms involved. It is due to the need for values still undetermined at this step, which is required for its computation. Fortunately, it seems that this problem can be bypassed, using some of the simple equations coming from the simplest model used in the hand-based process.

5. CONCLUSION

Using constrained optimization techniques for magnetic component industrial design has proved to dramatically decrease the design time, as well as to bring other benefits. Those results are a consequence of the sizing process iteration automation, which is the most time costing and boring task for a designer. The other steps have still to be carried out by the designer, whose knowledge and experience are strongly required to boost the speed of convergence toward a solution, as well as to analyze its industrial feasibility and to make fundamental technological choices.

REFERENCES

1. F. Wurtz, J. Bigeon, J.C. Poirson, "A Methodology and a tool for the Computer Aided Design with Constraints of Electrical Devices", *IEEE Transactions on Magnetics*, Volume 32, Issue 3, Part 1, May 1996, pp 1429-1432
2. E. Atienza, J. Bigeon, F. Wurtz, V. Leconte, "EDEN : A generic integrated component software for constraint optimization", *13th Conference on the Computation of Electromagnetic Fields (Compumag 2001)*, July 2-5, 2001, Evian, France, pp.232-233
3. B. Delinchant, F. Wurtz, E. Atienza, J. Bigeon "Components Approach - The integration way for electromagnetic simulation software", *COMPUMAG'01, 13th Conference on the Computation of Electromagnetics Fields*, Evian, France, July 2-5, 2001
4. A. Schellmanns, K. Berrouche, J.P. Keradec, "Multi-winding transformers: a successive refinement method to characterize a general equivalent circuit", *IEEE Transactions on Instrumentation and Measurement*, October 1998, Volume 47, N°5, pp 1316-1321
5. C.T Lawrence, J.L Zhou, A.L Tits, *User's guide for CFSQP version 2.5*, Electrical Engineering Department and Institute for Systems Research, University of Maryland, 1997
6. Harwell, *Routine Library Specification*, VF13, 10th October 1987

NUMERICAL OPTIMIZATION METHODS IN MAGNET DESIGN

Marek Rudnicki*, Arkadiusz Tomczyk, Krzysztof Sasiak
*Department of Artificial Intelligence and Informatic Systems, Academy of Humanities and Economics in Lodz, Poland
Institute of Computer Science, Technical University of Lodz, Poland

Abstract: This paper describes two optimization techniques applied to electromagnet pole-shape design along with numerical results. The aim was to find a shape, such that a desired magnetic field distribution is achieved in the air-gap. The chosen optimization techniques were simulated annealing and taboo search both applied to the aforementioned problem. For approximate solving of partial differential equations the finite element method was used. Calculations were performed with MATLAB and its PDE toolbox.

Key words: optimization, simulated annealing, taboo search, magnet shape, electromagnetism, potential equations, finite element method

1. INTRODUCTION

Magnetic fields are widely used in the science and industry of today. In practice, we aim to achieve desirable magnetic field distributions, suitable for all kinds of purposes. In electromagnets the distribution depends on the current density, the materials used to build the electromagnet and its construction (i.e. size and shape). In this work we focus on the last item and describe two optimization techniques used for optimal shape design. These methods are *simulated annealing* and *taboo search*. For testing purposes both methods were implemented in MATLAB. In order to approximately solve partial differential equations required by the problem the *finite element method* implemented in the MATLAB PDE toolbox was used.

The problem chosen for numerical simulation was that of achieving a prescribed magnetic flux density distribution on the magnet's axis, however

M. Rudnicki and S. Wiak (eds.), Optimization and Inverse Problems in Electomagnetism, 101–112.

these techniques have, of course, a wider range of application. The numerical results seem to be interesting and useful. This way of magnet design can be of use anywhere, where magnetic fields with particular properties are needed, e.g. medicine.

In the following chapter we will discus the general issues concerning magnet design. Chapter 3 deals with the specific problem used to test two optimization methods performance. Chapter 4 will give the reader some insight into MATLAB, the mathematical tool used in obtaining the results, described in this article. Chapters 5 and 6 describe the *simulated annealing* method and the results produced by it. Similarly, the next two chapters deal with *taboo search*. Last but not least, chapter 9 contains our conclusions as to these methods, their results and applicability

2. ELECTROMAGNET OPTIMIZATION

Neglecting hysteresis, the constitutive relation linking magnetic field strength **H** and magnetic flux density **B** can be written as follows

$$\begin{aligned} \mathbf{H} &= \nu(\mathbf{B})\mathbf{B} \\ \nu &= \nu_r \nu_0 \end{aligned} \tag{1}$$

Where ν is the reluctivity of the medium expressed as a product of the vacuum value $\nu_0 = \frac{1}{\mu_0}$ and the so-called relative reluctivity ν_r. From Maxwell equations

$$\begin{aligned} &\operatorname{curl}\mathbf{H} = \mathbf{J} \quad \text{in} \quad \Omega \subset \mathrm{R}^2 \\ &\operatorname{div}\mathbf{B} = 0 \end{aligned} \tag{2}$$

with Ω being a simply connected domain with a Lipschitz boundary ∂Ω, and

$$\begin{aligned} &\operatorname{curl}\mathbf{H} = \nabla \times \mathbf{H} = \left(\frac{\partial H_1}{\partial x_2} - \frac{\partial H_2}{\partial x_1} \right)\mathbf{k}, \quad \mathbf{H} = (H_1, H_2) \\ &\mathbf{B} = (B_1, B_2) \\ &\mathbf{J} = (0,0,J) \end{aligned} \tag{3}$$

We can deduce a non-linear, scalar and nondecreasing relationship of the form

$$\mathbf{H}(\mathbf{x}) = \nu\left(\|\mathbf{B}(\mathbf{x})\|^2\right)\mathbf{B}(\mathbf{x}) \qquad \mathbf{x} \in \Omega \tag{4}$$

According to second Maxwell's equation above there exists a function A (magnetic potential) such that

$$\mathbf{B} = \nabla \times A = \left(\frac{\partial A}{\partial x_2}; -\frac{\partial A}{\partial x_1} \right) \quad \text{up to a certain additive constant.} \tag{5}$$

Since $\nabla A = \left(\frac{\partial A}{\partial x_1}, \frac{\partial A}{\partial x_2} \right)$ there follows that $\|\mathbf{B}(\mathbf{x})\| = \|\nabla A(\mathbf{x})\|$. As a result we obtain the following non-linear elliptic problem

$$-\sum_{i=1}^{2} \frac{\partial}{\partial x_i} \left(\nu\left(\|\mathbf{B}\|^2\right) \frac{\partial A}{\partial x_i} \right) = J \qquad \text{in} \quad \Omega \tag{6}$$

Or, equivalently

$$-\operatorname{div}\left(\nu_r\left(\mathbf{x}, \|\nabla A\|^2\right) \nabla A\right) = \mu_0 J \tag{7}$$

where $\mu_0 = 4\pi 10^{-7}$ denotes permeability of the free space. For practical considerations we can apply, e.g. the following material relationships

$$\begin{aligned} \nu_r\left(\mathbf{x}, \|\nabla A\|^2\right) &= c + (f - e) \frac{\|\nabla A\|^{2\alpha}}{\|\nabla A\|^{2\alpha} + g} \\ \nu_r\left(\mathbf{x}, \|\nabla A\|^2\right) &= \frac{1}{\mu_0} \left(\alpha + (1 - \alpha) \frac{\|\nabla A\|^8}{\|\nabla A\|^8 + \beta} \right) \end{aligned} \tag{8}$$

With (first relation): α=5.419241, e=5.163619×10^{-4}, f=0.175775, g=8758.756 and (second relation): α=0.0003 and β=16000. All the above constants were obtained experimentally. However, for the present paper we assumed a linear dependence between magnetic fields, i.e. no saturation takes place.

The variational formulation of the non-linear elliptic problem reads:

$$\begin{aligned} &\int_\Omega \nu\left(\|\nabla A\|^2\right) \nabla A \cdot \nabla v dx = \int_\Omega J v dx, \qquad v \in V \\ &V = \left\{ \varphi \in \mathrm{H}^1(\Omega) \middle| \ \varphi|_\Gamma = 0 \right\} \\ &\mathrm{H}^m(\Omega) = \left\{ v \in L^2(\Omega) \middle| \ D^\alpha v \in L^2(\Omega), \ |\alpha| \le m \right\} \end{aligned} \tag{9}$$

Applying the finite element method of the first order (linear triangles) we arrive at the non-linear algebraic system

$$\mathbf{K}(\mathbf{u})\mathbf{u} = \mathbf{f} \tag{10}$$

where **u** is the vector of nodal potentials, **K**(**u**) is a stiffness matrix and **f** denotes the source term (current).

This is the starting point for the optimal shape design problem, which is as follows: design a part of the boundary (pole profile) to achieve prescribed distribution of the magnetic field in the air-gap. The objective function is given by:

$$F(\mathbf{x}) = \int_{\Omega_c} \|\nabla \times A - B_0\|^2 d\mathbf{x} \tag{11}$$

or equivalently

$$F(\mathbf{x}) = \sum_{j=1}^{N} \frac{(\mathbf{B}_j(\mathbf{x}) - \mathbf{B}_0)^2}{\mathbf{B}_0}, \tag{12}$$

where N is a number of checkpoints in the controlled subregion Ω_c and $\mathbf{B}_0$ denotes prescribed value of magnetic field in it.

3. THE PROBLEM

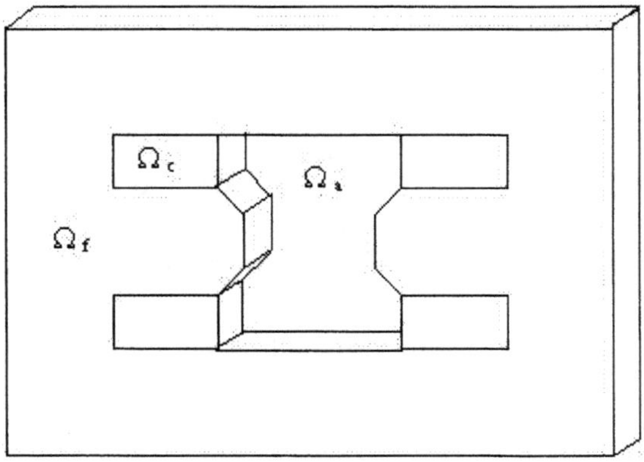

Figure 1. Optimized electromagnet ([1])

The problem to be optimised is the shape of the electromagnet. The desired result is a shape, which in effect gives a certain magnetic field distribution (calculated at specified points). The electromagnet we have chosen to optimise looks like this

Since the points we are interested in specifying field values in are situated in the centre and the magnet itself is symmetrical along both axes, it is sufficient to optimise only a quarter of the magnet. Hence, here we will optimise the shape of this section:

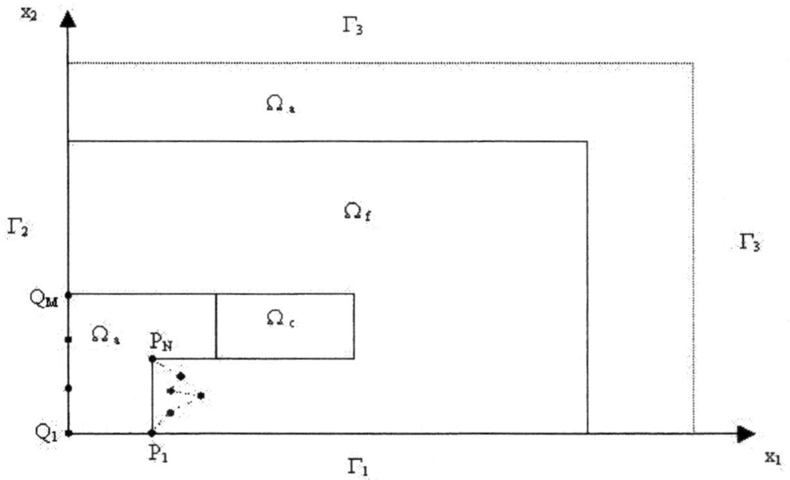

Figure 2. The quarter of the magnet, which shape will be optimized. P1,...,PN – points that will be changed during optimiza-tion process. Q1,...,QM – points where desired field distribution will be checked.

The points in which we specify desired field values lie along the vertical axis at regular distances between Q_1 and Q_M. The part of the magnet to be optimised is the segment between points P_1 and P_N. This edge will be divided up by N points, which will be moved left or right, thus modifying the field distribution.

The optimised function will be a least-squares deviation between the desired and actual magnetic flux density values at the specified points in the middle of the vertical axis of the magnet:

$$E(P_1,\dots,P_N)=\sum_{i=1}^{M}(B_i^{cal}(P_1,\dots,P_N)-B_i^{des})^2 \tag{13}$$

where B_i^{cal} represents the field values calculated (in point Q_i), based upon the current solution (shape), while B_i^{des} represents the desired values, and M is the number of these points, where field values are checked.

To calculate B_i^{cal} we have to refer to physical description of magnetic fields. Because we are considering problem in two dimensions the magnetic vector potential **A** can be represented by the scalar A. Therefore, Ampere's law can describe it as follows ([1]):

$$-\nabla\cdot(\nu\cdot\nabla A)=J \qquad A|_\infty=0 \tag{14}$$

where ν is magnetic reluctivity and J represents current density. In our problem we can decompose it into three equations (one for each medium):

$$\begin{aligned}&\nabla\cdot(\nu(\|\nabla\times A\|^2)\nabla A)=0 && \text{in iron}\\ &\nu_a\,\Delta A=0 && \text{in air}\\ &\nu_c\,\Delta A=J && \text{in copper}\end{aligned} \tag{15}$$

Magnetic reluctivity ν is different in each medium. It is constant in air and copper but in iron, which is a ferromagnetic, in general it depends on magnetic flux density (it is calculated as a curl of A). However in this work we assumed constant value of ν in iron because it has no influence on the optimization techniques described herein. Boundary conditions for our problem can be written as ([1]):

- $A=0$ on Γ_1 and Γ_3 (boundaries lie sufficiently far from the field source)
- $\partial_n A=0$ on Γ_2 (∂_n is the normal derivative, $\mathbf{n}\cdot\nabla A$, where $\mathbf{n}$ denotes the unit outward normal).

Solving the above system of partial differential equations we can find A. But to calculate B_i^{cal} we have to find curl of A. Fortunately in our case $\nabla\times A=(\partial_2 A,-\partial_1 A)$ and in consequence $\|\nabla\times A\|=\|\nabla A\|$.

In closing it is worth mentioning that this is constrained problem. Points cannot move in an arbitrary manner. For example they cannot move beyond the vertical axis of the electromagnet. Moreover for experimental purposes it was assumed that the desired magnetic flux density was to be constant but, of course, this is not the only possibility.

4. THE TOOL

In order to conduct mathematical computations, we used the MATLAB environment. As an interpreter for a math oriented language, MATLAB allows easy and effective coding of mathematical formulas, problems and algorithms. It is also equipped with a PDE toolbox, which can solve the partial differential equations required in this problem. It uses the finite element method and can be applied to solving, among many others, the elliptic equation with which we are dealing here. Its utilisation for our purposes was very simple.

5. SIMULATED ANNEALING

This algorithm searches the space of possible solutions in a random way, however it not only accepts better ones but sometimes also those of them that increase value of the optimised function E. The latter are accepted with the probability:

$$p = e^{-\frac{\Delta E}{T}} \tag{16}$$

where ΔE means increase of the optimised function E (in this algorithm this function is called an energy function or simply an energy) and T is a control parameter, called temperature, that changes during the search. This algorithm works in analogy to the physical process of annealing in metallurgy where slowly cooled metal freezes in an optimal state (has the best structure). It has two main advantages. Firstly teoretically it is able to find global minimum. Secondly it can be easily applied to problems with constraints.

Basic elements of this method, which must always be defined, are:

- a way of representing of possible solutions
- a generator of random changes in proposed solutions (movement generator) – it should meet two main assumptions, namely the changes should be small and allow all possible solutions to be achieved in a finite sequence of moves
- a way of evaluating the proposed solutions (target function) – it is worth mentioning that this evaluation depends only on value of the function and we do not need any gradient information as in many other optimization techniques
- annealing schedule – the way we decrease the temperature (it consists of: initial temperature, decreasing rules and the moment of ending the search); decreasing should not be too slow (it would take a very long time) and too fast (we would possibly get stuck in first local minimum)

In our problem solutions are represented as a vector of points (it is enough to describe changes in shape). Used movement generator moved one randomly chosen or all shape point in horizontal direction. The step which was used for that move was small and random (normal and uniform distributions where here tested). Because problem has, as previously stated natural geometric constraints those of the solutions that were not satisfying them were simply rejected. Proposed solutions are of course evaluated using the target function. The annealing schedule was standard. The initial temperature was calculated using an approach proposed by Kirkpatrick. In this method temperature should be chosen in such a way that initially worse solutions were accepted with a probability of 0.8. To choose it properly an initial search must be performed, during which all moves are accepted (except of course those that do not satisfy the constraints). We calculate the mean value of observed increases of optimised function ΔE^+ and then the initial temperature can be found using the following equation:

$$T_0 = -\frac{\Delta E^+}{\ln(0{,}8)} \tag{17}$$

The temperature was decreased using exponential annealing schedule:

$$T_{i+1} = \alpha \cdot T_i \tag{18}$$

where α was assumed equal to 0.95. Because method needs great number of iterations and in our case each iteration lasted a very long time (in each iteration system of partial differential equations was solved) the stop criterion was the author himself. He could observe the progress of the search and stop the algorithm if no better solutions have been found for a longer period of time.

For practical reasons it is very often worth using the other version of this algorithm with Markov chains, which can noticeably reduce time of calculations.

6. SIMULATED ANNEALING RESULTS

This chapter presents some results of applying simulated annealing to our optimization problem in such a sequence in which experiments were performed. The calculations for each case lasted usually 1 or 2 days on 500MHz machine so not many results could be investigated but these here can at least show what types of shape would give a desired magnetic field distribution. *Figures* from *3* to *8* presents shapes and magnetic potential A.

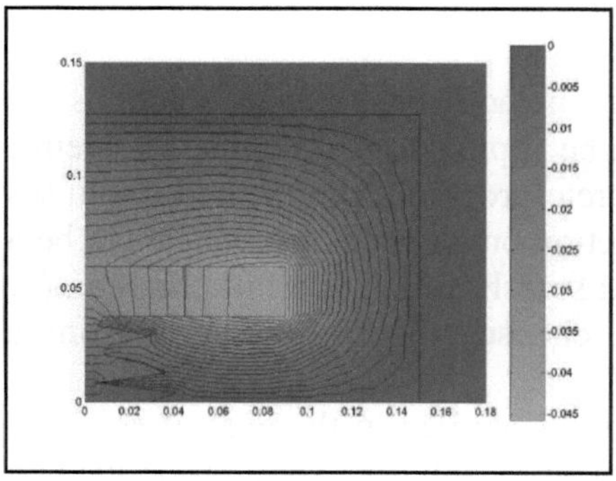

Figure 3.

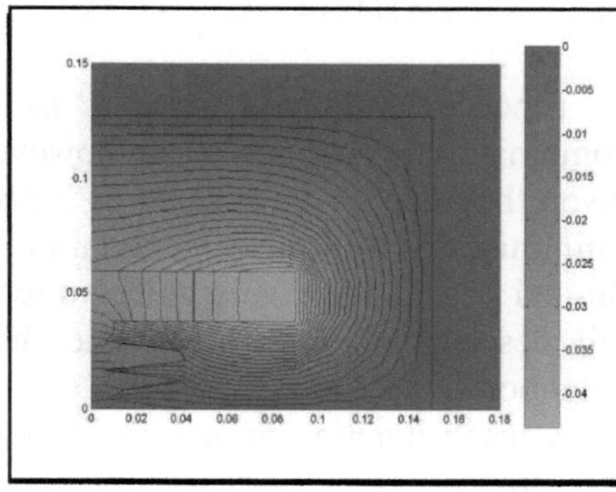

Figure 4.

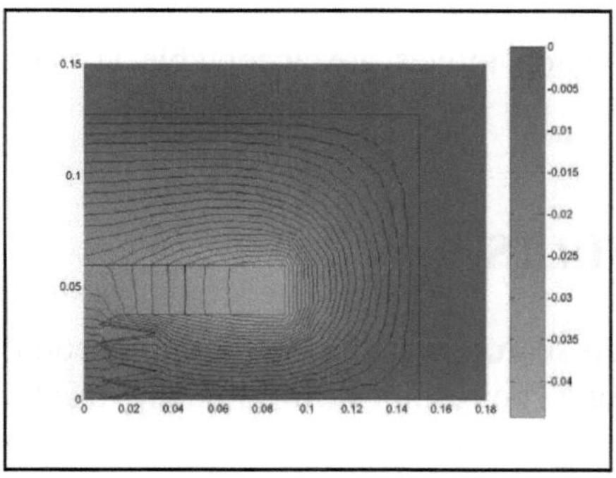

Figure 5.

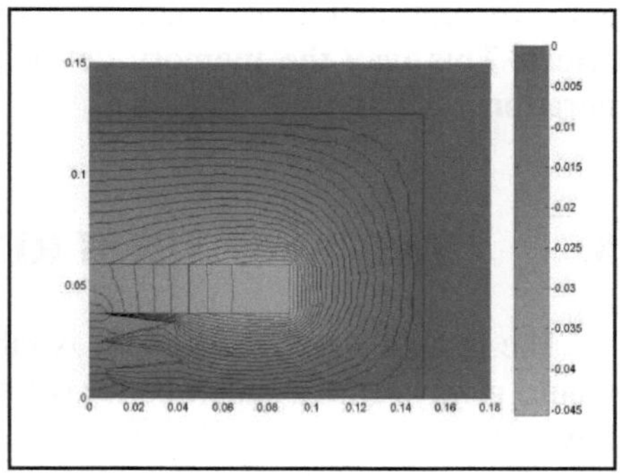

Figure 6.

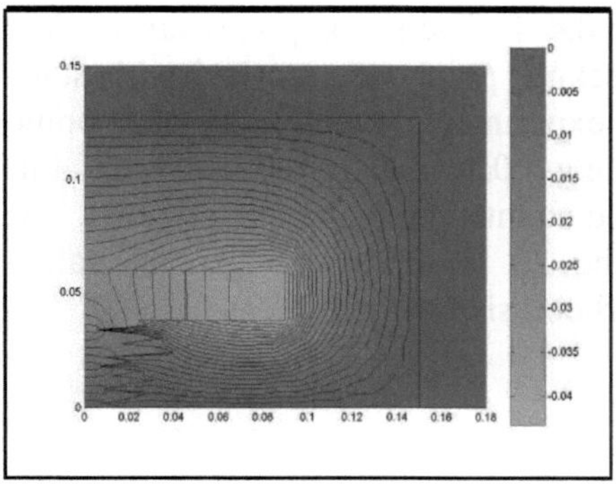

Figure 7.

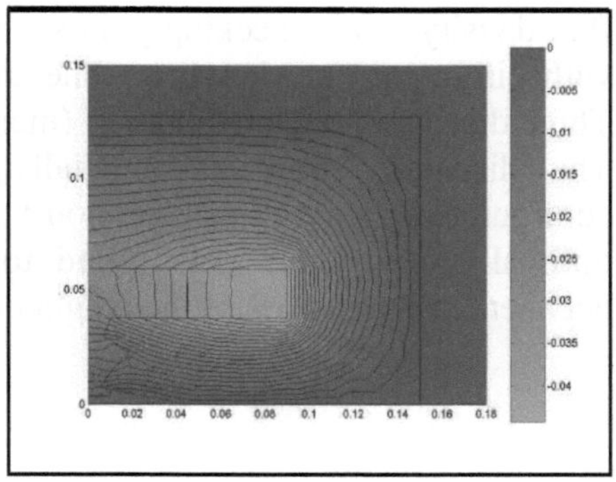

Figure 8.

7. TABOO SEARCH

Taboo search was created as a method of finding solutions to combinatorial problems, it can however be applied to real-number problems, given they can be formulated as a discrete problem. The method optimises (minimises or maximises) a certain function on a certain set, which can be a limited one. It adaptively guides a local search, which is a hill climbing (or hill descent) algorithm, i.e. one that chooses the best solution from a neighbourhood.

At each iteration an acceptable move is applied to the current solution, thus proceeding to the best of the neighbouring solutions (best, as defined by the function being optimised). Moves, which cause an inferior value of the target function, are acceptable. Some moves (e.g. a move reversing the previous one) are prohibited, or *taboo*. These movement limits are based on a movement memory, which defines how long these limitations stay in effect. This way the memory defines which moves are acceptable at each iteration.

8. TABOO SEARCH RESULTS

The results were quite good. A taboo search, starting at a random point, would always find a much better solution. The results showed much variety, however in most cases, the shape found was jagged and irregular, as might have been expected.

Most of the numerical experiments were conducted with the following settings: 10 shape points, 6 checking points evenly distributed along the central axis, from the centre to the level of the magnet's coil. The required flux density in all checking points was 0,4 T. The shape points are marked with circles on the drawings. The checking points are marked with stars. Three distances in each direction (most experiments) could move each point. The distances grew exponentially, e.g. 0.1, 0.01, 0.001. With this configuration, a solutions neighbourhood counted 60 elements (at most). The optimal taboo tenure was found to be 20 iterations (in general values between 25% and 50% of the neighbourhood size worked best).

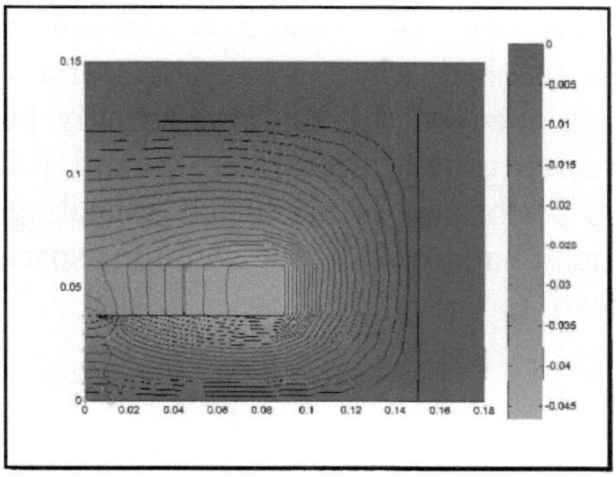

Figure 9. (I = 122, E = 0.00033282)

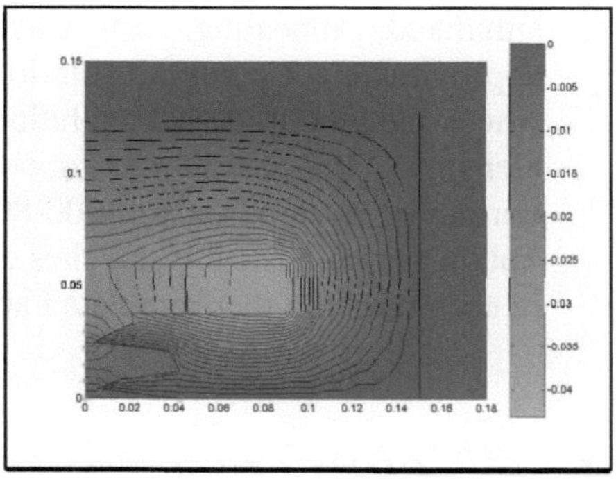

Figure 10. (I = 113, E = 0.0642345)

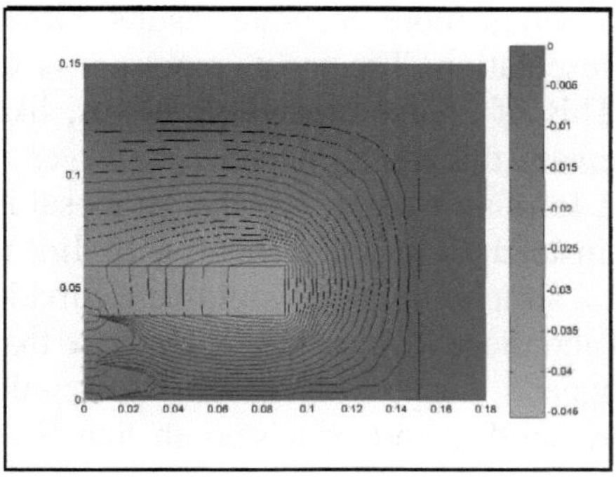

Figure 11. (I = 111, E = 0.00390493)

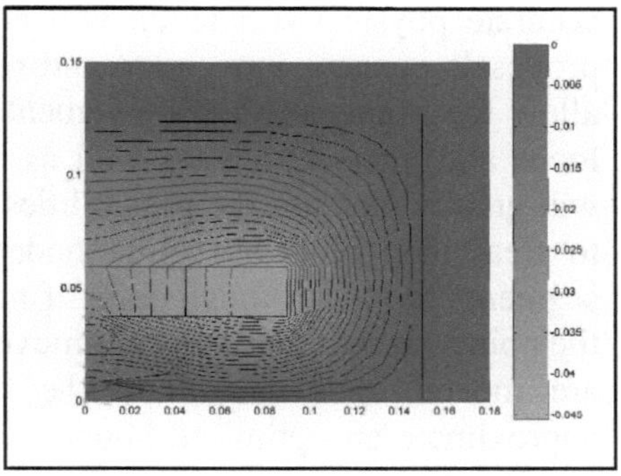

Figure 12. (I = 100, E = 0.000454669, 4 moves per point)

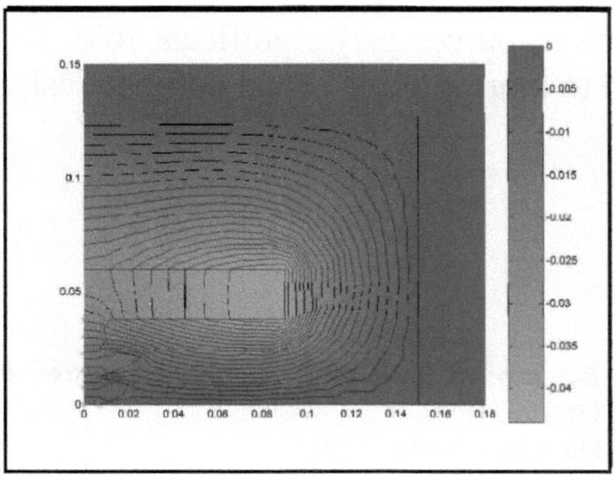

Figure 13. (I = 76, E = 0.00335, 4 moves per point)

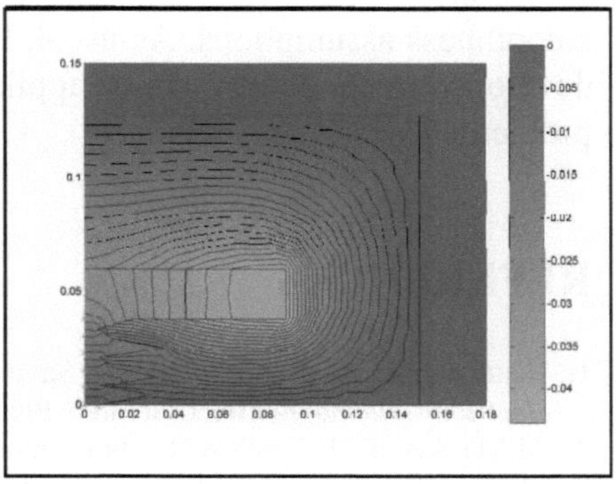

Figure 14. (I = 91, E = 0.004556, 8 moves per point)

An iteration of taboo search is not the same as a single iteration of simulated annealing. In each iteration of simulated annealing approximately 1 potential solution is checked, whereas in taboo search a whole neighbourhood of solutions is checked (in this case usually 60 elements). Because of this taboo search usually performed around 100 iterations, as opposed to 4000-8000 performed by simulated annealing. Taboo search basically searches a much larger part of the solutions space in each iteration then simulated annealing.

9. CONCLUSIONS

The quality of the solutions could be improved by increasing the complexity of the problem and giving it more possibilities. For one, a more accurate physical simulation will help obtain more accurate results. Other proposals concern movement and representation. The most important is to allow the points vertical movement. This of course creates problems, like loops and intersections, as well as element thickness concerns, however it will greatly increase the possibilities of forming shapes. Another proposal is to treat the points as spline nodes instead of joining them with linear segments to form a rough edge. If not, a simpler version would be to divide the points up into those that are moved and those that are pulled by those that are moved, thus smoothing the magnet's shape. Moreover, we could approximate an optimised boundary by another sort of a smooth line like, e.g. Bezier curve. This would facilitate a prototype construction. The results of the paper compare well with those given in [1,3]. The paper clearly demonstrates advantages of using derivative-free optimization algorithms for the solution of optimal shape design problems. Instead of time-consuming and prone to error sensitivity analysis we are led to the solution without any smoothness assumptions. As usual, time resources play significant role. The developed methodology can be applied to more general shape optimal design problems [3].

REFERENCES

1. Neittaanmäki P., Rudnicki M., & Savini, A.: *Inverse problems and optimal design in electricity and magnetism*, Clarendon Press, Oxford, 1996
2. MATLAB - PDE Toolbox Documentation, The MathWorks, 2001
3. M. Krizek and P. Neittaanmaki, *Mathematical and Numerical Modelling in Electrical Engineering: Theory and Applications,* Kluwer Academic Publishers, Dordrecht 1996.

THE USE OF QUASI-RED AND QUASI-YELLOW NONOBTUSE REFINEMENTS IN THE SOLUTION OF 2-D ELECTROMAGNETIC PDE'S

Jacek Stańdo*, Sergey Korotov**, Marek Rudnicki*** & Dorota Krawczyk-Stańdo*

*Institute of Mathematics, Technical University of Lodz, Poland, standoj@ck-sg.p.lodz.pl

**University of Jyväskylä, Dept. of Mathematical Information Technology, Finland korotov@mit.jyu.fi

*** Department of Artificial Intelligence and Informatic Systems, Academy of Humanities and Economics in Lodz, Poland, rudnicki@ics.p.lodz.pl

Abstract: In the paper, new refinement techniques called "quasi-red" and "quasi-yellow" are applied for the solution of a non-linear Poisson's equation describing magnetic flux density in 2-D. The equation is solved using the finite element method with non-obtuse triangulation. The goal is to modify initial mesh in some non-standard way to have solely non-obtuse triangles. Usefulness of the method is demonstrated on the problem of optimal shape design of an electromagnet.

Key words: nonobtuse triangulation, polygonal domain, finite element method, discrete maximum principle, quasi-red and quasi-yellow refinements, grid generation, non-linear Poisson's equation.

1. INTRODUCTION

In this paper, we consider conforming (planar) triangulation of a polygon into triangles, i.e., the union of all triangles of any particular triangulation is always equal to the polygon, and any two different triangles in any particular triangulation may only have a common edge, or a common vertex, or no common points [10]. In most of cases

M. Rudnicki and S. Wiak (eds.), Optimization and Inverse Problems in Electomagnetism, 113–124.

namely such conforming triangulation are used in finite element modeling and finite element analysis.

A planar triangulation is called *nonobtuse* if all interior angles in all triangles of the triangulation are less than or equal to $\pi/2$. This condition was first introduced in [4] and [13], where it was particularly used to prove the discrete maximum principle for the piecewise linear finite element approximations to the Poisson equation with the Dirichlet boundary condition in plane. Later this condition was imposed on the interior angles between faces of tetrahedra in tetrahedral partitions, and again used to prove the discrete maximum principle for the finite element solutions non-linear elliptic problems for which the continuous maximum principle to a class of holds, see [11].

Note that in nonobtuse triangulation, all triangles satisfy the maximum angle condition with the same constant, $\pi/2$. This condition is very important for the finite element analysis, since the optimal interpolation properties of the finite elements are preserved [10, Chapter 4].

a) red refinement b) yellow refinement

Figure 1.

The construction of a conforming triangulation of a polygon into (arbitrary) triangles is, obviously, simple; moreover, there are proofs of the existence of nonobtuse triangulation of polygons [1], [5]. Thus, having a nonobtuse initial triangulation, and then applying the (standard) red refinement [2] and/or the (standard) yellow refinement [7] (see Figure 1, where the same ``bar indices" are used for segments of the same length), we can easily build a family of globally refined triangulation where no obtuse angles occur.

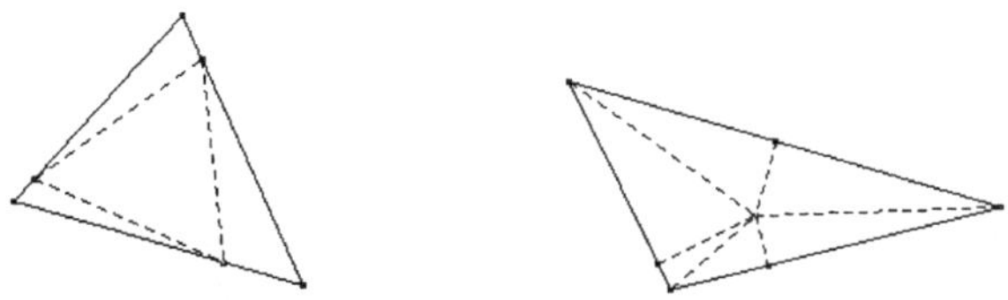

a) quasi- red refinement b) quasi-yellow refinement

Figure 2

Nevertheless, in practice, the initial (coarse) triangulation of the polygonal domain is performed by a grid generator, and it is most probably that such made initial triangulation is not a nonobtuse one. One way to change such a „undesirable" triangulation into a nonobtuse one is to move locally several vertices in it in order to get locally nonobtuse triangles. Another approach, first presented in [9] and also used in this paper, is to use special (local) refinement techniques for small polygonal zones (in the solution domain), where each such zone is formed by an isolated obtuse triangle and several nonobtuse triangles surrounding it, so that only midpoints of the sides forming such a zone and several additional points inside of it are used in the partition additionally to vertices of the zone. Thus, if we use standard red and/or yellow refinements for all the triangles outside of such „special" zones, after one global refinement step we get nonobtuse conforming triangulation of our domain. In [9], such zones are quadrilaterals made by isolated obtuse triangles and nonobtuse triangles adjacent to these obtuse triangles along their longest edges. In this paper, those zones are the unions of isolated obtuse triangles and triples of adjacent to them nonobtuse triangles. The conditions when the proposed way of partition for each such zone is possible are given in terms of co-ordinates of six vertices involved. To define the partition of each zone, we first introduce the ideas of quasi-red and quasi-yellow refinements (see Figure 2), which are, in fact, the natural generalisations of standard red and yellow refinement techniques. The quasi-red refinement is a refinement of the ``red-type" where three auxiliary points are not necessarily taken at midpoints of sides of the given triangle (cf. Figure 2a). The quasi-yellow refinement of a triangle is based on a choice of some point inside of the triangle so that three perpendiculars from this point onto three sides divide the triangle into six or four right triangles. We notice that the quasi-yellow refinement can be applied to any triangle (for example, always a proper choice is a centre of the inscribed sphere), at the same time the standard yellow refinement (where such point has to be a centre of the circumscribed sphere) is only applicable to nonobtuse triangles. Further, we analyse a possibility of refining globally the given triangulation, initially containing several isolated obtuse triangles (cf. Figure 3), into a conforming nonobtuse triangulation. To refine each obtuse triangle, we use the quasi-yellow refinement, and to refine the nonobtuse triangles, adjacent to those obtuse triangles along its sides, we use the quasi-red refinements. The other triangles are refined in some standard way (using the red and/or yellow refinements).

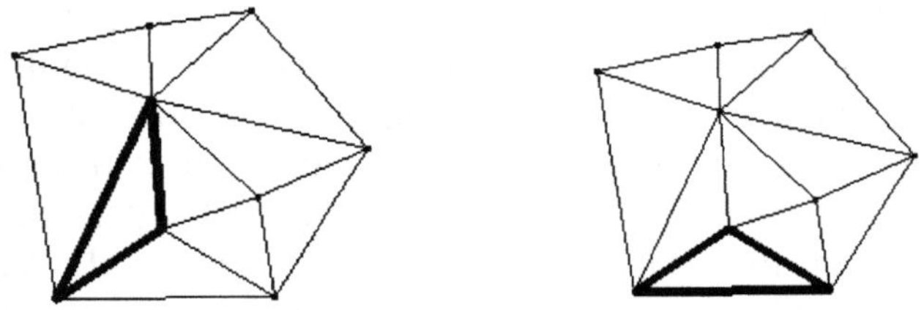

a) obtuse triangle is inside of solution domain b) obtuse triangle is adjacent to the boundary

Figure 3.

It is also worth to mention here, that the non-obtusity condition is only sufficient to prove the discrete maximum principle: it is possible to observe the validity of this principle also in situations when some of interior angles in the triangulation are slightly obtuse. This issue, leading to so-called weakened acute type conditions, is addressed in [12] and [8] in the two- and the three-dimensional cases, respectively.

Another known refinements of triangles are green [2], and blue [6], they are used for local refinement procedures. For a comprehensive survey of the results on the grid generation issues, we refer to monograph [3].

2. QUASI-RED NONOBTUSE REFINEMENTS OF A TRIANGLE

The standard red refinement is based on the choice of three auxiliary points on sides of the triangle at midpoints of these sides (see Figure 1a). Obviously, under certain circumstances, we might have a certain freedom in a choice of such auxiliary points: for example, one, or two, or even three points, can be taken on sides of the given triangle, but not necessarily at the midpoints; a refinement of „red-type" which is called *quasi-red refinement* can be applied to those cases as well, and it may still produce four nonobtuse subtriangles (cf. Figure 2a).

In our paper, we employ the idea of quasi-red refinement for the case when only one auxiliary point is allowed to be taken arbitrarily (and the other two are still midpoints of two sides of the triangle). Our purpose is to define the set of points from which this third auxiliary point can be taken, so that the quasi-red refinement, applied to a nonobtuse triangle, produces four nonobtuse (sub)triangles.

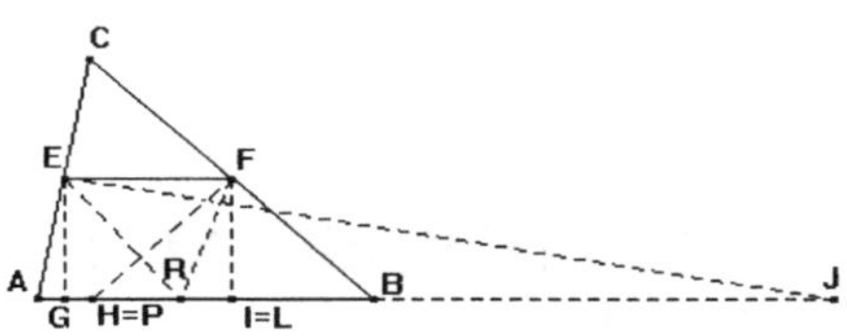

Figure 4

The main result, in this respect, is stated as follows.

Lemma 2.1 Let *ABC* be a nonobtuse triangle with vertices $A=(x_A,y_A)$, $B=(x_B,y_B)$, $C=(x_C,y_C)$ (see Figure 4), and let *E* and *F* be midpoints of *AC* and *CB*, respectively. We define four real numbers k_1, k_2, k^1 and k^2 as follows:

$$(2.1)\quad k_1=\frac{1}{2}\cdot\frac{(x_C-x_A)(x_B-x_A)+(y_C-y_A)(y_B-y_A)}{(x_B-x_A)^2+(y_B-y_A)^2}$$

$$(2.2)\quad k_2=\frac{1}{2}\cdot\frac{(x_B+x_C-2x_A)(x_B-x_C)+(y_B+y_C-2y_A)(y_B-y_C)}{(x_B-x_A)(x_B-x_C)+(y_B-y_A)(y_B-y_C)}$$

$$(2.3)\quad k^1=\frac{1}{2}\cdot\frac{(x_B+x_C-2x_A)(x_B-x_A)+(y_B+y_C-2y_A)(y_B-y_A)}{(x_B-x_A)^2+(y_B-y_A)^2}$$

$$(2.4)\quad k^2 = \frac{1}{2}\cdot\frac{(x_C - x_A)^2 + (y_C - y_A)^2}{(x_B - x_A)(x_C - x_A) + (y_B - y_A)(y_C - y_A)}$$

if $(x_B - x_A)(x_B - x_C) + (y_B - y_A)(y_B - y_C) = 0$, we set $k_2 = -\infty$,
if $(x_B - x_A)(x_C - x_A) + (y_B - y_A)(y_C - y_A) = 0$, we set $k^2 = +\infty$.

Let $k_* = \max\{k_1, k_2\}$ and $k^* = \max\{k^1, k^2\}$, and

$$(2.5)\quad P = (k_* x_B + (1-k_*)x_A, k_* y_B + (1-k_*)y_A),$$
$$(2.6)\quad L = (k^* x_B + (1-k^*)x_A, k^* y_B + (1-k^*)y_A).$$

Then, for any point $R \in [PL]$ $(\subset [AB])$, triangles *AER*, *ERF*, *FRB*, and *CEF* are nonobtuse (we note that it might occur the situation when *P=L*).

Proof. The proof is straightforward, since (in terms of Figure 4),

$$G = (k_1 x_B + (1-k_1)x_A, k_1 y_B + (1-k_1)y_A)\ ,$$
$$H = (k_2 x_B + (1-k_2)x_A, k_2 y_B + (1-k_2)y_A)\ ,$$
$$I = (k^1 x_B + (1-k^1)x_A, k^1 y_B + (1-k^1)y_A)\ ,$$
$$J = (k^2 x_B + (1-k^2)x_A, k^2 y_B + (1-k^2)y_A).$$

3. QUASI-YELLOW REFINEMENT OF A TRIANGLE

The idea of the yellow refinement of triangles was first introduced in [7]. The standard yellow refinement essentially uses the assumption that the centre of the circumscribed sphere of the given triangle lies in the closure of the triangle. Using the centre and the midpoints of sides of the triangle, we get four or six right sub-triangles, depending on the value of maximum angle in the original triangle (cf. Figure 1b).

Nevertheless, if we do not necessarily have to use namely midpoints for the refinement into right subtriangles, we can always employ some point in the triangle, for example, a centre of the inscribed sphere, and three projections of the centre onto the sides, and get again a partition of a triangle into six right subtriangles. This way of partitioning is called *quasi-yellow refinement*. The only difference with the standard yellow refinement is that additionally inserted nodes do not necessarily coincide with the midpoints of the sides of the triangle (cf. Figure 2b). In fact, we are able to define the set, called {\it admissible area of triangle}, of points in the triangle so that the above-described partition is possible. In case of nonobtuse triangle, this admissible area coincides with the whole triangle; if there is an obtuse angle in the triangle *ABC* (see Figure 5a for the denotations), then the admissible area is the subtriangle *AEF*, where the points *E* and *F* are taken on the segment *CB* in such a way that *AE* is perpendicular to *AB*, and *AF* is perpendicular to *AC*. As shown in figure 5b, for any point from this admissible area the „yellow-type" division into right triangles is possible.

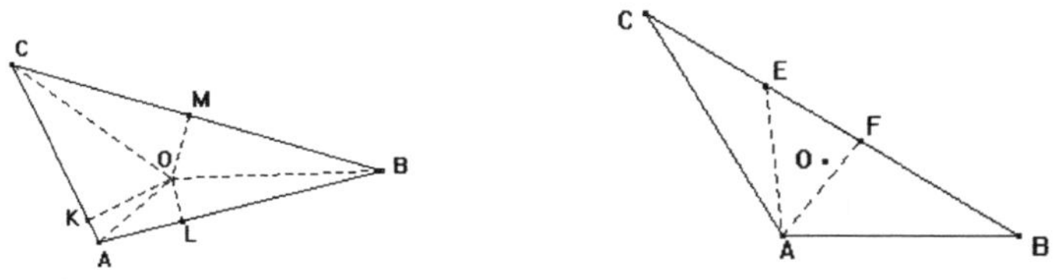

a) b)

Figure 5

Obviously, in case when ABC is nonobtuse and *O* is the centre of circumscribed sphere, the quasi-yellow refinement coincides with the standard yellow refinement.

4. MAIN RESULTS

Consider the situation of Figure 3. We assume that our initial triangulation is such that the obtuse triangles are „rarely" placed, i.e., surrounded only by nonobtuse triangles.

Let *BAG* be one of obtuse triangles in such an initial triangulation. We assume that it is surrounded by three nonobtuse triangles *ABC*, *AGE*, and *BGD*, adjacent to *BAG* along its sides (see Figure 6).

Using Lemma 2.1, we are able to find co-ordinates of points $L=(x_L, y_L)$ and $P=(x_P, y_P)$ (using co-ordinates of vertices *A*, *B*, *C*), so that for any point from segment [*LP*] and (mid)points *F* and *E*, the quasi-red refinement of *ABC* produces nonobtuse (sub)triangles.

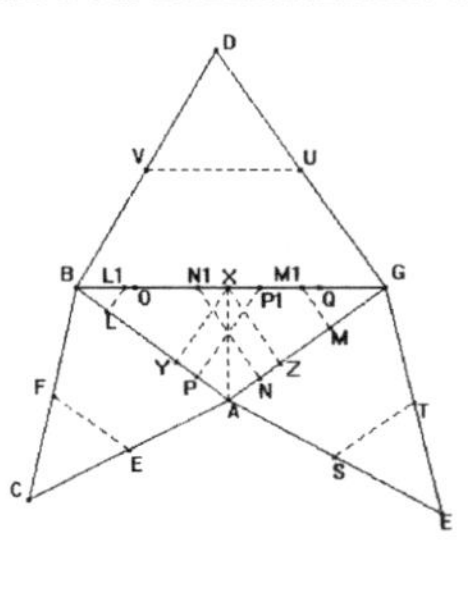

Figure 6

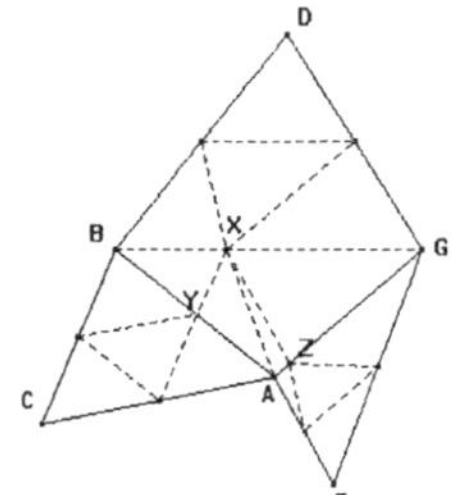

Figure 7

We can similarly find points $N=(x_N, y_N)$ and $M=(x_M, y_M)$ for triangle *AGE*; and points $O=(x_O, y_O)$ and $Q=(x_Q, y_Q)$ -for triangle *BGD*. Our goal is to find conditions on all the vertices *A*, *B*, *C*, *G*, *E*, and *D*, under which there exists a point $X \in [OQ]$ so that its orthogonal projection *Y* on side *AB* belongs to $[LP]$, and its orthogonal projection *Z* on side *AG* belongs to $[NM]$. It means then that we can conformably divide triangles *ABC*, *AGE*, and *BGD* (using the quasi-red refinement), and triangle *BAG* (using the quasi-yellow refinement), into nonobtuse (sub)triangles (see Figure 7).

To find conditions on the existence of such point *X*, we define point $\overline{L}$ as the intersection of perpendicular to *BA* at point *L* with side *BG*, and point $\overline{P}$ as the intersection of perpendicular to *BA* at point *P* with side *BG*. Similarly, let $\overline{N}$ denote the intersection of perpendicular to *AG* at point *N* with side *BG*, and $\overline{M}$ -the intersection of perpendicular to *AG* at point *M* with side *BG*.

In this way, we get six points on side *BG*: $\overline{L}, \overline{P}, \overline{N}, \overline{M}$, *O*, and *Q*. Using Lemma 2.1, we can easily find representations:

(4.1) $O=(q_1 x_G+(1-q_1)x_B, q_1 y_G+(1-q_1)y_B)$,
(4.2) $Q=(q^1 x_G+(1-q^1)x_B, q^1 y_G+(1-q^1)y_B)$,

where q_1 and q^1 are some real numbers. Analogously, we can assume that

(4.3) $\overline{L}=(q_2 x_G+(1-q_2)x_B, q_2 y_G+(1-q_2)y_B)$
(4.4) $\overline{N}=(q_3 x_G+(1-q_3)x_B, q_3 y_G+(1-q_3)y_B)$
(4.5) $\overline{P}=(q^2 x_G+(1-q^2)x_B, q^2 y_G+(1-q^2)y_B)$
(4.6) $\overline{M}=(q^3 x_G+(1-q^3)x_B, q^3 y_G+(1-q^3)y_B)$

where precise values of q_2, q_3, q^2, and q^3 are presented below in Lemma 4.2.

The following result is valid.

Lemma 4.1 Let $\underline{q}=\max\{q_1, q_2, q_3\}$ and $\overline{q}=\max\{q^1, q^2, q^3\}$. If $\underline{q} \le \overline{q}$, then point *X* (cf. Figure 7) always exists; for example, one proper choice is

(4.7) $X = (dx_G + (1-d)x_B, dy_G + (1-d)y_B)$,

where $d = \frac{(\underline{q} + \overline{q})}{2}$.

The proof is straightforward.

Lemma 4.2 The values of numbers in (4.3)--(4.6) are:

(4.8) $$q_2' = \frac{(x_L - x_B)(x_B - x_A) + (y_L - y_B)(y_B - y_A)}{(x_G - x_B)(x_B - x_A) + (y_G - y_B)(y_B - y_A)}$$

(4.9) $$q'^2 = \frac{(x_P - x_B)(x_B - x_A) + (y_P - y_B)(y_B - y_A)}{(x_G - x_B)(x_B - x_A) + (y_G - y_B)(y_B - y_A)}$$

(4.10) $$q_3' = \frac{(x_N - x_B)(x_G - x_A) + (y_N - y_B)(y_G - y_A)}{(x_G - x_B)(x_G - x_A) + (y_G - y_B)(y_G - y_A)}$$

(4.11) . $$q^3 = \frac{(x_M - x_B)(x_G - x_A) + (y_M - y_B)(y_G - y_A)}{(x_G - x_B)(x_G - x_A) + (y_G - y_B)(y_G - y_A)}$$

The proof is straightforward.

5. FINDING AN OPTIMAL SHAPE OF THE POLE OF AN ELECTROMAGNET.

The problem of practical interest is that of finding an optimal shape of the pole of an electromagnet such that magnetic flux density in the air-gap assumes prescribed value.

The general system of state equations has the form:

$$-\nabla \cdot (\nu \nabla A) = J \qquad A|_\infty = 0 .$$

In our case the above splits into three separate PDE's, one for each sub-region. The objective function to be minimised is given by the mean square deviation between prescribed and actual field distribution in the air-gap. $E = \sum_{i=1}^{M} (B_i^{obl} - B_i^{zal})^2$. The problem is analysed in [10] and [16].

In the paper, we demonstrate the use of a quasi-red and quasi-yellow non-obtuse refinement scheme for the solution of the above problem. A comparative study of these techniques is given. Related contribution [14] demonstrates further details of our approach.

6. NUMERICAL RESULTS

We present numerical example that demonstrates usefulness and efficiency of our approach. All numerical simulations were performed in Matlab 6.0 environment. The finite element tool was that of Matlab PDE Toolbox. To determine an unknown shape of the part of the pole boundary we select a number of moving nodes on the pole boundary equal to 8. The number of points at which the magnetic field is prescribed (0.4 Tesla) is fixed at 10. Our algorithm begins with checking the initial FEM mesh against obtuse triangles. The number of detected obtuse triangles at initial FEM mesh is 9. After first mesh refinement the number of obtuse triangles reduced to 4. At each analysis step we eliminate all obtuse triangles from the mesh to get better conditioning of the corresponding optimization problem. The results are summarized in table below.

Table 1

Coordinates of control points	Magnetic field value
v(0.000000,0.000000)	0.401146
v(0.000000,0.004167)	0.401147
v(0.000000,0.008333)	0.396453
v(0.000000,0.012500)	0.396453
v(0.000000,0.016667)	0.394602
v(0.000000,0.020833)	0.399188
v(0.000000,0.025000)	0.404835
v(0.000000,0.029167)	0.407663
v(0.000000,0.033333)	0.404314
v(0.000000,0.037500)	0.384658

Total mean square error: 0.000394. The following figures show the procedure of mesh rectification.

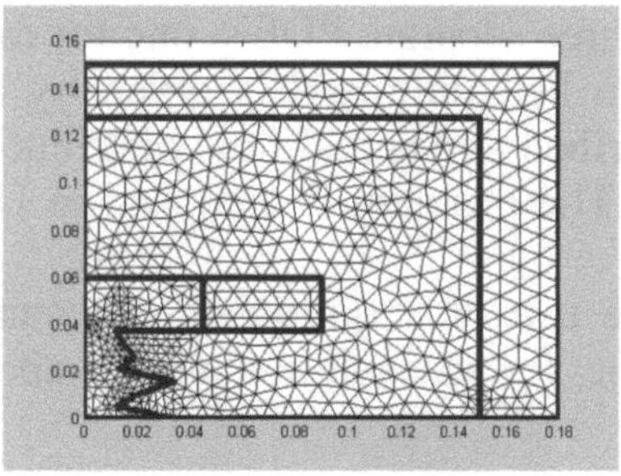

A finite-element refined mesh

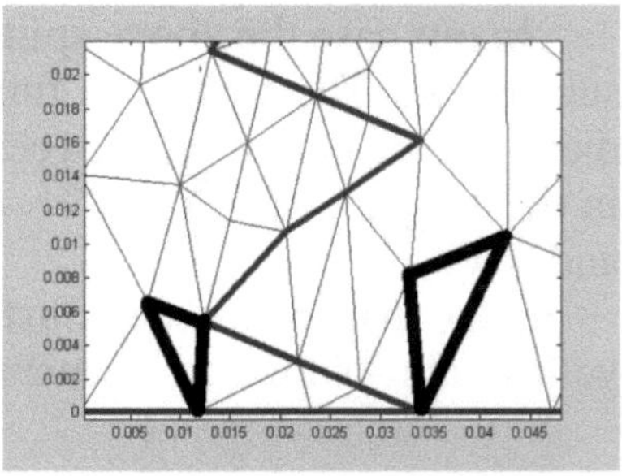

A finite-element initial mesh : obtuse triangles

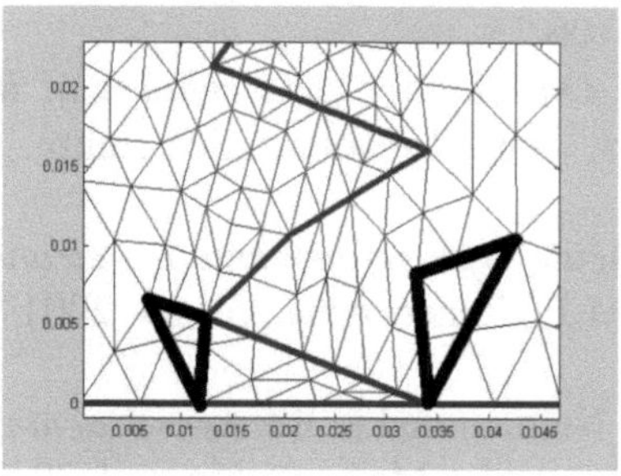

A finite-element refined mesh
(quasi-red and quasi-yellow)

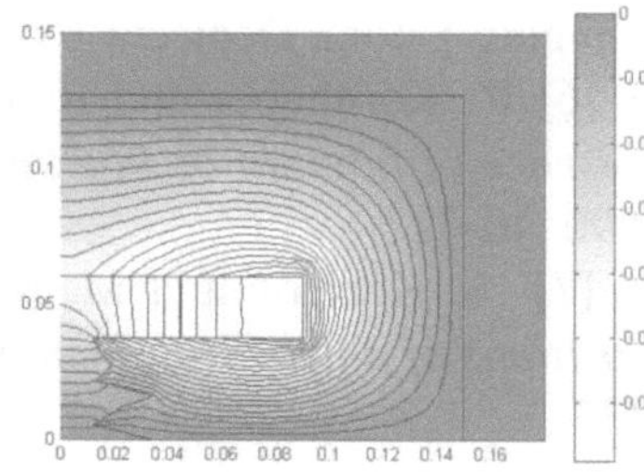

Field distribution for final shape

Conclusions

The results clearly demonstrate advantages of the elaborated algorithm. They compare well with those given in [15]. Further work is under way to extend the algorithm to 3-D case.

References

[1] B.S. Baker, E. Grosse and C.S. Rafferty, *Nonobtuse triangulation of polygons,* Discrete Comput. Geom vol 3 1988, 147-168.

[2] R.E. Bank, PLTMG*: A software package for solving partial differential equations: Users' Guide 7.0,* SIAM, Philadelphia 1994.

[3] G.F. Carey, *Computational grids. Generation, adaptation, and solution strategies,* Series in Computational and Physical Processes in Mechanics and Thermal Sciences. Taylor & Francis, Washington, DC 1997.

[4] P.G. Ciarlet and P.A. Raviart, *Maximum principle and uniform convergence for the finite element method*, Comput. Methods Appl. Mech. Engrg. vol 2 1973, 17-31.

[5] J.L. Gerver, *The dissection of a polygon into nearly equilateral triangles*, Geom. Dedicata 16 1984, 93-106.

[6] R. Kornhuber and R. Roitzsch, *On adaptive grid refinement in the presence of internal or boundary layers*, Impact Comput. Sci. Engrg. vol 2 1990 40-72.

[7] S. Korotov and M. Krizek, *Acute type refinements of tetrahedral partitions of polyhedral domains* , SIAM J. Numer. Anal. vol 39 2001 724--733 .

[8] S. Korotov, M. Krizek and P. Neittaanmaki, *Weakened acute type condition and the discrete maximum principle*, Math. Comp. vol 70 2000, 107-119.

[9] S. Korotov and J. Stando, *Yellow-red and nonobtuse refinements of planar triangulations*, (to appear in Math. Notes (Miskolc)).

[10] M. Krizek and P. Neittaanmäki, *Mathematical and Numerical Modelling in Electrical Engineering: Theory and Applications,* Kluwer Academic Publishers, Dordrecht 1996.

[11] M. Krizek and Qun Lin, *On diagonal dominance of stiffness matrices in 3D*, East-West J. Numer. Math. vol 3 (1) 1995, 59—69.

[12] V. Ruas Santos, *On the strong maximum principle for some piecewise linear finite element approximate problems of non-positive type*, J. Fac. Sci. Univ. Tokyo Sect. IA Math. l 29 1982, 473—491.

[13] G. Strang and G.J. Fix, *An analysis of the finite element method*, Prentice Hall, Englewood Cliffs, N. J. 1973

[14] J. Stańdo, M. Rudnicki and D. Krawczyk-Stańdo, *Novel numerical integration scheme for the solution of inverse magnetic problems*, presented at IGTE 2002, Graz, Austria, September 16-18

[15] M. Rudnicki, A. Tomczyk and K. Sasiak, *Numerical optimization methods for magnet design*, OIPE 2002, Lodz, Poland, September 12-14, 2002

[16] P.Neittaanmäki, M. Rudnicki and A. Savini, *Inverse Problems and Optimal Design in Electricity and Magnetism*, Oxford University Press,, Oxford, 1996.

2-D INVERSE PROBLEM: IDEAL CURRENT DISTRIBUTION GENERATING A MAGNETIC FORCE FIELD WITH GIVEN CONSTRAINTS ON HOMOGENEITY

Lionel Quettier, Alain Mailfert

GREEN, Groupe de Recherche en Electronique et Electrotechnique ,
2, av. de la Forêt de Haye, 54516 Vandoeuvre-les-Nancy, France
lionel.quettier@ensem.inpl-nancy.fr, amailfer@ensem.inpl-nancy.fr

Abstract: Generation of uniform field of magnetic forces in a given domain is a problem with various applications. The main idea is to compensate the gravity by means of a magnetic force acting on a small diamagnetic volume. This paper deals with a way of identifying ideal current distributions able to generate a quasi- uniform force field. Based on one particular expression of the complex magnetic potential, an inverse methodology, which is valid for structures with plane symmetry (2-D field), has been worked out. Currents are modeled by purely surface (or thin shell) current distributions.

Key words: magnetic levitation, inverse problem

1. INTRODUCTION

Generation of a quasi-uniform magnetic force field is a problem with several applications: magnetic separation of particles with small differential magnetic susceptibility, micro-gravity (magnetic levitation, crystal growth). Our paper deals mainly with magnetic levitation of diamagnetic material. The structure of magnets usually investigated to generate a quasi-uniform field of magnetic forces is an axisymmetric coil [1-3]. Unfortunately, this kind of designs leads necessarily to a restricted levitation domain, which can be an inconvenient in crystal growth experiments for example, where it is very interesting to operate in a volume as big as possible.

M. Rudnicki and S. Wiak (eds.), Optimization and Inverse Problems in Electomagnetism, 125–137.

In the first part, we will show why it is not possible to generate a constant magnetic force field. Then, we will focus on systems with plane symmetry (2-D field): the theoretical study presented here is a generalization of concepts previously approached for cylindrical structure with plane symmetry [4][5]. This leads to one particular expression of the complex magnetic potential creating a magnetic field with given properties, specially concerning constraints on the required homogeneity of the levitation force components. The proposed approach to identify current distributions solving the inverse problem is so based on the following criteria: the value of the vertical component of the magnetic force necessary to compensate gravity for a given diamagnetic material, the radius of the cylindrical working area, and finally the homogeneity of horizontal and vertical components of the residual force (sum of magnetic levitation force and gravity).

2. MAGNETIC FORCE FIELD

The magnetic force acting on a small volume of magnetic material is given by:

$$\mathbf{F_m} = \frac{1}{2} V \chi_m \mu_0 \, \mathbf{grad}\left(\mathbf{H}^2\right) \tag{1}$$

Here V is the volume to make levitated, χ_m is magnetic susceptibility (<0 for a diamagnetic material), μ_0 is magnetic permeability in vacuum and **H** is magnetic field.

Defining G as the vertical component of the vector μ_0^2 **grad(H²)**, the condition to compensate gravity for a diamagnetic material can be easily expressed as following:

$$G = \frac{\mu_0 \rho \|\mathbf{g}\|}{|\chi_m|} \tag{2}$$

with ρ volume density for the considered material and **g** gravity vector.

The aim is so to identify a current distribution leading to an area where the vector **grad(H²)** is constant. But unfortunately, as we are now going to show, it is not possible for a magnetic field to satisfy fully this condition.

Let us consider the vector space E = J^3 and its Euclidean inner product defined by:

$$\forall\, (\mathbf{x}, \mathbf{y}) \in J^3,\ <\mathbf{x},\mathbf{y}> = \sum_{i=1}^{3} x_i . y_i \tag{3}$$

Defining then $N(\mathbf{x})$ as the norm associated to this scalar product, we can write:

$$<\mathbf{H},\mathbf{H}>= N(\mathbf{H}) = \sum_{i=1}^{3} H_i^2$$

where H_i represents coordinates of **H** in the system $\mathbb{R}^3$

The final target is to get:

$$\mathbf{grad}(\mathbf{H}^2) = \mathbf{a} \tag{4}$$

where **a** is a constant vector defined by $\mathbf{a} = a_1\ \mathbf{i} + a_2\ \mathbf{j} + a_3\ \mathbf{k}$, with $(a_1,a_2,a_3) \in \mathbb{R}^3$ and (O, **i**, **j**, **k**) represents a rectangular coordinate system.

Step 1: $\mathbf{grad}(\mathbf{H}^2) = \mathbf{a} \Leftrightarrow \exists\, C \in \mathbb{R}\ /\ \mathbf{H}^2 = a_1.x + a_2.y + a_3.z + C$

Step 2: General form of solution of the previous equation is given by:

$$\mathbf{H} = \begin{pmatrix} \sqrt{\alpha_1 x + \beta_1 y + \chi_1 z + C_1}\,\mathbf{i} \\ \sqrt{\alpha_2 x + \beta_2 y + \chi_2 z + C_2}\,\mathbf{j} \\ \sqrt{\alpha_3 x + \beta_3 y + \chi_3 z + C_3}\,\mathbf{k} \end{pmatrix} \tag{5}$$

with $(\alpha_i, \beta_i, \chi_i, C_i)$, $(i, j, k) \in \{1,2,3\}$, such as

$$\sum_{n=1}^{3} \alpha_n = a_1 \quad \sum_{n=1}^{3} \beta_n = a_2 \quad \sum_{n=1}^{3} \chi_n = a_3 \quad \sum_{n=1}^{3} C_n = C$$

Step 3: If **H** represents magnetic field, according to Maxwell equations in stationary conditions, it must necessary satisfy:

$$\Delta\mathbf{H} = \mathbf{0} \tag{6}$$

Therefore, as $\Delta\mathbf{H} = \begin{pmatrix} \Delta H_x \mathbf{i} \\ \Delta H_y \mathbf{j} \\ \Delta H_z \mathbf{k} \end{pmatrix}$, we have $\Delta\mathbf{H} = \mathbf{0} \Leftrightarrow \begin{cases} \Delta H_x = 0 \\ \Delta H_y = 0 \\ \Delta H_z = 0 \end{cases}$

But starting from general form of **H** defined in step 2, we then get:

$$\Delta H_x = -\frac{1}{4} \frac{\alpha_1^2 + \beta_1^2 + \chi_1^2}{(\alpha_1 x + \beta_1 y + \chi_1 z + C_1)^{3/2}} \tag{7}$$

We can now easily see that the last expression leads necessarily to $\Delta H_x \neq 0$ (except if $\mathbf{a} = \mathbf{0}$, which is not an interesting case). Of course the same study is also possible for ΔH_y and ΔH_z by circular permutation of subscripts. For these reasons, the following conclusion can be drawn: magnetic field can not satisfy both equations (4) and (6), or in other words, it is not possible to generate a fully constant magnetic force field. That is why, we are going to study, in the following part, the problem of generation of quasi-uniform magnetic force field.

3. A PARTICULAR EXPRESSION OF THE COMPLEX MAGNETIC POTENTIAL

Let us now consider systems with plane symmetry (2-D magnetic field) where the currents are distributed along the z-axis. Using complex coordinates defined by $\underline{z} = x + iy$, the complex magnetic potential in the two-dimensional geometry considered, is defined by [6]:

$$\underline{W}(\underline{z}) = \frac{A_z}{\mu_0} + iV \tag{8}$$

where A_z is the only component of the vector potential $\mathbf{A}$
V is the scalar magnetic potential such as $\mathbf{H} = -\mathbf{grad}(V)$

Let us now study the following particular expression of $\underline{W}(\underline{z})$:

$$\underline{W}(z) = -\lambda\, \underline{z}^{\alpha},\ \alpha>0 \text{ and } \lambda \text{ is a real number} \tag{9}$$

In cylindrical coordinates, this leads to:

$$\mathbf{H} = \begin{cases} \alpha\lambda r^{\alpha-1} \sin(\alpha\theta)\mathbf{u}_r \\ \alpha\lambda r^{\alpha-1} \sin(\alpha\theta)\mathbf{u}_\theta \end{cases} \tag{10}$$

And one finally gets:

$$\mathbf{grad}(\mathbf{H}^2) = (\alpha\lambda)^2(2\alpha-2)r^{2\alpha-3}\mathbf{u}_r \tag{11}$$

This particular expression of $\underline{W}(\underline{z})$ gives a radial force, independent of θ and of constant modulus when α=3/2.

Let $R_1=(O,x,y,z)$ designate a rectangular coordinate system and let us consider $R_2=(O',x',y',z)$ as one local rectangular system as represented in figure 1.

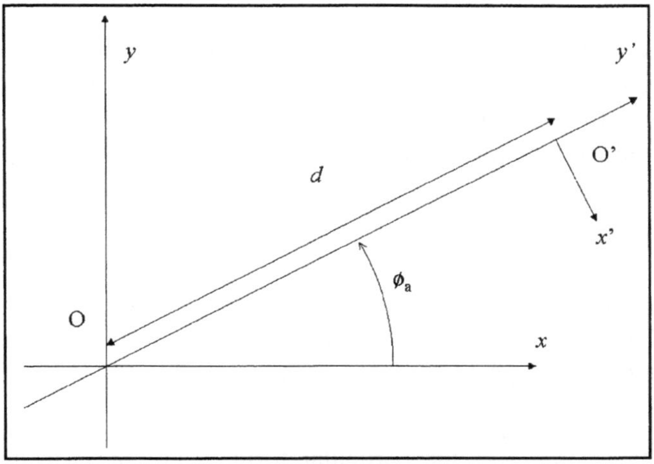

Figure 1. Coordinates systems definition

Thanks to two basic coordinate system transformations, the complex magnetic flux density created by the particular complex magnetic potential form previously defined can be expressed in the local system R_2 as following:

$$\underline{H}(\underline{z}'') = \alpha\lambda \underline{z}_0^{\alpha-1}\left(1+\frac{\underline{z}''}{\underline{z}_0}\right)^{\alpha-1} e^{i\alpha\left(\varphi_a-\frac{\pi}{2}\right)} \tag{12}$$

where: $\underline{z}''$ designates the complex variable in system R_2.

and $\underline{z}_0 = d\,e^{i\frac{\pi}{2}}$ designates the change of coordinate system center

Since $\underline{H}$ is analytic, it can be expanded into a Taylor's series around the disk of center O' as following:

$$\underline{H}(\underline{z}'') = \alpha\lambda \sum_{n=0}^{\infty} N_n \underline{z}_0^{\alpha-1-n} (\underline{z}'')^n e^{i\alpha\left(\varphi_a-\frac{\pi}{2}\right)} \tag{13}$$

with $N_n = \dfrac{1}{\alpha n!}\displaystyle\prod_{i=0}^{n}(\alpha-i)$ for $n \geq 0$

Depending on the value of ϕ_a, the chosen working area moves, thus enabling the angular offset between the directions of the magnetic field and of the magnetic force field to be varied.

When $\varphi_a = -\dfrac{\pi}{2\alpha}$, the magnetic field vector and the **grad**($\mathbf{H}^2$) vector are collinear but opposite way in the local coordinate system R_2.

4. STABILITY OF MAGNETIC LEVITATION

Considering the method used in [7] to determine the stability conditions for a system with axial symmetry, it is possible to express the stability conditions for the two-dimensional geometry invariant by translation here considered with the following formula:

Horizontal stability: $H'^2 + H.H'' > 0$ (14)

Vertical stability: $H'^2 - H.H'' > 0$ (15)

where H designates the magnetic flux density on the vertical axis ($\theta = \phi_a$), H' and H'' are respectively the first and the second derivatives of the function H with respect to r, position on this vertical axis.

For $\theta = \varphi_a = -\dfrac{\pi}{2\alpha}$, the particular form of complex magnetic potential defined above, by combination with (14) and (15) yields the conditions:

Horizontal stability: $(\alpha\lambda)^2(\alpha - 1)(2\alpha + 3)r^{2(\alpha-2)} > 0$ (16)

Vertical stability: $(\alpha\lambda)^2(\alpha - 1)(2\alpha - 3)r^{2(\alpha-2)} > 0$ (17)

A quick study of stability conditions as functions of α shows that α must be chosen strictly higher than 3/2 to satisfy both conditions (16) and (17).

5. FORMULATION OF THE "INVERSE SOURCE SYNTHESIS PROBLEM"

The aim is to research a current distribution creating a circular working area of center O' and of radius *R*, where G, value of the vertical component of the vector μ_0^2 **grad(H²)**, will at least allow to compensate the gravity.

Moreover homogeneity on both horizontal and vertical residual components is given constraint and has to be taken account.

Re-writing (11) in the local system R_2=(O,'*x'*,*y'*), the vector $\mathbf{grad}(\mathbf{H}^2)$ can be expressed at given points O', A1, A2, A3, and A4 as following:

$$\mathbf{grad}\left(\mathbf{H}^2\right)_{|O'} = \left((\alpha\lambda)^2(2\alpha-2)d^{2\alpha-3}\right)\mathbf{u}_{y'}$$

$$\mathbf{grad}\left(\mathbf{H}^2\right)_{|A_1} = \left((\alpha\lambda)^2(2\alpha-2)d^{2\alpha-3}\right)\left(sin(\delta)\mathbf{u}_{x'} + cos(\delta)\mathbf{u}_{y'}\right)$$

$$\mathbf{grad}\left(\mathbf{H}^2\right)_{|A_2} = \left((\alpha\lambda)^2(2\alpha-2)d^{2\alpha-3}\right)\left(-sin(\delta)\mathbf{u}_{x'} + cos(\delta)\mathbf{u}_{y'}\right) \quad (18)$$

$$\mathbf{grad}\left(\mathbf{H}^2\right)_{|A_3} = \left((\alpha\lambda)^2(2\alpha-2)(d-\mathrm{R})^{2\alpha-3}\right)\mathbf{u}_{y'}$$
$$\mathbf{grad}\left(\mathbf{H}^2\right)_{|A_4} = \left((\alpha\lambda)^2(2\alpha-2)(d+\mathrm{R})^{2\alpha-3}\right)\mathbf{u}_{y'}$$

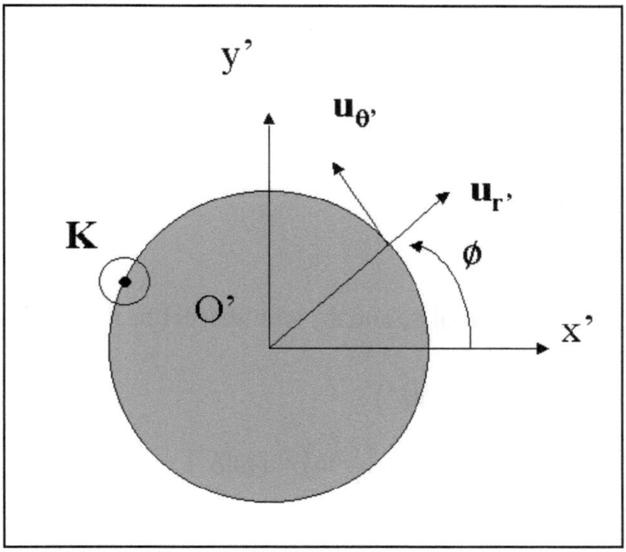

Figure 2. Circular working area

In this section, the assumption $\alpha > 3/2$ is made: it ensures stability conditions (16) and (17) in the case of a diamagnetic material as previously studied and in the same times, it satisfies $\left\|\mathbf{grad}\left(\mathbf{H}^2\right)\right\|_{|A4} > \left\|\mathbf{grad}\left(\mathbf{H}^2\right)\right\|_{|A3}$

Furthermore let ε_x and ε_y respectively define the homogeneity on horizontal and vertical components as following:

$$\varepsilon_x = \frac{\left\|\mathbf{grad}\left(\mathbf{H}^2\right)_{|A1x'}\right\|}{\left\|\mathbf{grad}\left(\mathbf{H}^2\right)_{|A3y'}\right\|}$$, homogeneity of the residual force on Ox'

Then $$\varepsilon_{y1} = \frac{\left\|\mathbf{grad}\left(\mathbf{H}^2\right)_{|A4y'} - \mathbf{grad}\left(\mathbf{H}^2\right)_{|A3y'}\right\|}{\left\|\mathbf{grad}\left(\mathbf{H}^2\right)_{|A3y'}\right\|}$$

and $$\varepsilon_{y2} = \frac{\left\|\mathbf{grad}\left(\mathbf{H}^2\right)_{|A1y'} - \mathbf{grad}\left(\mathbf{H}^2\right)_{|A3y'}\right\|}{\left\|\mathbf{grad}\left(\mathbf{H}^2\right)_{|A3y'}\right\|}$$

The homogeneity of the residual force on Oy' is therefore defined as $\varepsilon_y = \max(\varepsilon_{y1}\,;\,\varepsilon_{y2})$,

Where $\varepsilon_{y1} = \left(\dfrac{1+sin(\delta)}{1-sin(\delta)}\right)^{2\alpha-3} - 1$ and $\varepsilon_{y2} = \dfrac{cos(\delta)}{(1-sin(\delta))^{2\alpha-3}} - 1$

which gives $\varepsilon_x = \dfrac{sin(\delta)}{(1-sin(\delta))^{2\alpha-3}} - 1$ (19)

and $\varepsilon_y = \varepsilon_{y1} = \left(\dfrac{1+sin(\delta)}{1-sin(\delta)}\right)^{2\alpha-3} - 1$ because $\varepsilon_{y1} > \varepsilon_{y2}$ for $\alpha > 3/2$ (20)

A series expansion of ε_x and ε_y, with respect to the variable δ, about the point 0 yields:

$$\varepsilon_x = \delta + (2\alpha - 3)\delta^2 + o(\delta^3) \tag{21}$$

$$\varepsilon_y = (4\alpha - 6)\delta + (8\alpha^2 - 24\alpha - 36)\delta^2 + o(\delta^3) \tag{22}$$

Defining $\xi = \dfrac{\varepsilon_x}{\varepsilon_y}$, the above equations give at the first order:

$$\alpha = \frac{1+6\xi}{4\xi} \tag{23}$$

Table 1. Evolution of α against ξ

ξ	0.5	1	1.5	2	2.5	5	10	$\to \infty$
α	2	7/4	5/3	13/8	8/5	31/20	61/40	$\to 3/2$

The proposed approach allows identifying the value of α, depending on required constraints on homogeneity.

<u>Remarks:</u>

The case $\alpha = 3/2$ has always been studied in [4] and gives a constant modulus radial force.

The case $\alpha < 3/2$ leads to an unstable equilibrium but it can be nevertheless studied with the same method as previously and yields $\alpha = \dfrac{3}{2} - \dfrac{1}{4\xi}$.

6. EXAMPLE OF IDENTIFICATION OF A MAGNETIC FIELD DISTRIBUTION WITH GIVEN CONSTRAINTS

Let us consider in this last part an example to point out the pertinence of the proposed method to identify a two-dimensional magnetic field distribution. Let us now assume the aim is to make liquid hydrogen levitate (diamagnetic substance).

How identify parameters to identify the right form for the complex magnetic potential $\underline{W}(\underline{z}) = -\lambda\,\underline{z}^{\alpha}$, with $\alpha>0$?

Using (2) and as $\chi_m=-1.78.10^{-6}$ m^3/kg and ρ = 70.78 kg/m^3 for liquid hydrogen, one gets that G has to reach a value of 982 T²/m to compensate gravity in this case. Moreover, the following assumption will be made: *R* radius of the working area will be 5 mm, $\varepsilon_x = 2\%$, and $\xi = 1$.

Starting from (11), we can evaluate an approximated value of the magnetic flux density with the following formula:

$$\|\mathbf{B}\| \approx \sqrt{\frac{\mathrm{G}\,d}{2\alpha-2}} \tag{24}$$

That is why, the assumption $\alpha > 3/2$ is made, which ensures a stable levitation and in the same time allows decreasing the value of the magnetic field.

δ is the solution (in radian) from equation $\varepsilon_x = \dfrac{sin(\delta)}{(1-sin(\delta))^{2\alpha-3}} - 1$ on the interval [0 ;1].

But as δ is small, it is possible to consider $\varepsilon_x \approx \delta$ at the first order.

$d = R$ / sin(δ) is the distance between both global and local coordinates systems and $d \approx R / \delta$ following the last assumption on δ.

This yields $\alpha = 7/4$ because $\xi = 1$, and $\lambda = \dfrac{\sqrt{\mathrm{G}}}{\mu_0\alpha\sqrt{2\alpha-2}(d-R)^{\alpha-3/2}}$ (25)

It gives: d = 0.252 m, $\|\mathbf{B}\| \approx 13\mathrm{T}$, $\lambda = 1.649\ 10^7\ \mathrm{T.m}^{-3/2}$

Once a magnetic field distribution having the required properties has been identified, the next stage to solve completely this inverse problem is to use a method developed at G.R.E.E.N. [8]. Its goal is to identify an ideal superconducting coil structure (ideal as the currents are modeled by purely superficial densities) generating the above magnetic field distribution $\underline{W}(\underline{z})$.

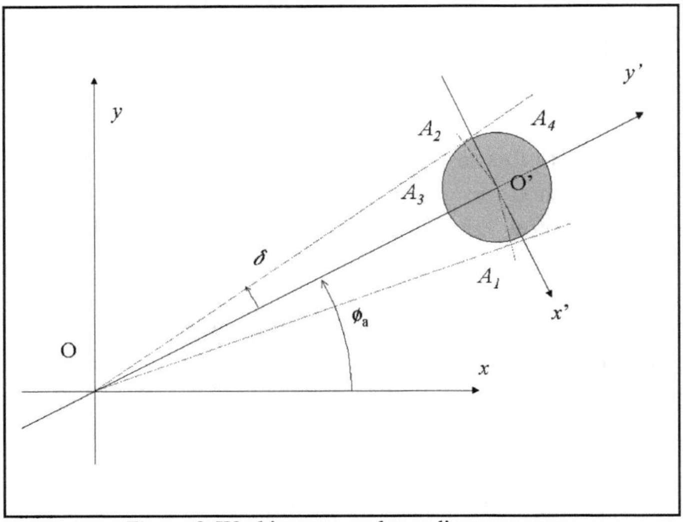

Figure 3. Working area and coordinates system

Let us define H_t as the tangential field on the circle of center O' and of radius *R* created by the previously identified complex magnetic potential.

Let then define **n**, normal vector to the cylindrical working area.

The magnetic field distribution generated by a current density $\mathbf{K} = 2H_t \times \mathbf{n} = K_z \mathbf{u_z}$ distributed over the circle will exactly have the wished properties of $\underline{W}(\underline{z}) = -\lambda\, \underline{z}^{\alpha}$.

We have:

$$\begin{cases} \mathbf{u_{r'}} = cos(\varphi)\mathbf{u_{x'}} + sin(\varphi)\mathbf{u_{y'}} \\ \mathbf{u_{\theta'}} = -sin(\varphi)\mathbf{u_{x'}} + cos(\varphi)\mathbf{u_{y'}} \end{cases} \tag{26}$$

Furthermore,

$$K_z = 2H_{\theta'} \tag{27}$$

As $\underline{H} = H_{y'} + iH_{x'}$, by combining (26) and (27), we get:

$$H_{\theta'} = -sin(\varphi).Imag(\underline{H}) + cos(\varphi).Re(\underline{H}) \tag{28}$$

where $Re(\underline{z})$ and $Imag(\underline{z})$ respectively design real part and imaginary part of the complex number $\underline{z}$.

This superficial current density can therefore be expressed as a function of G, *R*, ε_x and ε_y as following:

$$K_z = \frac{8(1-\varepsilon_x)^{-\left(\frac{\varepsilon_x}{4\varepsilon_y}\right)}}{\mu_0(6\varepsilon_x+\varepsilon_y)} \sqrt{\frac{2\varepsilon_x G R}{2\varepsilon_x+\varepsilon_y}} \sum_{n=1}^{\infty}\left(\left(\prod_{u=0}^{n-1}\left(\frac{6\varepsilon_x+\varepsilon_y}{4\varepsilon_y}-u\right)\right)\frac{\sin\left(n\left(\varphi-\pi/2\right)\right)}{(n-1)!\,\varepsilon_x^{1/2-n}}\right)$$

Figures 4. represents respectively the superficial density K_z against ϕ (solid line), and the function $50(K_z - \max(K_z).\cos(\phi))$ (dotted line) in the case of liquid hydrogen levitation.

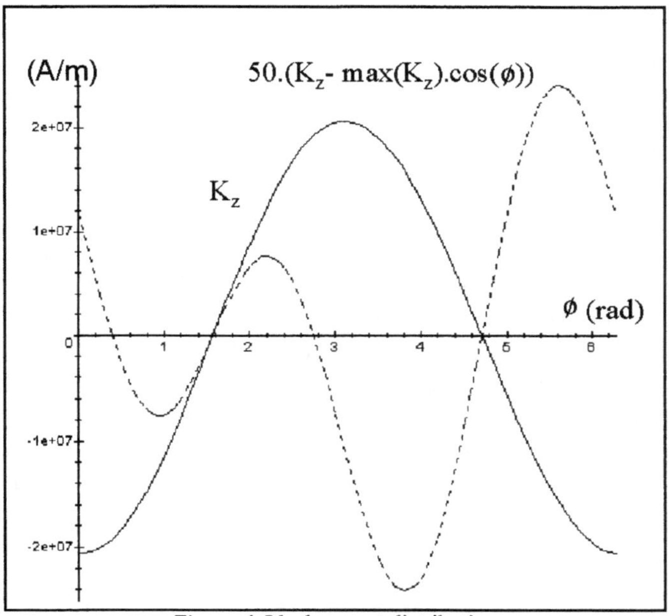

Figure 4. Ideal current distribution

Figure 5. represents the field distribution in the local coordinates system.

This utopian solution is almost the same as in the one generating a uniform magnetic field where it is a purely $\cos(\phi)$ distribution.

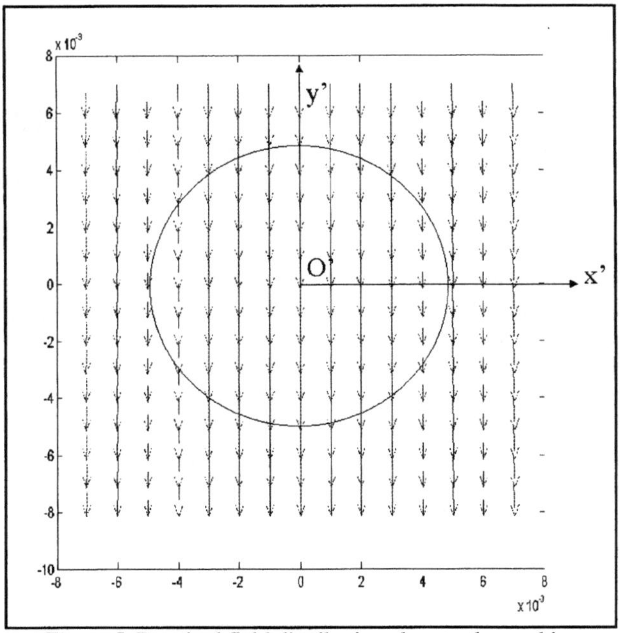

Figure 5. Required field distribution close to the working area

Finally, this theoretical structure will have to be discretized into a 2-D real design whose currents will be modeled by volume densities as in [5]. Due to the order of magnitude of the required magnetic field, the magnetic device will be achieved using superconducting cables. This cross-section will then have to be optimized to reach more precisely the required magnetic properties by changing variables like number of blocks, blocks position, number of superconducting conductors per block, etc.

7. CONCLUSION

In this paper, one particular expression of the complex magnetic potential has been studied to solve a two-dimensional magnetic levitation problem. The proposed method allows to identity a theoretical magnetic field distribution answering the posed problem. The most important point is the possibility to determinate an ideal current distribution generating a magnetic force field with given constraints on homogeneity. One can also point out the fact that the complex formulation used for magnetic calculation can be easily used to include some aspects of stability in the particular case of a geometry invariant by translation. This fully analytical method can be used as a first step to perform the calculation of real design of magnetic device for magnetic levitation.

ACKNOWLEDGEMENTS

The authors gratefully acknowledge Dr. A. Colteu, working with GREEN, for his helpful contribution to the first part of this work concerning the demonstration on impossibility of constant magnetic force field generation.

REFERENCES

1. Beaugnon E., Tournier R., *"Levitation of Water and Inorganic Substances in High Static Magnetic Field"*, Journal of Physics III, France, 1, 1991.
2. Berry M. V., Geim A. K., *"Of flying frog and levitrons"*, Eur. J. Phys. Vol. 18, pp 307-313, 1997.
3. Ozaki O., Kiyoshi T., Matsumoto S., Koyanagi K., Fujihira J., Nakayama H., Wada H.,*"Design Study of Superconducting Magnets for Uniform and High Magnetic Force Field Generation"*, IEEE Transactions on Applied Superconductivity, Vol. 11, No 1, pp 2252-2255, March 2001.
4. Vincent-Viry O., Mailfert A., Colteu A., Dael A., Gourdin C., Quettier L.,*" Micro-gravity: current distributions creating a uniform force field"*, The European Physical J.Appl. Phys., Vol. 13, No 2, pp 121-123, 2001.
5. Quettier L., Mailfert A., *"New design of Superconducting Magnet for Generation of Uniform Magnetic Force Field"*, presented in ASC 2002, Aug 2002, Houston, Texas, USA.
6. Beth R.A., *"Complex Representation and Computation of two-dimensional Magnetic Fields,"* J. Appl. Phys., Vol. 37, No. 7, pp 2568-2571, 1966.
7. Holmes L. M., "Stability of magnetic levitation", J. Appl. Phys. Vol. 49, No 6, pp 3102-3109, June 1978.
8. Mailfert A., Vincent-Viry O.,*" 2-D inverse problems with applications to micro-gravity and energy storage "*, International Journal for Computation and Mathematics in Electrical and Electronic Engineering COMPEL, vol. 20, n°3, pp 869-878, 2001.

S.M.E.S. OPTIMIZATION USING THE EXPERIMENTAL DESIGN METHOD

S. VIVIER, M. HECQUET, F. GILLON, P. BROCHET
L2EP - Ecole Centrale de Lille, Cité Scientifique
B.P. 48, 59651 Villeneuve d'Ascq, France

Abstract: Some results on optimization techniques and the Experimental Design Method (EDM) are presented in this paper and applied to a superconducting magnetic energy storage (SMES) device. The optimization process is realized through two different approaches.

In order to solve easily this problem, an optimization manager based on numerical electromagnetic simulations is employed. In addition, a PC network has been used so as to manage the numerous and heavy calculations required.

Key words: Experimental Design Method, Optimization techniques, Electromagnetic simulations

1. INTRODUCTION

The studied problem deals with the optimization of the main dimensions of a superconducting magnetic energy storage system. This is a relatively well-known problem which is now considered as a benchmark for optimization methods. The test case is the discrete one, where only three factors are considered (r2, h22 and d2). It is described in detail in [1,2]. The goal is to minimize the objective function *OF*. This function is also called the response.

Heuristic techniques based on the Experimental Design Method (EDM) are applied [3, 4, 5]. Generally speaking, this approach is characterized by the use of models of the studied responses, and by the implicit aim to minimize the global cost of optimization procedures.

M. Rudnicki and S. Wiak (eds.), Optimization and Inverse Problems in Electomagnetism, 139–150.

Here, all the experiments are computer calculations based on axi-symetrical and electromagnetic finite element models. The OPERA 2D software [6] is used for this purpose and all the calculation processes are driven by an optimization manager: Sophemis [7, 8].

Two different approaches have been considered in order to solve the present optimization problem.

The first one cuts the studied domain into subset. In each of them are calculated the effects of each factor on the response *OF*. The comparison of these effects, from a subset to an other gives the possibility to detect non-influent factors. The definition of the problem can then be strongly simplified.

The second approach does not use simplifications to treat the problem. Experimental Designs give one more time models of the objective function, but they link in the present case the response to the 3 factors r2, h22 and d2. This way is more difficult, because it demands more experiments to be done, and graphical interpretation becomes less easy and useful, but it gives a more complete overview of the design space.

2. S.M.E.S. OPTIMIZATION BENCHMARK

A graphical representation of the S.M.E.S. device is given by Figure 1. This device consists of two coaxial tori with rectangular cross sections. These rings are made of superconducting materials; current densities are defined for each of them.

Two formulations of this optimization problem exist: they differ about the number and the type of the factors. We will consider the three discrete factor case. The main characteristics of these three variables are summarized by the table hereafter.

Table 1. Factor characteristics

Name	Unit	Lower bound	Upper bound	Step
R2	m	2.6	3.4	0.01
h22	m	0.204	1.1	0.007
d2	m	0.1	0.4	0.003

These three factors are related to the outer ring.

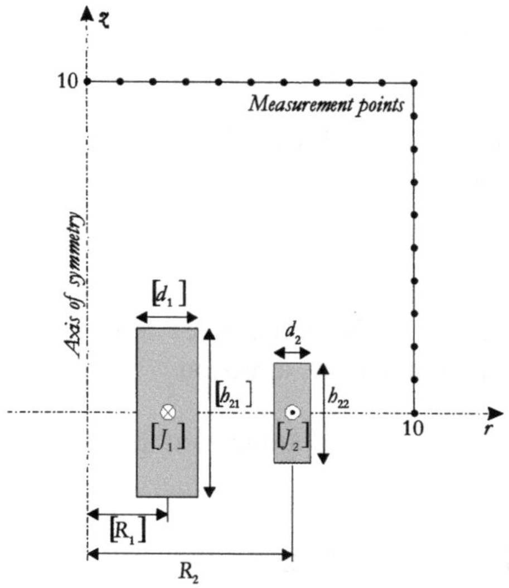

Figure 1. Right part of the cross section of the S.M.E.S. device

The optimization problem consists in finding the minimal value of the objective function defined as follows:

$$OF = \frac{B_{stray}^2}{B_{norm}^2} + \frac{|E - E_{ref}|}{E_{ref}}$$

Where $E_{ref} = 180$ MJ, $B_{norm} = 3.10^{-3}$ T and $B_{stray}^2 = \frac{1}{22} \cdot \sum_{i=1}^{2} |B_i|^2$ (B_{stray}^2 is evaluated along 22 equidistant measurement points).

A constraint function is defined: it involves the maximum induction value induced by the S.M.E.S. system: $B_{max} < 4.92$ T.

3. OPTIMIZATION MANAGER

3.1 Experimental Design Method

The Experimental Design Method (EDM) is based on the exploitation of statistical and algebraic properties of samples carefully chosen in the investigation domain. This remains an iterative method; however, at each iteration, *N* experiments must be carried out so as to deduce information. In a general way, the experimental design method requires *N* (*N*>1) simulations to be done, from which an objective function can be built. Subsequent simulations can then be deduced, leading to optimization techniques.

Its main features are:

- the identification of the significant factors acting on the variations of a response (Screening technique) [5];
- the building of simple but reliable mathematical models of the response (Response Surface Methodology (RSM)) [9]. In addition, a polynomial expression for the response function is then calculated over the validity domain of the design. This model gives reliable information about the optimum location, or at least its direction.

It is possible to drive an optimization process by the use of these fundamental facilities.

3.2 Optimization Manager tool

The optimization manager, called Sophemis, has been set up mainly to handle the Experimental Design Method.

Sophemis is an easy-to-use software, based on classical OS graphical standards like dialog boxes and menus. The main menu takes in charge the optimization processes and the application of the Experimental Design Method.

The very efficient matrix manipulation functions and the graphical functionalities included in the Matlab® environment are fully exploited. Hence, time for implementation has been saved, and the reliability of matrix calculations increased. An help using the HTML format has been written so as to give assistance to beginners.

Several optimization methods are available, such as:

- direct methods (Conjugated Gradient, BFGS, ...);
- stochastic methods (genetic algorithms, simulated annealing, ...);
- heuristic methods using Experimental Designs, acting with different strategies.

3.3 Distributed calculations

The main asset of Sophemis is certainly the possibility to prepare batches of simulations before submitting them to several computers linked up with a network, and gathering their results. Therefore, calculation costs can be strongly reduced.

Indeed, electromagnetic simulations are often time consuming, and their number may be high in optimization processes.

Figure 2 summarize this technological solution.

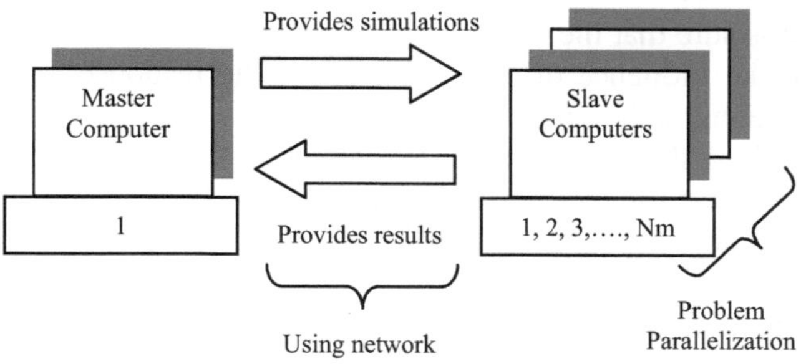

Figure 2. Principle of distributed computing

The EDM suits well to simultaneous evaluations on computer networks. Indeed, the final results are obtained after the computation of *N* (virtual) experiments, and not before. Since the *N* experiments are mutually independent, they can be calculated at the same time by different machines.

Using several computers to carry out these simulations is the easiest way. The sole condition demands to have a sufficient number of machines linked up with a network, that is often the case in today organizations.

The present configuration uses the network of five identical computers (PC Pentium 550MHz).

The parallel evaluation is based on a master-slave relationship. One computer is designed to be the server. It manages:

- the application of calculations (experimental designs or optimization algorithms);
- the distribution of the simulations among the remote machines;
- the dispatching of the files describing the experiments to be done to each slave computer;
- the gathering of the results sent back by each remote machine.

All those operations are fully automated.

The master computer does not perform any electromagnetic simulation. Hence, at least 3 machines are required to operate the distributed computing.

4. OPTIMIZATION: FIRST APPROACH

As it was described in the previous part, 3 factors are considered: r2, h22 and d2. These variables are defined over intervals that limit their variations. This design space is then a 3-dimensionnal. Each point inside this domain is the graphical representation of a combination of the possible values for the 3 factors. Hence, at least one of these points is the answer to the optimization problem, since it corresponds to a configuration where the objective function *OF* is minimal.

Since it is sure that the response *OF* varies inside the studied domain, it follows that the influence of each factor over the response variations also changes inside this same space.

The goal of this approach is to obtain the best estimation of these influences, with the use of a minimum of experiments.

The first step is to cut the studied space into identical subset. In the present case, the interval of the 3 factors have been cut in 4. Therefore, the whole studied domain is cut into 4×4×4=64 subset.

Each of them can be identified with a vector $\boldsymbol{v}$ composed of 3 elements, indicating respectively the division number according to the factor r2, h22 and d2. For example, the vector $\boldsymbol{v}$=[2 1 3] stands for the under-domain where r2∈[2.8 ; 3], h22∈[0.204 ; 0.428] and d2∈[0.25 ; 0.335].

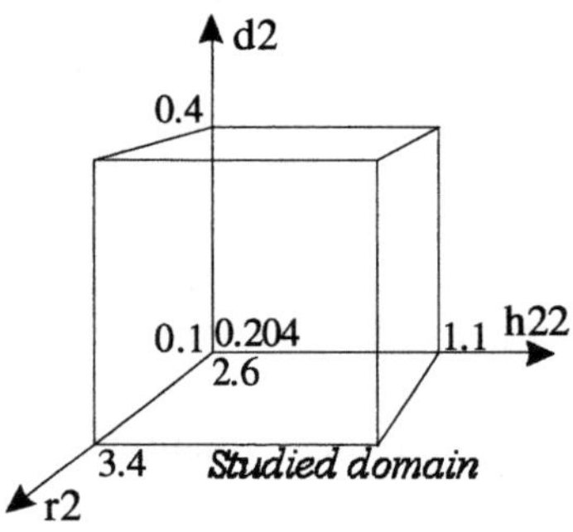

Figure 3. Graphical representation of the studied domain

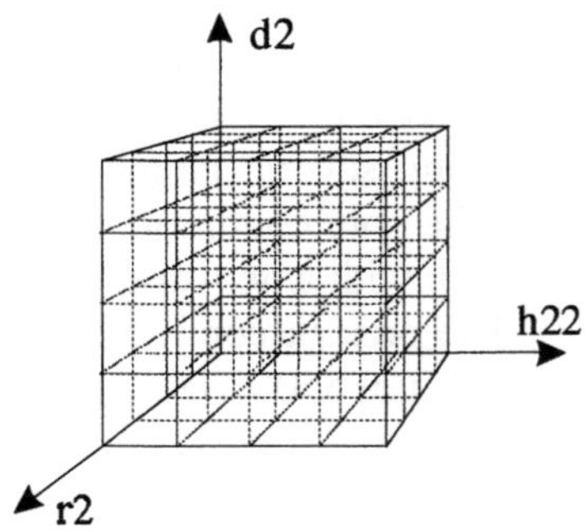

Figure 4. Splitting up into 64 under-spaces of the original studied domain

A design of experiments is realized inside each of the 64 subsets. The same type of DoE is used, that is a fractional factorial design which notation is 2^{3-1}. It is in fact a fraction of a full factorial design 2^3. The figure 5 shows a graphical view of such a design. Experiments are represented by black dots.

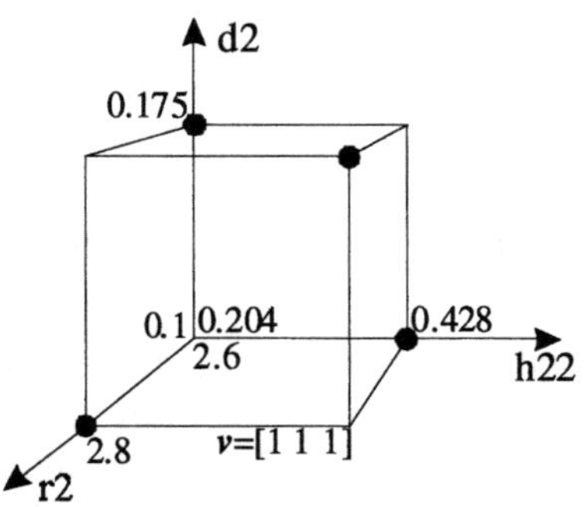

Figure 5. Graphical representation of the base design (fractional factorial design 2^{3-1})

In order to minimize the number of experiments, adjacent subsets do not use the same experiment positioning, but a symmetry is applied so as to share all the experiments located on subset faces. This is illustrated by Figure 6, where it is easy to see that 2 experiments are shared between the 2 designs defined in the under-domains ν=[1 1 1] and ν=[1 2 1].

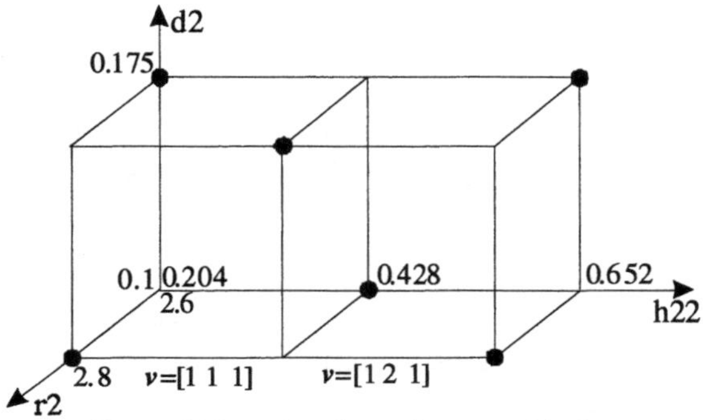

Figure 6. Example of experiment positioning
Experiments where h22=0.428 are shared between the 2 designs

This technique is used to place the experiments inside the whole studied domain. This particular way to locate experiments inside the studied domain implicitly defines a new type of designs: the *trellis designs* [8].

With this method, only 63 experiments are necessary to cover the totality of this volume.

The characteristics of the best point among the 63 experiments are given in Table 2 (first approach).

After calculating these 63 simulations, it is possible to deduce the effect of each factor on the objective function variations *OF*, for each subset. It can be shown that the effect vector **E** can be calculated by:

$$\boldsymbol{E} = \frac{1}{N}.\boldsymbol{X}'.\boldsymbol{y}$$

Where:

- N is the number of experiments for each design (that is 4);
- $\boldsymbol{y}$ is the response vector composed of the response values at the N=4 experiments defined by the present design;
- $\boldsymbol{X}$ is the experimental matrix; it has N lines and N columns. The first column is composed of 1s exclusively; the N-1 other columns give the coordinates of the N experiments defined by the present design;
- $\boldsymbol{E}$ is a N-element vector where the N-1 last values give the effects (or influences) of the factors on the response variations.

At this stage, it is interesting to plot the effects of the 3 factors versus the design positioning inside the studied domain, thanks to the values of vector ***v***. This graphical representation is given by Figure 7.

The upper plot shows the effect variations for the 3 factors. The other plots give all the possible combinations of the elements of ***v***.

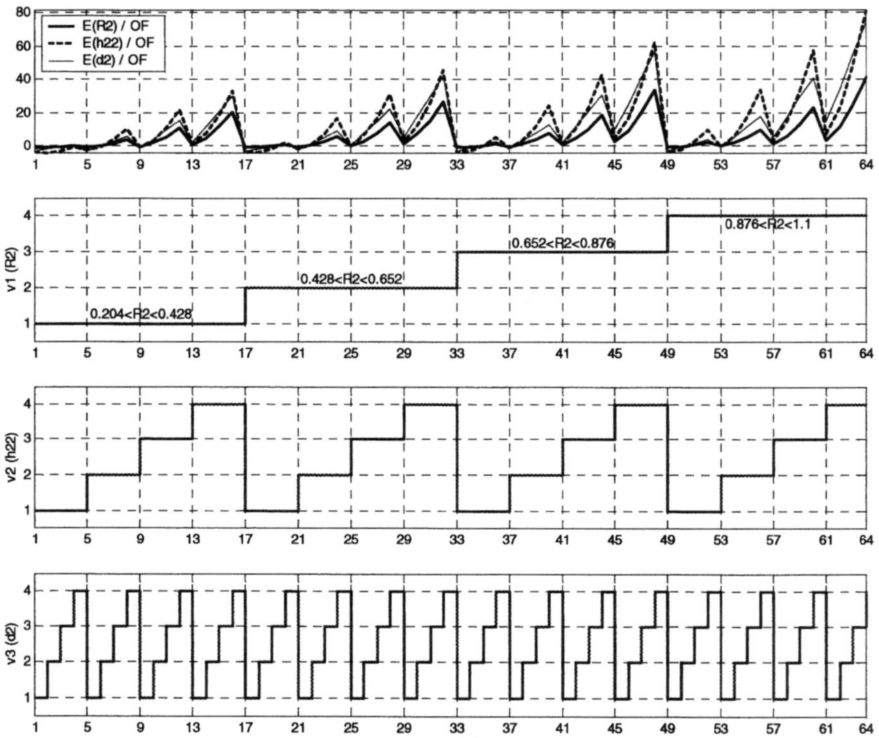

Figure 7. Effects of the 3 factors versus design positioning

The previous figure shows that the effect of the factor r2 does not significantly changes when ***v***(2)=1 and ***v***(2)=2, that is when h22 is between 0.204 and 0.652, whatever the values of r2 and d2. Inside this zone, the effect of r2 equals approximately 1.41.

The previous considerations are important because they lead to the simplification of the optimization problem. The variable r2 can be partially removed, and considered as an analytical term. The variations of the objective function *OF* can then be evaluated according to the following relation:

OF=*ymod*(h22,d2)+3.50268.(r2-3)

Where *ymod*(h22,d2) stands for an unknown relation linking *OF*, h22 and d2 together. Establishing this relation is the last stage of the current first approach.

For this purpose, a Design of Experiments is calculated. Each of the two remaining factors (h22 and d2) takes 11 different level equally distributed among their lower and upper bounds. r2 is set to 3. The DoE is therefore composed of 11×11=121 experiments; however 15 experiments can be recycled from the previous calculation stage described above.

Thanks to this new DoE, *ymod*(h22,d2) can be deduced and interpolations built. In this context, graphical representations are made:

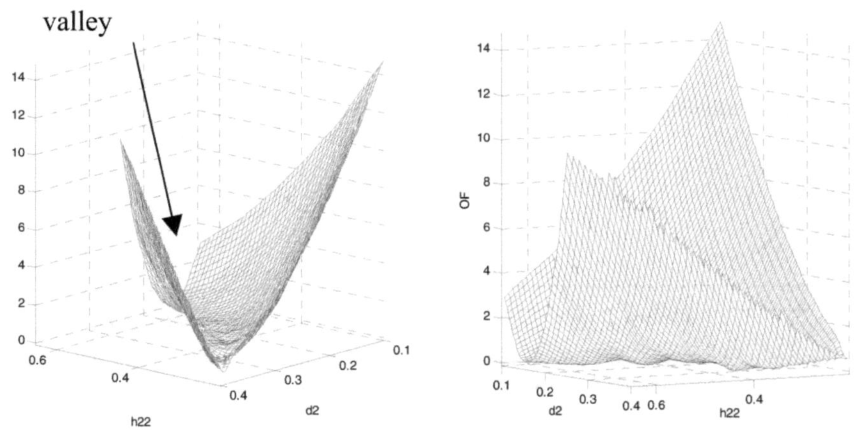

Figure 8. Variations of objective function *OF* versus h22 et d2 (r2=3)

It can be seen that minimal values for *OF* are located along a curved valley with a small slope. Successive multi minima can also be seen in this area.

Therefore, the research of the optimal conditions can be obtained exclusively by means of a systematic approach: all the experiments situated in the valley must be calculated. By this way, when r2 equals 3, the optimal point is found to be h22=0.239 and d2=0.394.

5. OPTIMIZATION: SECOND APPROACH

The second approach does not simplify the problem before treating it. The 3 factors are always considered. It consists of a global modeling algorithm of the function *OF* over the whole studied domain.

This procedure tries to fit locally second order models with *OF* variations.

At each iteration, a polynomial model is calculated from results of an Experimental Design. The quality of this modeling is then checked: it includes its representativeness and validity. Statistical measures are used for that, such as R^2 and Q^2 coefficients. If these measurements do not satisfy user limits, the domain under study is then cut into equal parts. The same process is then reapplied for each of these divisions.

This algorithm works in an iterative way. It explores the studied domain in an exhaustive manner.

The following figures illustrate this procedure on a one-dimensional domain. The objective function is plotted versus x the unique factor. We consider its variations inside the studied domain.

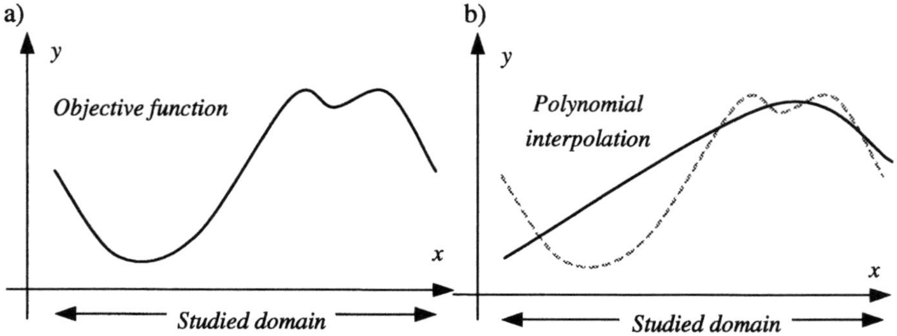

Figure 9. Illustrative example; a) variations of the objective function *y* versus the unique factor *x*; b) First polynomial interpolation of the objective function variations, over the entire studied domain

Let's assume that the first polynomial model does not verify the minimum levels for R^2 imposed by the experimenter. The studied domain is then cut in half. In these two subset, the same process is applied. The algorithm stops when no more domain need to be split up, or when the maximum number of cuts is reached.

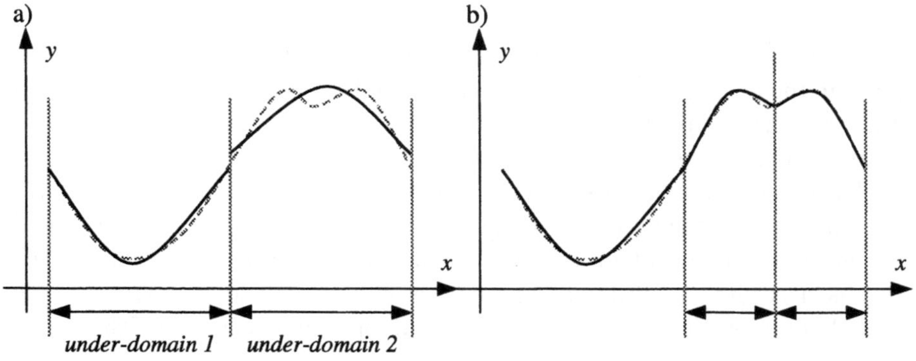

Figure 10. Illustrative example; a) First division of the studied domain in 2 parts; b) Second division: the under-domain 2 is cut in two parts

This method is generalized and applied to the current three factors optimization problem.

2247 experiments only are required; it takes 10h 19min on a PC computer.

As shown in Table 2 (2nd approach (1)), this best point is not the global optimum given by IGTE [1]: the difference equals approximately 12%.

Nevertheless, an other discrete optimization algorithms, equally based on EDM, is started. This method starting is simple and robust and uses a predefined grid of possible experiments. From an initial point, the algorithm checks the points of the grid next to it, according to certain directions. Among these points, if a better point exists, that is an experiment where the objective function *OF* is smaller than the initial one, then this point becomes the initial point, and the algorithm restarts. It stops when no better point can be found.

This method has been used here; the initial point has been set with the coordinates of the best experiment found previously by the exhaustive technique. A new best point, where *OF* equals 0.08876 (Table 2), is found by this mean. The difference with the global optimum is decreased, and becomes 2%.

Table 2. Best results according to different approaches

Variable	r2	h22	d2	*OF*
Best point (IGTE)	*3.08*	*0.239*	*0.394*	*0.08646*
1st approach	3.2	0.876	0.1	0.15204
2nd approach (1)	3.04	0.316	0.306	*0.09889*
2nd approach (2)	3.096	0.295	0.315	*0.08876*

6. CONCLUSION

The Experimental Design Method combined with numerical simulations gives to the designer the ability to understand the influence of each factor. In addition, the Response Surface Method is straightforward to localize an optimum taking into account its sensitivity.

Different optimization strategies and different experimental designs have been presented. Due to the low number of simulations required, EDM should be an appreciable part of any electromagnetic optimization package. The results according to different approaches are given and compared to the global optimum (IGTE). The difference with optimum is equal to 2%.

REFERENCES

1. SMES Optimization benchmark; definition of the 3 parameter problem, discrete case – IGTE Internet site: http://www-igte.tu-graz.ac.at/team.
2. Alotto P., Brandstätter B., Fuerntratt G., Magele C., Molinari G., Nervi M., Repetto M., Richter K. R.: Some Results On a SMES Device Optimization Benchmark Problem.
3. Box G. E. P., Draper N. R.: Empirical model-building and response surfaces, John Wiley & Sons, New York, 1987.
4. A. Garcia-Diaz, Don T. Philips, "Principles of experimental design and analysis", Chapman & Hall, 1995.
5. M. Pillet, "Introduction aux plans d'expériences par la méthode Taguchi", Les Editions d'Organisation, 1994.
6. OPERA 2D User's Guide, Vector Fields Ltd, Oxford, UK
7. Vivier Stéphane, Gillon Frédéric, Hecquet Michel, Brochet Pascal : A Design Optimization Manager, Proceedings Compumag 2001, Evian, France, June 2001, pp. 228-229.
8. Vivier Stéphane, Stratégies d'optimisation par la méthode des plans d'expériences et Application aux dispositifs électrotechniques, Thèse de doctorat, Université des Sciences et Techniques de Lille, July 2002.
9. Goupy J., Plans d'expériences pour surfaces de réponse", Dunod, 1999.

OPTIMAL DESIGN OF AXIAL FIELD SILICON MICROMOTOR BY USE OF FIELD/CIRCUIT MODELS

Sławomir Wiak
Department of Artificial Intelligence and Informatic Systems, Academy of Humanities and Economics in Lodz, ul. Rewolucji 1905 r. nr 64, 90-222 Lodz, Poland.

e-mail: wiakslaw@p.lodz.pl

Abstract: The paper deals with the problems of field and circuit modeling of silicon disc-type micromotors. 3-D motor structure is proposed to be studied. The field analysis and torque calculation results have given the knowledge about the changes of the motor structure in order to increase unit motor torque. Comparative analysis leads to the choice of optimal design of the structure.

Key words:

1. INTRODUCTION

Actually, the task of the designer is complicated by the evidence that different energy domains, in general being mutually coupled, are involved when modeling micro-electro-mechanical devices (MEMD), in the frame of more general micro-electro-mechanical systems (MEMS). Basically, electrical and mechanical domains interact to realize the main performance of the micractuator. In addition, other energy domains can either contribute to the main effect or give rise to second-order effects in the operation of the device. Moreover, the designer has to take into account the constraints imposed by the process technology, that limit the feasibility of innovative devices.

M. Rudnicki and S. Wiak (eds.), Optimization and Inverse Problems in Electomagnetism, 151–159.

In principle, modelling tools are widely available to simulate electromagnetic fields in both 2D and 3D geometries. On the other hand, algorithms applicable for the automated optimal design of devices and systems come from soft computing, i.e. artificial neural networks, evolution strategies, genetic algorithms, hybrid algorithms and expert systems.
Even more, artificial intelligence tools are likely to become a vital part of complex electromechanical systems of different size. In particular, applications of artificial-neural-network driven electromechanical devices in robotics, medicine, electronics and space technology are easily predictable.

As far as numerical tools are concerned, it should be noted that nowadays many simulation codes are available both commercially and from various public domains as well for supporting the designer of MEMD. For instance, user-friendly environments based on finite-element codes can simulate the electrostatic behaviour of complex two- and three-dimensional geometries of microdevices, given field source and boundary conditions.

In principle, evaluating both local quantities (field map, force distribution) and global quantities (induced charge, total force) becomes straightforward a task. In particular, detailed field analyses can be developed for investigating local effects in the steady state operation of the device. Even more, a first approach to dynamic simulation of a microdevice can be exploited: the basic idea is to extract a set of global parameters from field analysis and then attribute it to an equivalent circuit that undergoes the time-varying source.

MicroElectromechanical Systems (MEMS) technology has generated a significant interest of academic, research and business sectors. The above is caused by very promising technology market and as well cost reduction connected with microscale devices fabricated by use of silicon technology. Due to rapid grow of this market, year after year, the market is just segmented. This is also true in the reference to accelerometer market. Efficiency of the device development is also due to design flexibility, developing of simulation tools and CAD systems, integration of mircodevices and microelectronics, etc. In general, multi-layer surface mircoactuator fabrication process is enough flexible to produce advanced device structures [3,6]. MEMS structure development requires reliable fabrication processes and flexible CAD/(analysis and optimal design) tools. We could treat mask layout tools as complementary and CAD/(analysis and optimal design) tools. The device geometry is, in general, defined in the mask layout tool, then transferred to integrated circuit layout package, while specific surface micromachining process ahs been employed. Some specific codes based on Finite Element Method could be successfully applied to MEMS designing while microdomain physics are also taken into account. Such a sophisticated software could allow creating 3D structure of the device, full model analysis (structural mechanics, electrostatics, fluid flow, vibration, etc.).

It is easy to realize that literature is rather poor of advanced methodologies for both computer aided design (CAD) and automated optimal design (AOD) at a device level; this aspect seems to be paradoxical if one considers the large amount of experimental data that need to be interpreted and justified by means of field models. In other words, there is still lack of accurate numerical tools for understanding the influence of geometry and material properties on the electromechanical response of the microdevice, in both static and dynamic conditions.

These remarks become even more important when passing from CAD to AOD. In fact, the ultimate goal of the latter is to set up a computational environment able to optimise, in a completely automatic way, an objective function defined by the designer. The objective function represents the design criterion and depends simultaneously on several degrees of freedom, that in turn are subject to bounds and constraints. Often, more than on objective function should be considered at a time.

The strongest impulse to the development of such advanced tools should come mainly from industrial companies; in fact, virtual prototyping of optimal devices would determine a twofold benefit:

- decreasing the total time lasting from the device ideation to its implementation as a marketable product;
- reducing the number of fabrication experiments on real prototypes;
- identifying the device of minimum cost, given the performance or, conversely, the device of maximum performance, given the cost;
- increasing the global quality of the industrial product.

By exploiting the mechanical properties of Silicon - as extraordinary as the electrical ones, using the technology already perfected for integrated circuit manufacturing, and integrating power supply, motion actuation and control functions, it is possible, nowadays, to make MEMS smaller than one millimeter, for several applications. Actually, pressure and acceleration microsensors, as well as microswitches, microvalves, micromirrors, microresonators, linear and rotating microactuators and microrobots, have already been developed, for application in medicine, space and automobile industry, and precise mechanical industry.

Electrostatic motors in the microscopic scale thanks to the impressive developments in the technology for integrated circuits has attracted the scientists. The micromotor is investigated at steady state and particular attention is paid to calculation of the torque-angle characteristics. Two selected motor geometries are simulated, as well as design considerations are drawn about their influence on the electromechanical performances. Both 3-

D field model and circuit model of a synchronous electrostatic micromotor are set up.

2. SUBJECT OF THE ANALYSIS

The subject of investigation is top-drive structure. A proposed new construction of the motor is shown in Figure 1. This motor is a variable capacitance (VC) stepping motor with its stator electrodes (poles) underneath the rotor (top drive). Field lines in the air gap are globally axially oriented, and the planar geometry leads to larger VC facing areas. This results in an increased capacitance change rate and hence an increased motor torque. Due to electrostatic attraction the rotor teeth will be attracted to the stator teeth and even contact the base plane if a certain (high) 'pull in' voltage is applied, thus preventing any rotation from taking place. In this structure the bearing, stator and rotor consist of polysilicon. The stator is connected by means of terminals to a voltage source. The rotor has a potential of zero volts, due to its contact via the bushings and bearing to the plane.

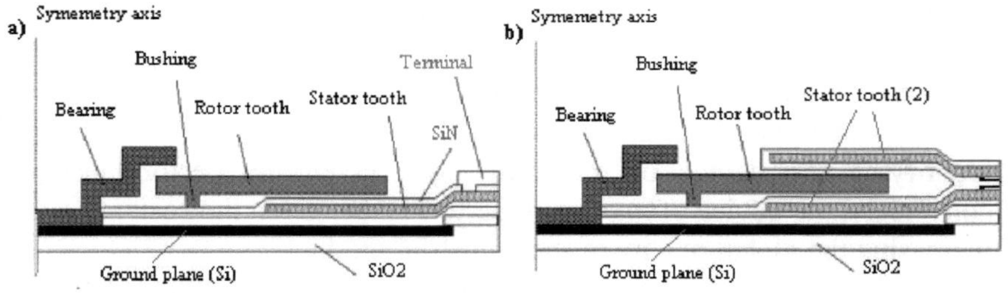

Fig. 1. a) Simplified cross-sectional sketch of the motor with axial field, b) Cross sectional view of micromotor with double axial field.

This motor is a variable capacitance (VC) stepping motor with its stator electrodes (poles) underneath the rotor (top drive). Field lines in the air gap are globally axially oriented, and the planar geometry leads to larger VC facing areas. This results in an increased capacitance change rate and hence an increased motor torque. Due to electrostatic attraction the rotor teeth will be attracted to the stator teeth and even contact the base plane if a certain (high) 'pull in' voltage is applied, thus preventing any rotation from taking place. In this structure the bearing, stator and rotor consist of polysilicon. The stator is connected by means of terminals to a voltage source. The rotor has a potential of zero volts, due to its contact via the bushings and bearing to the plane. As is shown in Figure1b, stator consists of two teeth stators above each other in full alignment.

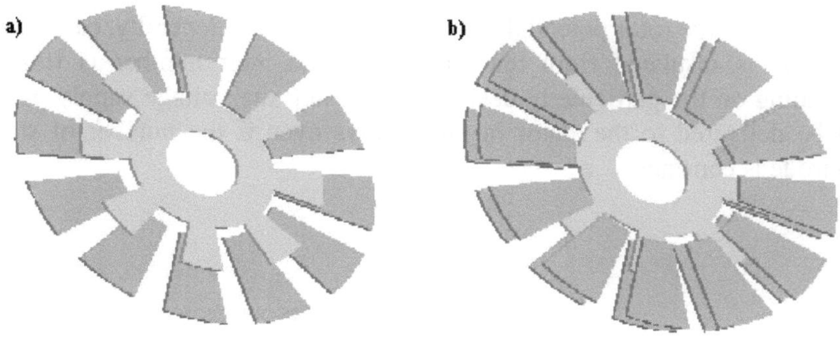

Fig. 2. New motor structure by field models a) motor 1 with 12 stator electrodes and 8 rotor electrodes, b) motor 2 with 12 stator electrodes and 8 rotor electrodes [5]

3. MANUFACTURED DEVICES (ITE TECHNOLOGY)

The technology adequate for both considered micromotors was developed by Institute of Electron Technology (ITE) in Warsaw, then examined. The sequence of steps leading to fabrication of the top drive micromotor presented in Fig.1 and 2 are shown in Fig.3, whereas Fig.4 covers the photo of the manufactured devices. At the very beginning, silicon wafers were thermally oxidised to create thin pad oxide layer and low stress, 200 nm thick silicon nitride layer was deposited using LPCVD method. Next, first polysilicon layer, 300nm thick, was LPCVD deposited and patterned to form anchor and stator (Fig.3a). First 1500 nm thick sacrificial PSG oxide layer was then deposited with CVD and 900 nm holes were wet etched to form dimples for rotor bushing (Fig.3b). 2000 nm thick second polysilicon layer was then deposited using LPCVD in the process that ensured low residual mechanical stress in the layer. On the layer, an aluminium hard mask was deposited to define the rotor shape for the RIE etching process in SF_6/Ar plasma resulting in the rotor formation (Fig.3c). Next, 700 nm thick, PSG oxide was deposited to form second sacrificial layer. During photo-litography the holes were then opened to the first polysilicon layer to anchor the bearings (Fig.3d). The third polysilicon layer, 1500 nm thick, was then deposited and covered by aluminium to form hard mask against RIE plasma etching during formation of bearings anchored to base polysilicon (Fig.3e). Contact holes were then opened and aluminium was deposited and patterned to form electric connections to stator poles (Fig.3f). The last processing step was etching away all sacrificial oxide to release rotor (Fig.3g). The final operation lubricating of bearings was. In the presented micromotor the supply voltage is delivered to the stator electrodes only whereas the rotor potential usually equals to zero. Generally, each stator electrode is insulated from the other ones and their number is flexible depending on the particularly design. As a result, the voltage supply pulses can be delivered to the stator in many ways even if the number of stator electrode is constant.

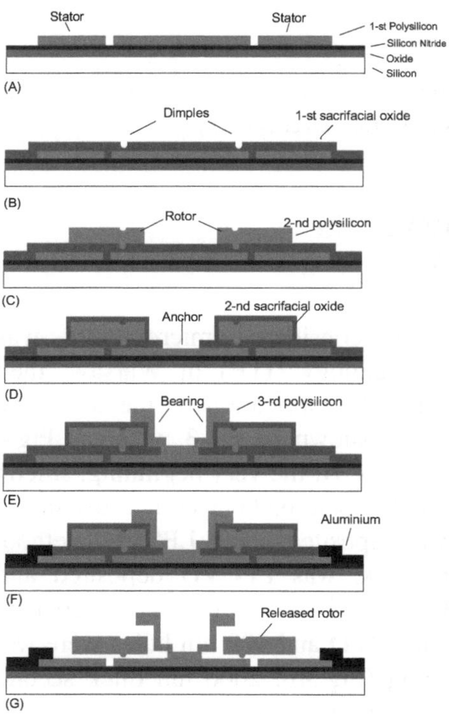

Fig.3 [5]. Three – level polysilicon process for axial field micromotor fabrication.

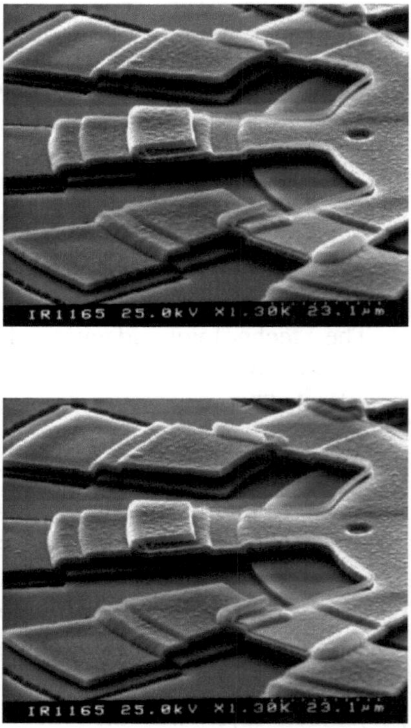

Fig.4 [5]. Microscopy photographs of the manufacture micromotor presenting the rotary of released rotor.

4. CIRCUIT MODEL (CAPACITANCE MODEL)

In order to define the objective function (with the set of variables) the equivalent capacitance model should be elaborated. Author proposes the improved model (see Fig. 5) to be verified by field model.
Analysed positions of rotor electrode against stator electrode are as follows:

- Both electrodes overlaps, (φ=0),
- Rotor electrode moves negligibly small against stator electrode (φ>0),
- Rotor electrode moves against stator electrode (φ>>0),
- Analysed rotor electrode leaves energised stator electrode, φ is greater than stator electrode width
- Analysed rotor electrode left energised stator electrode, and φ is much greater than stator electrode width (*in this case the capacitance is negligibly small*).

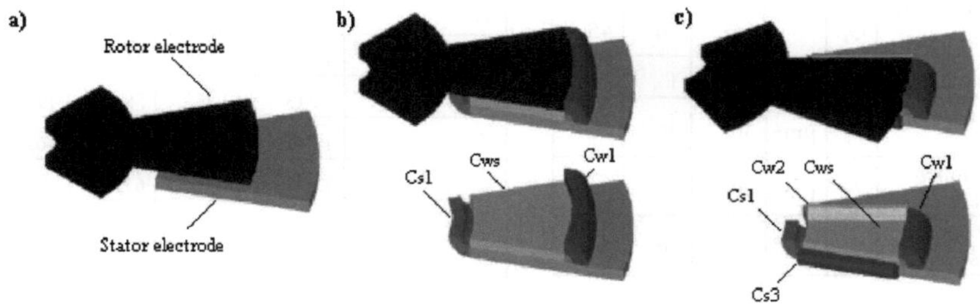

Fig. 5. Capacitances for rotor electrode: b) position (φ = 0), c) position (φ > 0)

Thanks to this model it is possible to define the formula for equivalent capacitance of the motor for each rotor electrode position. The net-drive torque defined as the objective function strongly depends on motor structure (geometrical optimisation variables)

5. OPTIMAL DESIGN BY FIELD/CIRCUIT MODELING

To design stepping motors for micro positioning, large starting torque, small frictional force and fine angle resolution is desirable. The design of a rotating motor begins with an estimation of the torque exerted by the electric field. This can be expressed in terms of the derivative of the stored electrical co-energy which, for a given bias V between rotor and stator, is conveniently represented as 1/2 TV^2, where C represents the capacitance across the driving electrodes that have voltage V across them. In the paper top-drive motor structures are proposed, then investigated by use of field method leading to resultant capacitance characteristics. Thanks to C curve obtained from the field method equivalent circuit is then modified.

The optimization technique based on the circuit model is fully satisfactory in order to simulate the electromechanical characteristics of such type of micromotors. The electrostatic drive torque (Tdrive), and friction drag torque terms (Tfric) are calculated using this model. Table 1 shows the list of micromotor basic variables (for motor 1- Fig.1) to be introduced into optimization algorithm for selected: pole configurations, air gap length and geometric parameter variations (see results in Tab. 2).

TABLE 1. OPTIMISATION VARIABLES

Voltage	U	V	50	50	100
Angular width of rotor pole	Q_r	°	25	10	45
Air gap length	d	μm	2	1	20
Bushing radius	R_a	μm	30	30	60
Stator inner radius	R_i	μm	70	60	90
Rotor outer radius	R_c	μm	90	70	100

TABLE 2. RESULTS OF OPTIMISATION PROCESS

6/4	30	3,141	15,491	30,98	15,49
8/6	15	1,570	30,982	31,96	30,98
10/6	24	2,513	19,363	38,27	19,36
10/8	10	1,047	46,474	92,94	46,47
12/8	15	1,570	30,982	61,96	30,98
12/10	10	1,047	46,474	92,94	46,47
14/8	19,28	2,019	24,097	48,19	24,09
14/10	10,28	1,077	45,183	90,36	45,18
16/12	10	1,047	46,474	92,94	46,47

REFERENCES

1. Di Barba P., Savini A., and Wiak S., *2-D numerical simulation of electrostatic micromotor torque*, IEE Second International Conference on Computation in Electromagnetics, Nothingham, UK, 1994.
2. Price R.H., Wood J.E., and Jacobsen S.C., *Modelling considerations for electrostatic forces in electrostatic microactuators*, Sensors and Actuators, 20, 107-114, 1989.
3. Trimmer W. and Jebens R., *Harmonic electrostatic motors*, Sensors and Actuators, 20, 17 –24, 1989.
4. Van Dessel M., Johansson T.B., Belmans R., and W. Geysen, *An optimisation scheme of electrostatic micromotors based on an equivalent circuit-finite element approach*, ICEM'92, UK, Manchester, Conference Proceedings, 1157-1160, 1992.
5. Wiak S., *Optimal Design of Silicon Micromotors by Use of Field/Circuit Modeling*, COMPUMAG, PE1-13, vol. III, pp.38-39, Evian, July 2-5, 2001.
6. Wiak S., Dumania P., Lisik Z., Szczepaniak Z., *Silicon Electrostatic Micromotor*, EPEMC'2002, Dubrovnik, Croatia, September 2002.

EXPONENTIAL FAMILY AND INVERSE PROBLEMS

On Metamorphoses of Probability Theory

Aleš Gottvald

Institute of Scientific Instruments, Academy of Sciences of the CR, Královopolská 147, 612 64 Brno, Czech Republic

Abstract: We associate Inverse Problems with a structure called Exponential Family. This structure is pivotal in many fundamental theories (Bayesian Probability Theory, Maximum Entropy Principle, Fourier-Laplace Transforms, Generatingfunctionology, Sufficient Statistics, Lie Groups, ...), and offers extensive applications to Physical Theories and Complex Systems.

Key words: Exponential family, inverse problems, Bayes' theorem, MaxEnt principle, sufficient statistics, entropic functionals, Lie groups, complex systems

1. INTRODUCTION

Inverse Problems (IP) may naturally be associated with a structure called *Exponential Family*. Suppose that a solution of IP is represented by a conditional Probability Density Function (PDF), say $p(x|\lambda)$. To put this PDF to a form of the exponential family often requires nothing more than a simple "exp-log trick", resulting in a structure of the type:

$$p(x|\lambda) = \frac{1}{Z(\lambda)} e^{-S(x|\lambda)} q(x) \tag{1}$$

where the quantities $Z(\lambda)$, $S(x|\lambda)$, and $q(x)$ may possess some amazingly diverse interpretations and names. This is just a topic of this

M. Rudnicki and S. Wiak (eds.), Optimization and Inverse Problems in Electomagnetism, 161–180.

work. It follows that the innocent-looking exponential structure is intimately linked with many concepts and methods of fundamental importance, which unlock its hidden meanings. [1, 2, 6-10, 12-18, 20, 23, 27]

The idea of investigating phenomena of nature in the exponential format have emerged historically in many scattered areas. Its systematic roots go back to *Laplace's probabilistic legacy* (Generating and Characteristic Functions, Laplace's Transform, error distributions) [3, 4]. A fundamental importance of this idea has later been recognized in *Statistical Thermodynamics* (J. W. Gibbs) [9, 10]. Its information-theoretic generalizations, in terms of *Maximum Entropy Principle* (E. T. Jaynes), are particularly inspiring for *Information Geometry* and *Information Physics*. [1, 2, 9, 10, 18, 20, 27] A geometric and algebraic view of the exponentiation idea is instrumental in *Symmetry Groups* (S. Lie, É. Cartan) [5, 22-25]. Fundamental role of the exponentiation is particularly stressed in some formulations of *Quantum Mechanics* (von Neumann, Dirac, Feynman, Wigner and Moyal), *Statistics* (Pitman and Koopman, Kullback, Grendár), etc. [6, 7, 12, 13, 20] The very idea of *Complex Numbers* (Euler) and *Analytic Functions* (Cauchy, Green, Riemann) obtains new meanings when being interpreted in terms of the exponential family [19, 21].

Exponential Family shows many metamorphoses. This concept sets a stage for Bayes' Theorem [8, 1], MaxEnt-Principle [9, 10], Fourier-Laplace Transforms [14], Characteristic and Generating Functions [15, 16], Sufficient Statistics (Pitman-Koopman theorem) [7], Multinomial Asymptotics (Grendárs' theorems) [12], Symmetries and Lie Groups [5, 22-25], etc. This also brings our treatment of IP closer to some fundamental ideas of modern physics, probability theory, and information theory. [1, 2, 18, 20, 21, 25]

In *Electromagnetics*, many quantities naturally come in the exponential format, or may simply be adapted to it. E.g., Fourier spectral representations of electromagnetic laws may naturally be interpreted in terms of the exponential family and associated partition sums. It supports a view-angle from which the electromagnetics features as an intrinsically spectral theory, closely linked to statistical thermodynamics [1, 27]. However, the idea of exponentiation is much more general. Its multifaceted nature allows many applications to IPs in physics (e.g., a novel treatment of Planck's law [1, 27]), signal processing (NMR), complex systems (spiral geometries, power laws (Zipf), fractal patterns, phyllotaxis, chaos theory, etc.) [1, 2, 26]. Consequently, the very concept of Inverse Problems addresses the same fundamental situation as the information-theoretic extensions of Statistical Thermodynamics, and namely -- a *reproducible logical inference from incomplete information*.

2. INVERSE PROBLEMS VS EXPONENTIATION

We emphasize that the fundamental difference between the concepts of *Inverse Problems* and *Direct Problems* just corresponds to different levels of description and different information available. In IPs, we construct our solutions using some *incomplete (partial)* information only; consequently, our sharp δ-type probability assignments are "smeared" into some general PDFs, say $p(x\,|\,\lambda)$. This is analogical to a difference between a *microscopic* and *macroscopic* levels of description in Statistical Physics or Quantum Mechanics:

$\delta(x - x_0)$		$p(x\,|\,\lambda)$
Direct	<----->	Inverse
Complete info'	<----->	Incomplete info'
"Microscopic" description	<----->	"Macroscopic" description

Many aspects of *Inverse Problems* may be investigated with the help of *Exponential Family*. The very formulation of IPs in probabilistic terms leads to *Bayes' Theorem* and *MaxEnt Principle*, which are intimately related to Exponential Family. *Uncertainty of solutions*, which is inherent to IPs, naturally calls for some probabilistic analysis in an exponential format. Particularly involved are questions about optimal representations of data, which shows the Exponential Family in a context of *Sufficient Statistics*. These concepts are also instrumental in stochastic optimizations (evolutionary, genetic, etc.) which are often applied to IPs. [1, 9, 6, 7]

Concerning the underlying *Direct (or Forward) Problems*, the Exponential Family may also come very naturally. It might be surprising, as some form of deterministic equations is still a common starting point. E.g., in electromagnetics we usually start with some form of Maxwell's equations. However, in vast domain of statistical physics, the Exponential Family (dressed as a Gibbs' "grand canonical ensemble") is the very central structure. Here, the electromagnetic quantities feature as the parameters of some "most entropic" probabilistic distributions, taking place in the exponent (and a normalizing coefficient) of the exponential model. For another example consider representations of Maxwell's equations in Fourier spectral domain. This takes ubiquitous convolutionary relationships to a multiplicative form, which is closely related to Bayes' Theorem and Exponential Family. Thus, even the underlying forward problems may often come in a format of Exponential Family, or may advantageously be adapted to this form.

Other fundamental aspects of IPs correspond to some changes of *topological structures*, which in turn correspond to *symmetries*, *conservation laws*, and many associated topological, algebraic and geometric questions. Here we meet the Exponential Family in a context of *Lie groups* and *Lie Algebras*, at the crossroads of algebra, topology, geometry, and probability theory. [21-25, 17, 11]

Thus, let us investigate an *exponential format* for analyzing conditional probabilities $p(x\,|\,\lambda)$. Here, the quantity λ may typically be interpreted as some "macroscopic parameters" which however possess various physical, statistical and mathematical meanings depending on a context.

3. BAYES' THEOREM

The first metamorphosis of the Exponential Family arises in a context of *Bayesian Probability Theory* (BPT) [8, 7]. Recall that the BPT - which is formally just the calculus of conditional probabilities - is governed by two fundamental rules - the *product rule* and the *sum rule*:

$$P(XY\,|\,I)=P(X\,|\,YI)\,P(Y\,|\,I)=P(Y\,|\,XI)\,P(X\,|\,I) \tag{2a}$$

$$P(X+Y\,|\,I)=P(X\,|\,I)+P(Y\,|\,I)-P(XY\,|\,I) \tag{2b}$$

The sum rule may equivalently be stated as the *resolution of unity*:

$$P(X\,|\,I)+P(\bar{X}\,|\,I)=1 \tag{3}$$

A simple adaptation of the product rule leads to Laplace's "theorem on inverse probability", a fundamental probabilistic law for which H. Poincaré (around 1898) suggested a name *Bayes' Theorem*:

$$P(X\,|\,YI)=\frac{P(Y\,|\,XI)}{P(Y\,|\,I)}\,P(X\,|\,I) \tag{4}$$

3.1 Beyond Bayes' Theorem

Bayes' Theorem involves many hidden and often overlooked messages [1]. In terms of Probability Density Functions (PDF), the Bayes Theorem may be percepted as a (normalized) product of two factors $f(.,.)$ and $q(.)$:

$$p(x|\lambda)=\frac{1}{Z(\lambda)} f(\lambda|x)\ q(x) \tag{5}$$

Posterior PDF = (Sampling PDF) x (Prior PDF)

To obtain explicitly a link with Exponential Family, let us apply the "exp-log trick" to the sampling PDF in Bayes' formula:

$$p(x|\lambda) = \frac{1}{Z(\lambda)} e^{\ln f(\lambda|x)}\ q(x) \tag{6}$$

This multiplicative structure may be interpreted as a statement about:

-the fundamental law for learning from new data ...
-the Theorem between Theories and Things ...
-Sampling PDF and Prior PDF are mutually independent ...
-Epistemological and Ontological layers are mutually independent ...
-data in "Memory" and data in "Sensors" are mutually independent ...
-Prior PDF is equivalent to "more data" (from a "Memory") ...
-Future knowledge comes from the Memory and Sensors ...

Note 1: Applying the "exp-log trick" directly to the posterior PDF produces:

$$p(x|\lambda) = e^{\ln p(x|\lambda)} = \exp[-\ln Z(\lambda)+\ln f(\lambda|x)+\ln q(x)] \tag{7}$$

where the individual terms may physically be related to some vacuum, kinetic and potential forms of energy, respectively.

Note 2: The fundamental structure of Bayesian Probability Theory involves 6 conditional probabilities,
$P(X|I), P(Y|I), P(X+Y|I), P(XY|I), P(X|YI), P(Y|XI)$,
which are constrained by two equations - the sum rule and the product rule. Consequently, the BPT immediately suggests a geometry of a 4-dimensional space-time continuum, as a subspace of the 6-D world of conditional probabilities. We do are cruising a "space of hypotheses" ... [18]

Note 3: Today we know with mathematical certainty that both the factors (i.e., the sampling PDF and the prior PDF) are absolutely inevitable in Bayes' Theorem [8]. In particular, there is no logically consistent way how to "delete" the prior factor from Bayes' Theorem. Corresponding frequentists' claims are just illusions which can be refuted on many fundamental grounds. Every probabilistic reasoning inevitably starts from some priors, and the opposite would be literally like an analytical geometry without Euclidean spaces, a group theory without Abelian groups, or a theory of electromagnetic potentials without Laplace's equation.

3.2 Arrow of Time from Bayes' Theorem

Bayes' Theorem may naturally be associated with three sets of data and intrinsically involves the concept of *Arrow of Time* [1]. Every action of Bayes' Theorem actually imposes a topological structure on the time-axis, splitting the time-axis into three intervals, thus breaking the symmetry $t \rightarrow -t$ of both time directions.

Prior PDF: data in Memory, $\{t < 0\}$ ...
Sampling PDF: data in Sensors $\{0 < t < \varepsilon\}$ ...
Posterior PDF: Predicted data $\{t > \varepsilon\}$ ...

$\{t < 0\}$ $\{0 < t < \varepsilon\}$ $\{t > \varepsilon\}$

Memory ---> Sensors ---> Prediction
Past ---> Present ---> Future

Consequently, the concept of *Arrow of Time* is intrinsically captured by the fundamental probabilistic law -- the *Bayes Theorem*! This point of view is equivalent to a physically more common suggestion -- "The Arrow of Time from MaxEnt Principle" -- but offers perhaps yet more elementary and tangible conceptual rationale for the logical origin of time and its direction.

4. THE MAXIMUM ENTROPY PRINCIPLE

In this metamorphosis, the Exponential Family arises as the maximizer of some unique information-theoretic (or entropic) functionals [9, 6, 12, 13]. Most important of these entropic functionals are the following three:

$$S(p) := -\int p(x|\lambda) \ln p(x|\lambda)\, dx \tag{8a}$$

(Shannon-Jaynes) [9, 10]

$$H(p|q) := -\int p(x|\lambda) \ln \frac{p(x|\lambda)}{q(x)}\, dx \tag{8b}$$

(Kullback-Leibler) [6, 18]

$$J(p|q) := -\int [p(x|\lambda) - q(x)] \ln \frac{p(x|\lambda)}{q(x)}\, dx \tag{8c}$$

(Jeffreys-Grendár) [7, 12]

The Exponential Family is most closely related to the maximization of the second functional - the Kullback-Leibler entropy (also called "KL-distance", "KL-number" or "relative entropy"). The role of other two entropies will be indicated in a chapter on multinomial asymptotics (see below).

Let the posterior PDF $p(x|\lambda)$ be inferred on the basis of the following information: (i) given *prior* PDF $q(x)$; (ii) given *average values*:

$$\bar{u}(\lambda) = \int u(x)\; p(x|\lambda)\, dx := \langle u(x) \rangle \tag{9}$$

where both the *"generalized potentials"* $u(x)$ and associated parameters λ may be vectorial quantities of the same dimension. Define the *MaxEnt* task:

maximize: $H(p|q)$ (p varied, q fixed)

subject to: $\bar{u}(\lambda) := \langle u(x) \rangle$

Following Lagrange, Gibbs and Jaynes [9], every individual generalized potential gives rise to a corresponding lagrangean multiplier and the solution of the constrained variational problem - the most probable posterior PDF - leads to a form of the Exponential Family:

$$p(x|\lambda) = \frac{1}{Z(\lambda)} e^{-\lambda \bullet u(x)} q(x) \tag{10}$$

where $\lambda \bullet u(x)$ is a scalar product which, in a fairly general case of a discrete parameter *k* and a continuous parameter *t*, may be promoted to the form

$$\lambda \bullet u(x) := \sum_k \int \lambda(t)\, u_k(x,\ t)\, dt \tag{11}$$

and the normalizing coefficient $Z(\lambda)$, called a *partition function*, is:

$$Z(\lambda) = \int e^{-\lambda \bullet u(x)}\, q(x)\, dx \tag{12}$$

The partition function $Z(\lambda)$ depends only on the lagrangean multipliers λ (macroscopic parameters) and captures the complete information about the MaxEnt solution - the most entropic posterior PDF $p(x\,|\,\lambda)$.

4.1 Examples of Generalized Potentials

The "most entropic" PDF represents an optimal inference about a "microsystem", given some global macroscopic description (partial information) only. Let us show several examples of the generalized potential functions $u(\bullet)$ which may stand inside the averaging bracket $<...>$. These examples indicate a tremendous flexibility and ubiquity of the MaxEnt scheme:

$<1>$ ---> induces a uniform PDF, Laplace's "Insufficient Reason" ...

$<x>$ ---> first-order potentials/moments, Gibbsian statistical mechanics

$<x^k>$ ---> k-th order potentials/moments ...

$<u(x)>$ ---> generalized potentials/moments ...

$<u(x,\ t)>$ ---> t-parametrized potentials/moments ...

$<e^{tx}>$ ---> characteristic function of probability theory (t real) ...

$<e^{-i\,2\pi t x}>$ ---> characteristic function (Fourier Transform case)

$<e^{px}>$ ---> characteristic function (Laplace Transform case, p complex)

$<\cos(xt) + \sin(xt)>$ ---> characteristic function (Hartley Transform case)

$\ln < e^{tx} >$ ---> cepstrum, cumulant-generating function

$- < \ln \frac{p(x|\lambda)}{q(x)} >$ ---> entropy

Particularly important are those cases which involve some forms of parametrized exponential functions. These encompass not only the ubiquitous concepts of Fourier-Laplace transforms, but also that of characteristic functions, which represent key-objects of probability theory. The game need not stop here. Roughly speaking, with a proper choice of the averaged quantities (generalized potentials) the scheme also naturally leads to cepstra and cumulants (in signal theory), Feynman's path integrals (in quantum mechanics), string theories (in particle physics), and other interesting structures of modern information physics. Consequently, all these theories may be cast in a unified frame - as some sorts of MaxEnt principle or statistical physics - with a central unifying role of the Exponential Family.

5. FOURIER-LAPLACE TRANSFORMS

This metamorphosis may be a "royal road" to the MaxEnt world for many physicists and engineers, as many of them are more familiar with Fourier Analysis [14] than with Statistical Physics.

A key-point is very simple: The partition function $Z(\lambda)$, which captures the whole information about the Exponential Family, may be treated as a (properly generalized) Fourier-Laplace Transform (FL) of some "structure function" $w(u)$:

$$Z(\lambda) = \int e^{-\lambda \bullet u(x)} q(x)\ dx = \int w(u) e^{-\lambda \bullet u} du = FL[w(u)] \tag{13}$$

The structure function $w(u)$ involves information about priors and topology, and, more precisely, may exist only in an abstract sense of a distribution (these mathematical subtleties may be faced by Fourier-Stieltjes integral, etc.).

Equivalently we may say that the partition function $Z(\lambda)$ is actually a Characteristic Function (ChF) of Probability Theory [15, 8]. For many good reasons, these ChFs are major objects rather than mere tools in Probability Theory. (i) the ChFs are well defined objects even for some ill-defined PDFs; (ii) local characteristics (derivatives) of the CHFs translate as the global characteristics (moments) of associated PDFs; (iii) convolutions of PDFs translate as products of CHFs, which may easily be associated with Bayes' Theorem and Exponential Family.

More generally, all theorems of Fourier-Laplace analysis may be interpreted in probabilistic terms. In particular, Borel's *Convolution Theorem.* Some useful probabilistic metamorphoses of a (multi)convolutionary relationship are briefly listed below, as a sequence of interrelated formulae:

$$X = \sum_k X_k \qquad \text{(a sum of independent random variables)}$$

$$f(x) = \bigotimes_k f_k(x) \qquad \text{(a multiconvolution of PDFs)}$$

$$F(t) = \prod_k F_k(t) \qquad \text{(characteristic function)}$$

$$\ln F(t) = \sum_k \ln F_k(t) \qquad \text{(cepstrum)}$$

$$\ln F(t) = \sum_{n=-\infty}^{\infty} c_n t^n \qquad \text{(cepstrum, cumulant GF)}$$

$$e^{-\ln F(t)} = \exp[-\sum_k \ln F_k(t)] \qquad \text{(exponential family)}$$

$$e^{-\ln F(t)} = \exp[-\sum_k c_k t^k] \qquad \text{(exponential family)}$$

Notice a close link between the concepts of partition functions, cepstra, and cumulant expansions [8], as indicated by the last four formulae. These links are formally simple, but surprisingly rarely mentioned in literature on statistical physics and signal processing.

Note that full understanding of these relationships and analogies implies a non-trivial question about coding probabilities into the *complex* rather than real numbers [19]. This question is particularly important for Quantum Physics. Many traces show a fundamental relevance of Lie groups and Octonions [21] to this question, e.g.: (i) The Probability Theory may be promoted to the realm of *Normed Division Algebras*: as we know from a Hurwitz Theorem, there are exactly four of them, in dimensions 1, 2, 4 and 8: reals **R**, complexes **C**, quaternions **Q**, and octonions **O**. (ii) The questions of a complexification, quaternification or even octonification of Probability Theory are intimately connected with the theory of Lie groups, their role in Probability Theory, and especially with the mystery of five exceptional Lie groups [21, 24] (see also Ch. 9 below).

6. GENERATING FUNCTIONS

This metamorphosis touches another of Laplace's great inventions or improvements [3] which is now of paramount importance in various branches of *Analytical Combinatorics* (the art of counting discrete decomposable structures) [16].

To make a long story short, a *Generating Function* (GF) is just a formal power series, say $A(t)$, associated with a given sequence, say (a_k), and some index set, $k \in K$:

$$A(t) = \sum_{k \in K} a_k \, t^k \tag{14}$$

Having a GF, the discrete sequence (a_k) is actually "shrinked" into a compact analytical object. Many "difficult" operations under (a_k) translate as much easier and transparent operations under corresponding GFs. And yet much more: there are some "dictionaries" where many operations under GFs are interpreted as some explicit combinatorial constructions; there are also some smart ways how to link singularities of GFs with asymptotic features of the k-th order coefficient, etc. The rationale holds its importance even for divergent series (formal power series). For a nice introduction to this vast realm, see Generatingfunctionology by Wilf [16].

Putting $t^k \to e^{kt}$, it is clear that every GF may be linked to an equivalent Fourier serier, which in turn may be interpreted as a Characteristic Function of Probability Theory (or a Partition Function in the Exponential Family). Consequently, we see that there is an intimate link between the concepts of Generating Functions and the Exponetial Family. In this way, the world of Analytical Combinatorics may be related to the world beyond the Exponential Family (MaxEnt Principle, etc.).

The theory of Generating Functions has a particularly rich vain of applications in counting problems. In this context, Wilf [16] describes the Exponential Formula in somewhat gambling terms of "cards", "hands" and "decks", as the following

Theorem (The Exponential Formula): Let F be an exponential family whose deck and hand enumerators are $D(x)$ and $H(x, y)$, respectively. Then

$$H(x,\ y) = e^{y\,D(x)} \tag{15}$$

In detail, the number of hands of weight *n* and *k* cards is

$$h(n,\ k) = [\frac{x^n}{n!}]\ \{\frac{D(x\ )^k}{k!}\} \tag{16}$$

Here, the expression in brackets [.] denotes the coefficient at the n-th order term of a power expansion of the quntity given in {.}. For more details about this terminology, formula, its proof and aplications, see Wilf [16]. Notice that in this counting problem, the information in question is actually topological and thus it is captured by a sampling PDF, whilst the prior PDF is proportional to a "hidden" unity. Interestingly enough, in this combinatorial/counting context, the Exponential Formula first appeared in the Thesis by Riddell (1953), co-authored by a celebrated statistical physicist G. E. Uhlenbeck.

7. SUFFICIENT STATISTICS

Any grouping, condensation or transformation of observables will in general result in a loss of information. In particular, projecting data by some functions (called "*statistics*") on a (smaller) number of parameters cannot, in general, represent a corresponding sampling PDF correctly. The only exception is an important case when the data are represented by *sufficient statistics* [6, 7]. This fundamental concept was introduced in statistics by R. A. Fisher around 1922. An example: For a Gaussian PDF, the parameters μ (mean) and σ (variance) are the sufficient statistics.

Let all information about a sampling PDF be captured by some functions of the data only ("sufficient statistics"). Exponential family is fundamental to the theory of sufficiency, as follows from

Pitman-Koopman Theorem (1936): A necessary and sufficient condition that a PDF may possess sufficient statistics is that it can be put in the form:

$$p(x\,|\,\lambda\,) = \frac{1}{Z(\,\lambda\,)}\ e^{\lambda \bullet u(x)}\,q(x) \tag{17}$$

where

$$\lambda \bullet u(x) = \sum_k \lambda_k u_k(x) \tag{18}$$

Thus, the concept of Sufficient Statistics implies the Exponential Family, and vice versa. Pay attention to the individual factors of Bayes' Theorem in (17), where $q(x)$ may again be interpreted as the prior PDF. However, even with the concept of sufficiency one cannot agree with Fisher's maxim, *"Let data speak for themselves"*, as we can never strip our data (and Bayes' Theorem) from its contextual prior factor [8]. Notice also that the concept of sufficiency suggests a rationale for wave-particle duality, and actually B. O. Koopman also authored a paper on Quantum Mechanics in 1931.

8. MULTINOMIAL ASYMTOTICS

Now we apply the "exp-log trick" to a *multinomial distribution* - the most general discrete PDF. The entropy functionals naturally emerge as an asymptotic case of the multinomial PDF. To have an underlying statistical picture, we can imagine a simple statistical experiment with a die, as recently elaborated in detail by M. Grendár [12] (improving a celebrated "Brandeis die" example of E. T. Jaynes [9]) . This offers a tangible example guiding us how to "read" a posterior probability of a discrete experimental "scene", using the entropic functionals.

Consider a multinomial PDF:

$$p(n|q) = \frac{N!}{n_1! \ \dots \ n_K!} q_1^{n_1} \ \dots \ q_K^{n_K} \tag{19}$$

where $n := (n_1, \dots, n_K)$ is a sampling distribution ("scene"), $\sum n_k = N$, and $q := (q_1, \dots, q_K)$ is a prior distribution ("generator"), $\sum q_k = 1$. Of course, for Jaynes' "Brandeis die" example, K=6.

We can recognize both factors of Bayes' Theorem (sampling PDF, prior PDF) in the multinimial PDF (19). To put the multinomial formula to the form of the Exponential Family, one shall apply the "exp-log trick" and investigate the logarithm of posterior probability $p(n|q)$.

Thus, let us interpret solution of the MaxEnt principle as a maximizer of the posterior log-probability (or multiplicity) of a scene n, given a prior generator q. Let us investigate an asymptotic behaviour of ln $p(n|q)$, for $N \to \infty$,

$n_k \to \infty$, putting $p_k := (n_k / N)$. With the help of Stirling's asymptotic formula [12] (or Laplace's Saddle-Point Method), the asymptotic analysis elucidates answers to the following three questions:

(i) What do is entropy??? Entropy is just a log-scaled posterior probability, defined on a space of PDFs.

(ii) Which functionals on a PDF asymptotically capture this log-probability???

Case 1: The asymptotic behaviour of $\ln\ p(n|q)$, given the uniform prior PDF $q = (1/K,\ \dots\ ,\ 1/K)$, is captured by the *Shannon-Jaynes entropy*:

$$S(p) = -\sum_k p_k \ln p_k \tag{20a}$$

Case 2: The asymptotic behaviour for $\ln\ p(n|q)$, given a prior PDF $q = (q_1,\ \dots\ ,\ q_K)$, is captured by the *Kullback-Leibler relative entropy*:

$$H(p|q) = -\sum_k p_k \ln \frac{p_k}{q_k} \tag{20b}$$

Case 3: The asymptotic behaviour for

$$\ln\ [p(n|q)\ p(\ qN|\frac{n}{N}\ )]$$

(prior and sampling PDFs "symmetrized") is captured by the Jeffreys-Grendar entropy:

$$J(p|q) = -\sum_k p_k \ln \frac{p_k}{q_k} + q_k \ln \frac{q_k}{p_k} \tag{20c}$$

(iii) What does the MaxEnt principle accomplish??? When maximizing the entropy, we are just seeking the most probable posterior PDF, subject to given constraints.

Consequenly, the above three entropic functionals correspond to *three different tasks* [12]. This result offers a clue to a puzzling question often asked

-- "which entropy is the correct one". Some historical remarks: A quantity equivalent to the Shannon-Jaynes entropy, and interpreted as a "measure of the probabilities of states", appears already in Boltzmann's "Lectures on Gas Theory" around 1896-98 (and in his papers even before). However, the MaxEnt rationale based on seeking the "most probable probability distribution" may be traced back even to Laplace's probabilistic papers around 1777 [3].

9. LIE GROUPS

The exponential mapping is of paramount importance in the theory of Lie groups (and Lie algebras) [21-25]. How the Lie groups enter the stage? Recall that the "most entropic" distributions are induced by conserved quantities. However, every conserved quantity also corresponds to a *symmetry* of a physical system, which in turn corresponds to a group of symmetries. Schematically,

$p(x\,|\,\lambda)$ <-----> MaxEnt <-----> Conserved Quantities <----->
<-----> Symmetries <-----> Groups

There are either discrete or continuous symmetries, and associated groups. Here we particularly focus on the continuous finite groups - called in mathematics the *Lie groups.*

In the theory of Lie groups, the exponential mapping *defines* a relationship between the *Lie groups* and *Lie algebras* [21-24]:

{Lie Group} = exp {Lie Algebra}

$$G = e^{tX} \tag{21}$$

In mathematics or theoretical physics, the exponential map is mostly understood as an algebraic and/or geometric concept. It is remarkable that even the best references on Lie groups are practically silent about any relationships to probability theory or statistical physics, as strongly suggested by the Exponential Mapping/Family. In this connection, see also [11, 17].

9.1 Probability vs General Decomposition of Lie Algebras

Taking the above analogies seriously, one naturally expects that some basic facts in the theory of the Lie groups may be associated with some fundamental probabilistic concepts. And really, it appears to me that a general decomposition of Lie algebras [22] can be made easy when interpreted in terms of Bayes' Theorem and/or Exponential Family. Let us define the following correspondences:

$p(x|\lambda)$ <-----> exp {Lie Algebra}

$q(x)$ <-----> exp {Abelian Lie Algebra}

$\exp[\sum \lambda \bullet u(x)]$ <-----> exp { $\sum$ Simple Lie Algebra} =

= exp {Semi-Simple Lie Algebra}

To motivate a rationale for the above, recall that any *Abelian Lie Algebra* is commutative and its commutator bracket [.,.] equals to zero by definition. Consequently, this concept is associated with *homeomorphic mappings* of Euclidean spaces, where no topological effects take place. Obviously, topology enters the stage via *Simple Lie Algebras*, and the commutator [.,.] just isolates the underlying topological information [22, 23].

Example: Let the symmetry operations be the *translations* and *rotations*, acting on a 3-D Euclidean space $\mathbf{R}^3$; these symmetry operations correspond to the following groups and their algebras:

Translation group T(3) ... Abelian Lie Algebra t(3)
Rotation group SO(3) ... Simple Lie Algebra so(3)

Note that for so(3), the topological information is extracted by the commutator [.,.] which is equivalent to a vector product in the 3-D euclidean space. In probabilistic terms, the translation and rotation groups correspond to the prior and sampling PDFs, respectively.

It is remarkable that the exponential link between a (posterior) *probability* and *entropy* appears to be fully analogical to the exponential link between a *Lie group* and a corresponding (semi)simple *Lie Algebra*. Consequently, essentially probabilistic readings of the Exponential Family may naturally be translated in terms of algebraic, geometric and topological facets of Lie group theory.

9.2 Probability vs Cartan's Classification of Simple Lie Algebras

One of the major results of 20th century mathematics, with many physical implications, was a *classification of simple Lie algebras* [5, 22]. This is of paramount importance for modern physics, as it closely corresponds with the question what are the fundamental symmetries of physical laws.

As discovered by *Sophus Lie, Wilhelm Killing,* and ultimately by *Élie Cartan* (1869-1951) [5, 21-25], there exist just four infinite *"classical" types*:

A_N ... SU(N+1), B_N ... SO(2N+1), C_N ... Sp(2N), D_N ... SO(2N)

plus exactly five finite *exceptional types* [5, 21-24]:

G_2, F_4, E_6, E_7, E_8.

Concerning the four *classical* types, A_N, B_N, C_N and D_N, these correspond to "external" symmetries of our physical laws, with respect to orthogonal rotations described by the above-indicated matrices with some real (B_N, D_N), complex (A_N), and quaternionic (C_N) entries. Corresponding conserved quantity may be interpreted as a generalized momentum.

However, what is the role of the mysterious exceptional Lie groups in physics, and what are the symmetries associated with them?

As for the geometric, algebraic and topological roles of the exceptional Lie groups, these follows from their intimate connection with the *Octonions* [21, 22] (the last member of the only four existing *normed division algebras* - reals **R**, complexes **C**, quaternions **Q**, and octonions **O**). They also correspond to five Platonic solids in 3-D space [5].

I propose the idea that *the exceptional Lie groups are just inner symmetries of the Probability Theory itself.* More specifically, particularly involved for this probabilistic interpretation is the so-called *A-D-E sequence* [25, 21, 24]:

A_1, A_2, A_3=D_3, D_4, D_5, E_6, E_7, E_8

A probabilistic rationale for the above "exceptional" A-D-E sequence may already be supported by several arguments, one of them being the fact that this sequence arises as a unique solution of a probabilistic (MaxEnt) problem, as indicated by the following

Conjecture: Let us reinterpret the *Sphere Packing Problem* [25] in N = 1, 2, 3, 4, 5, 6, 7 and 8 dimensions, respectively, as *the most probable (most entropic) distribution of the N-dimensional spheres*. The solutions define corresponding optimal *lattices*, which in turn define corresponding *root systems*, *Coxeter-Dynkin diagrams* and associated A-D-E *Lie algebras* uniquely. (Note: The root vectors for G_2 and F_4 are subsets of their A_2 and D_4 counterparts; only for A_n, D_n and E_n, the root vectors are of the same length. See [25], p. 98-99.)

Note: A particularly interesting probabilistic role emerges for the Lie group D_4, which stands as a bridge between the "classical" and the "exceptional" family [21-25]. This group shows a unique symmetry known as *Triality* [21, 24]. I conjecture that the same inner symmetry is inherent to Bayes' Theorem!

10. THE EXPONENTIAL FAMILY EVERYWHERE ...

The laplacean idea of exponentiation is truly ubiquitous and may be applied as a unifying rationale for many problems which are often treated under an umbrella of *Complex Nonlinear Systems* [26, 1, 2]. Missing both time and space, we mention just some typical situations. These involve: (i) ubiquitous *Spiral Geometries*, which are so common in all scales of nature: galaxies, weather patterns, biological structures (shells, pine cones, DNA, ...), NMR-signals, etc; (ii) ubiquitous *Power-Laws* phenomena which are often associated with the names of Zipf, Benford, Pareto, Mandelbrot, etc.); (iii) ubiquitous *Phyllotaxis* phenomena and some number-theoretic manifestations in biology, etc. [1, 2, 26]

In particular, Spiral Geometries in nature are straighforward manifestations of the Exponential Family and associated MaxEnt principle. Every *logarithmic spiral* may be treated as a special case of the Exponential Family.

Exponential Family <---> Logarithmic Spirals

Two great pioneers of probabilistic ideas have some logarithmic spirals on their tombs: Amazing invariance features of the logarithmic spiral brought so much pleasure to *Jacob Bernoulli* (1654-1705) that he asked the "Spira Mirabilis" be put on his tomb in Basel, together with the words *"Eadem mutata resurgo"* (*"I shall arise the same, though changed"*). And the second one is *Ludwig Boltzmann* (1844-1906), whose most famous formula $S = k \log W$ is engraved on his tomb in Wien.

11. CONCLUSIONS

Ubiquitous Exponential Family takes many metamorphoses and roles, which might be vital in context of Inverse Problems and many other applications. The very concept of Inverse Problems addresses the same fundamental situation as the information-theoretic extensions of Statistical Thermodynamics, and namely - the task of an optimal logical inference from given incomplete information.

Probability Theory appears as a very unique "design" among all our theories. It inherits most of its unique features just from the Exponential Family, which not only unifies some fine probabilistic, statistico-physical, spectral, combinatorial, statistical and asymptotic properties, but even locks them with some unique algebraic, geometric and topological features of Lie groups. As a suggestion for future investigations, I conjecture that the mysterious exceptional Lie groups might be actually the inner symmetries of the probability theory itself. Even such fundamental physical concepts, as the dimension of physical space-time continuum and the arrow of time, obviously show some place and rationale along this line of thought.

In this work, some analogies between the exponential structures have been taken very seriously, advocating that they might be not just mere "random coincidencies". Well, "*Though analogy is often misleading, it is the least misleading thing we have.*" (Samuel Butler (1835-1902)).

ACKNOWLEDGEMENT

This work was supported in part by the project ÚPT AV ČR #3151 and MŠMT-KONTAKT ME-526.

REFERENCES

1. A. Gottvald, "Information, Physics, Computing", unpubl. Lecture Notes presented at Kyoto University (Japan), Dec. 13-15, 2001 (to appear)
2. A. Gottvald, "Beyond the MaxEnt Principle: Bayes, Fourier, Zipf and Spirals in Metamorphoses", in Energy and Information in Non-linear Systems, A. Gottvald (ed.), CSAEM, Brno, 2001, pp. 48-51
3. Ch. C. Gillispie, "Pierre-Simon Laplace, 1749-1827, A Life in Exact Science," Princeton Univ. Press, Princeton, 1997
4. A. Gottvald, "Laplace's Demon and Maxwell's Demon: From Historical Myths to Topological Evolution," in Proc. of 8th MENDEL Conf. on Soft Computing, Brno (CR), 2002, pp. 139-146
5. M. A. Akivis and B. A. Rosenfeld, "Élie Cartan (1869-1951)," American Mathematical Society, Providence, 1993
6. S. Kullback, "Information Theory and Statistics," Dover Publ. Inc., Mineola, 1997 (first published 1959)
7. H. Jeffreys, "Theory of Probability," Clarendon Press, Oxford, 1998 (first published 1939)
8. E. T. Jaynes, "Probability Theory - the Logic of Science", Cambridge, to appear in April 2003; on-line: http://omega.albany.edu:8008/JaynesBook

9. E. T. Jaynes, "Where do we Stand on Maximum Entropy?" in The Maximum Entropy Formalism, R. D. Levine and M. Tribus (eds.), M. I. T. Press, Cambridge, MA, p. 15; on-line: http://bayes.wustl.edu
10. E. T. Jaynes, "The Evolution of Carnot's Principle," in Maximum-Entropy and Bayesian Methods in Science and Engineering, G. J. Erickson and C. R. Smith (eds.), Kluwer, Dordrecht, p. 267; on-line: http://bayes.wustl.edu
11. E. T. Jaynes, "Prior Probabilities and Transformation Groups," around 1965, unpublished manuscrips; on-line: http://bayes.wustl.edu/etj/groups.pdf
12. M. Grendár, jr., and M. Grendár, "What is the Question that MaxEnt Answers? A Probabilistic Interpretation," in Bayesian Inference and Maximum Entropy Methods in Science and Engineering, A. Mohammad-Djafari (ed.), AIP Press, New York, pp. 83-94; on-line: http://xxx.lanl.gov/abs/math-ph/009020
13. M. Grendár, jr., and M. Grendár, "MiniMax Entropy and Maximum Likelihood," in Bayesian Inference and Maximum Entropy Methods in Science and Engineering, A. Mohammad-Djafari (ed.), AIP Press, New York, pp. 49-61; on-line: http://xxx.lanl.gov/abs/math.PR/0009129
14. R. N. Bracewell, "The Fourier Transform and Its Applications," McGraw-Hill, Boston, 2000 (first published 1965)
15. A. Rényi, "Probability Theory," (Czech version: Academia, Praha, 1972; German original: Wahrscheinlichkeitsrechnung mit einem Anhang über Informationstheorie," VEB deutscher Verlag der Wissenschaften, Berlin, 1962
16. H. S. Wilf, "Generatingfunctionology," Academic Press, San Diego, 1994
17. C. C. Rodriguez, "Entropic Priors," on-line: http://omega.albany.edu:8008/entpriors.ps
18. C. C. Rodriguez, "Are We Cruising a Hypotheses Space,", in Maximum Entropy and Bayesian Methods, W. von der Linden, V. Dose, R. Fischer and R. Preuss (eds.), Kluver Academic Publishers, 1998, pp. 131-139; on-line: http://omega.albany.edu:8008/
19. C. C. Rodriguez, "Unreal Probabilities: Partial Truth with Clifford Numbers," in Maximum Entropy and Bayesian Methods, W. von der Linden, V. Dose, R. Fischer and R. Preuss (eds.), Kluver Academic Publishers, 1998, pp. 247-270; on-line: http://omega.albany.edu:8008/
20. A. Caticha, "Probability and Entropy in Quantum Theory," in Maximum Entropy and Bayesian Methods, W. von der Linden, V. Dose, R. Fischer and R. Preuss (eds.), Kluver Academic Publishers, 1998, pp. 237-246
21. J. Baez, "The Octonions," Bull. Amer. Math. Soc. 39 (2002), pp. 145-205; on-line: http://math.ucr.edu/home/baez/Octonions/octonions.htm
22. L. Frappat, A. Sciarrino and P. Sorba, "Dictionary on Lie Algebras and Superalgebras," Academic Press, San Diego, 2000
23. N. Bourbaki, "Lie Groups and Lie Algebras," Springer, Berlin, 1989 (first published 1971, 1972)
24. J. F. Adams, "Lectures on Exceptional Lie Groups," The University of Chicago Press, Chicago, 1996
25. J. H. Conway and N. J. A. Sloane, "Sphere Packings, Lattices and Groups," Springer, New York, 1999
26. M. Schroeder, "Fractals, Chaos, Power Laws," Freeman and Comp., New York, 1991
27. A. Gottvald, "Planck's Law and the Rise of Information Physics," in Energy and Information in Non-linear Systems, A. Gottvald (ed.), CSAEM, Brno, 2001, pp. 44-47

TOPOLOGICAL EVOLUTION AND INVERSE PROBLEMS

On Probability between Discrete and Continuous

Aleš Gottvald

Institute of Scientific Instruments, Academy of Sciences of the CR, Královopolská 147, 612 64 Brno, Czech Republic

Abstract: Topology is manifest in Inverse Problems, Electromagnetic Theory and many other fields. Cartan's exterior calculus of differential forms, Lie groups and Exponential Family are central tools for investigating phenomena where some topological changes take place. A fundamental law of topological evolution is equivalent to the First Law of Statistical Thermodynamics, involving a topological interpretation of heat. We link Bayes' Theorem with Triality, D4-group, and Octonions. We propose that the inner symmetries of Probability Theory itself correspond to an exceptional A-D-E sequence of Lie groups, with vital implications for fundamental physical laws.

Key words: Topological evolution, inverse problems, Cartan's exterior calculus, Lie groups, exponential family, triality, A-D-E sequence

1. INTRODUCTION

Topology is dealing with those fundamental features of nature which are preserved under homeomorphic mappings [1]. We argue that Inverse Problems, Electromagnetic Theory and many other domains are rich with topological phenomena, a fact still often neglected. E.g., Maxwell's integral equations or Planck's radiation law are of intrinsically topological nature.

Consider a mapping X → Y which characterizes an Inverse Problem (IP). A fundamental feature of IP is that the domain X and the range Y show *different*

M. Rudnicki and S. Wiak (eds.), Optimization and Inverse Problems in Electomagnetism, 181–192.

topological structures. A topology on the range Y is finer and requires more information than a topology on the domain X. Consequently, the *topological evolution* (induced by IP) allows a *retrodiction*, but not a *prediction* [1]. (For intuitive examples, consider a topological difference between "cutting - pasting", "cards shuffling - cards ordering", or "laminar - turbulent".) This implies a *topological definition of IPs* -- as the problems where some topological information is missing. Recall also Hadamard's classical definition of IP where the idea of topological changes is clearly involved, too.

To capture the topological phenomena, three closely related concepts appear as very central: (i) *differential forms* [1, 2]; (ii) *simple Lie groups and Lie algebras* [8-10]; (iii) *Exponential Family* [3]. First two of these concepts are particularly connected with the legacy of *Élie Cartan* (1869-1951) [6]. However, it is often overlooked that also Probability Theory (and Exponential Family) carries a topological information in its inner structure. No wonder that physical quanta was first unveiled as an application of the Second Law of Statistical Thermodynamics (Planck's struggle with a radiation law) [16]. The present paper shows an informal list of concepts and phenomena associated with some *topological* information, and puts them in a contrast with intrinsically *euclido-geometrical* phenomena where no topological changes take place.

2. EUCLIDEAN VS TOPOLOGICAL PHENOMENA

There are two principally distinguished kinds of phenomena (and mappings) in nature: Some phenomena may be "eliminated" or "removed" by a re-parametrization and, consequently, may be interpreted as mere coordinate artifacts. Typical examples are potential fields (gravitational, electric, etc.). A typical stage for this "topologically trivial" situation is the concept of an *Euclidean space* and *Euclidean geometry*. These phenomena represent an *epistemic* (logical) layer in our theories.

In contrast, many phenomena of nature are invariant with respect to the changes of coordinates. These are of intrinsically *topological* (and *ontological*) nature. Typical examples are shape- and material-independent phenomena (e.g., Planck's radiation law). Also "true" gravitational and electric fields cannot be *globally* eliminated by any continuous differentiable mapping. A typical stage for this "topologically non-trivial" situation is the concept of a differentiable *manifold* (an atlas of locally homeomorphic maps).

2.1 A Depository of Euclido-Geometric Concepts

There is typically no change of topological structure associated with in the concepts and phenomena listed below. These *indicate* those effects which can be (at least in principle) eliminated by the action of some homeomorphic coordinate mappings, and thus represent some parametrization artifacts:

-potential fields
-Laplace's equation
-closed loop integrals equal to zero, total differentials, exact forms
-Cauchy-Riemann conditions
-conformal mappings, their quaternionic and octonionic generalizations
-commuting diagrams
-compact geometry, continuum
-laminar flows
-actions of Abelian symmetry groups
-Hamiltonian evolution, unitary evolution
-jacobians (regular)
-tensors
-regular/forward/direct problems
-space-components of the space-time manifold ("Minkowski fluid") (??)
-antisymmetric quantities (??)
-Fermi-Dirac statistics, fermionic particles (??)

2.2 A Depository of Topological Concepts

The intrinsically *topological* world cannot be eliminated by mere re-parametrizations (diffeo-maps) and stands behind the following concepts:

-integer number relationships (e.g., $1^2 + 2^2 + 3^2 + \ldots + 24^2 = 70^2$)
-graphs
-non-commuting diagrams
-singularities, residua
-commutators, associators
-disconnected geometry
-cutting vs pasting
-cards shuffling vs cards ordering
-phase transitions
-heat changes
-chemical reactions
-particle "chemistry"

-charges, currents
-turbulence, chaos
-lightnings, clouds, rains, snowflakes, foams
-atoms, molecules, quanta, singularities
-genetic code (DNA)
-actions of simple Lie groups and associated symmetries;
-differential forms, Pfaff forms
-inverse problems
-time component of the space-time manifold ("Minkowski fluid") (??)
-symmetric quantities (??)
-Bose-Einstein statistics, bosonic particles (??)

2.3 Link to Bayes' Theorem

Where to "dig" in the omnipotent Bayes' Theorem (or Exponential Family) for the above euclido-geometric and topological phenomena, respectively? The intrinsically *euclidean* phenomena, being epistemic, correspond to the prior factor in Bayes' Theorem [3]. *Example:* Potential fields in different coordinates lead to different priors, while the sampling PDF is constant.

The intrinsically *topological* phenomena, being ontological, correspond to the exponential "sampling" factor of Bayes' Theorem. *Examples*:

Vertices - Edges + Faces = 2 (*Euler's formula of graph theory*)
Freedoms - Components + Phases = 2 (*Gibbs' phase rule*)

Both topological and euclido-geometric aspects are involved in the normalizing factor of Bayes' Theorem - the partition function of Exponential Family. *Examples:* Kirchhoff's and Planck's black-body radiation laws [16].

3. CARTAN'S EXTERIOR CALCULUS

Cartan's exterior calculus [1, 2] is an important tool for studying a *topological evolution*, associated with non-homeomorphic maps. A stage of this game is a linear space of *differential forms*, endowed with three operations (*exterior derivative (d)*, *exterior product (^)*, and *inner product* (symbols "<.,.>" or dot "." or "i(.)")). With these operations, one can constructs a *Cartan-Pfaff sequence* of (higher-order) differential forms, which terminates at a finite N (the Pfaff dimension, usually N=4). Let us emphasize that differential forms are not just another word of fancy. Being global quantities (actually the integrands), they incode a *topological information*, whilst a more common tensor calculus (as used in relativity) is limited to homeomorphic (=topology preserving) mappings [1, 2].

Cartan's method offers a unified treatment of classical deterministic systems (including unholonomic constraints).

Examples of 1-forms of actions from classical mechanics: [1]

(i) *Lagrange's 1-form* stands in the integrand of the action:

$$S = \int_a^b L(q, \dot{q}, t)\, dt \tag{1}$$

(ii) *Hamilton's 1-form* stands in the integrand of the action:

$$S = \int p\, dx - H\, dt \tag{2}$$

(iii) *Cartan-Hilbert 1-form* stands in the integrand of the action:

$$S = \int L(q, \dot{q}, t)\, dt + p\,(dq - \dot{q}dt) \tag{3}$$

where the difference term in (3), being not necessarily zero, is an unholonomic constraint ("Cartan's trick") which takes into account a friction [1].

4. DIFFERENTIAL FORMS IN ELECTROMAGNETISM

It is illuminating to classify the differential forms in electromagnetic theory in terms of the Cartan-Pfaff sequence. It shows that the *electromagnetic equations are intrinsically topological* laws [1], independent on coordinate transformations (homeomorphic maps). Moreover, these topological laws are *universal* and not restricted to electromagnetics or 4-dimensional spaces. Obviously, the top Pfaffian is reached for N=4 (the Cartan-Pfaff dimension).

The *Cartan-Pfaff sequence* for the *Maxwell system*:

$$\{\, A,\ F = dA,\ G,\ J = dG,\ A\wedge F,\ A\wedge G,\ A\wedge J,\ F\wedge F,\ G\wedge G \,\} \tag{4}$$

A (1-form of electromagnetic potentials, postulated)

$F = dA$ (2-form of fields, exact)

G (2-form of excitation fields, postulated)

$J = dG$ (3-form of charge-current densities, exact)

$A \wedge F,\ A \wedge G$ (3 forms)

$A \wedge J,\ F \wedge F,\ G \wedge G$ (4-forms)

Maxwell's equations, in terms of the differential forms, take the form:

$$F - dA = 0 \tag{5a}$$

$$J - dG = 0 \tag{5b}$$

Differential forms shown below serve as the *counters of singularities/ quanta* in quantum electrodynamics [1], proviso that associated field quantities equal to zero:

$$F = 0 : \qquad \oint A = n\phi \qquad \rightarrow \text{flux quanta [h/e]}$$

$$J = 0 : \qquad \oint\!\!\oint G = n\, e \qquad \rightarrow \text{charge quanta [e]}$$

$$F \wedge F = 0 : \qquad \oint\!\!\oint\!\!\oint A \wedge F = n\, t \qquad \rightarrow \text{torsion quanta } [(\text{h/e})^2]$$

$$F \wedge G - A \wedge J = 0 : \qquad \oint\!\!\oint\!\!\oint A \wedge G = n\, s \qquad \rightarrow \text{spin quanta [h]}$$

For an example of the above flux quanta counter, recall the Aharonov-Bohm effect [15].

5. TOPOLOGICAL EVOLUTION VS THE FIRST LAW OF THERMODYNAMICS

Given a differential form, a fundamental *law of topological evolution* follows from an action of *Lie derivative.* This concept shows a striking analogy with the *First Law of Thermodynamics*, a new insight emphasized by Kiehn [1]. We illustrate ubiquity of this analogy on three metamorphoses: (i) *generalized conservation laws* (statistical physics); (ii) *Cartan's magic formula* (differential geometry) [1]; (iii) the *Cauchy-Green formula* (theory of complex variables) [5]. A new interpretation for the concept of *heat* arises -- as a measure of topological changes in an arbitrary system.

5.1 The First Law of Thermodynamics

Given any *conserved quantity* $\overline{A}(\lambda)$:

$$\overline{A}(\lambda)=\int A(x)\, p(x\,|\,\lambda)\, dx \tag{7}$$

its variation is:

$$\delta\overline{A}(\lambda) = \int \delta A(x)\; p(x\,|\,\lambda)\, dx + \int A(x)\; \delta p(x\,|\,\lambda)\, dx \tag{8}$$

We can interpret this universal expression as a *generalized First Law of Thermodynamics*:

$$d\overline{A}(\lambda) = \delta W(\lambda) + \delta Q(\lambda) \tag{9}$$

$$d\ \{\text{Any Quantity}\} = \delta\,\text{Work} + \delta\,\text{Heat}$$

This is actually the *Law of Continuous Topological Evolution.* The concept of *heat*, which is usually given a kinetic energy interpretation (Maxwell, Boltzmann), is now percepted as a *measure of topological changes* in a system.

δ {Heat} ---> Topological changes ...

Along this line of analogies, the euclidean "work-like" and and the topological "heat-like" terms feature together in other metamorphoses of the First Law. Two examples:

(i) In differential geometry: Cartan's Magic Formula (Marsden's term):

$$L_{(V)} A = i(V)\, dA + d\,(i(V)\, A) \tag{10}$$

This is a *Lie differential* along a vector field V and an expression of *cohomology* (R. M. Kiehn, [1]). It may be "thermodynamically" interpreted as

$$\delta Q = \delta W + dE \tag{11}$$

(ii) In complex functions: Cauchy-Green formula: (f continuous, D compact)

$$f(z) = \frac{1}{2\pi i} \int_{\partial D} \frac{f(\zeta)}{\zeta - z}\, d\zeta + \frac{1}{2\pi i} \int_{D} \frac{\partial f}{\partial \bar{\zeta}}(\zeta) \frac{1}{\zeta - z}\, d\bar{\zeta} \wedge d\zeta \tag{12}$$

Holomorphic	+	Non-Holomorphic
Conformal	+	Non-Conformal
δ Work	+	δ Heat

5.2 From Differential Forms to Exponential Family

Differential forms actually generalize that constraint conditions which we considered in MaxEnt formalism (see [3]). Treating all members of a Cartan-Pfaff sequence as topology-carrying geometric objects (actually they form a Clifford's "geometric algebra"), one can establish a bridge between the *"deterministic"* and the *"probabilistic"* worlds (obscured in physics since Boltzmann's times ...). Symbolically:

conserved quantities ←→ differential forms $X := (X_k(x))$:

implies the Exponential Family:

$$p(x\,|\,\lambda) = \exp\{\lambda \bullet X\} \tag{13}$$

Note: Particularly important is a differential form which plays a central role in statistical physics - the *partition sum*, associated with the Exponential Family. A link between a corresponding cumulant expansion and a moment expansion (for the asssociated PDF) is instrumental in this context; topology complicates this link above the 4-th order.

6. FROM LIE GROUPS TO BAYES' THEOREM AND EXPONENTIAL FAMILY

How many topologically different symmetries are hidden in the Exponential Family? For continuous symmetries, the answer is provided by *Cartan's classification of simple finite Lie Algebras* [6, 8-10, 3]. There exist:

-Four *"classical"* infinite types:

A_N, B_N, C_N, D_N

-plus five (and only five!) finite *exceptional* types:

E_6, E_7, E_8, F_4, G_2

A "tangible" representation of these topology-carrying objects is provided by their root systems, lattices, and (particularly conveniently) their *Coxeter/Dynkin diagrams* [6, 8-11].

The Dynkin diagrams are invariant with respect to actions of Abelian groups (for which the Dynkin diagram is empty). So they carry just the *topological* content of the individual Lie groups. Actually this is the way how relativistic ideas enter the stage (recall that Élie Cartan served as a not-quite-properly acknowledged critic and adviser to Einstein).

6.1 Triality

Triality [10, 7, 14] is an exceptional symmetry of the *simple Lie Algebra D_4*, reflected by the corresponding Dynkin diagram (whilst duality is a common feature of the classical A-B-C-D Lie groups). A remarkable fact is that the same unique Triality symmetry is inherent to *Bayes' Theorem* (*Exponential Family*). We thus suggest a link which "locks" the algebraic and geometric features of the Lie group D_4 with a fundamental *probabilistic meaning*.

D_4 ←→ Exponential Family (Bayes' Theorem):

⊗	$p(\lambda \mid x)$
|	|
⊗---⊗---⊗	$p(x)$---$Z(\lambda)$---$p(x \mid \lambda)$

I suggest that the individual 4 representations of D_4 (see the above Dynkin diagram) be interpreted as the corresponding 3 conditional probabilities, plus the adjoint representation interpreted as the partition function.

Classical probabilities are represented by real numbers. However, I propose that also the three higher *normed division algebras*, and namely *complex numbers* **C**, *quaternions* **Q** (**H**), and *octonions* **O**, [7, 8, 13, 14], may be used for coding probabilities. A Cayley-Dickson "doubling" process [7] naturally leads to an "octonified" version of Bayes' Theorem (Exponential Family), with a Spin(8)-type of inner symmetry -- a full *Triality* symmetry unique to D_4. This implies a huge amount of remarkable algebraic and geometric facts, which specialists on exceptional Lie groups and particle physics know better than me [14, 10, 7, 8, 13]. Now these facts are also unique to Probability Theory, and vice versa.

Note: The Triality symmetry of D_4 (and Bayes' Theorem) obviously captures our *visual perception of colours* (the R, G, B colours map onto 3 conditional probabilities, and an intensity information maps on the partition function).

6.2 A-D-E Symmetries, Probability and Particles

Which are the *internal* symmetries of our fundamental physical laws? Which principle can guide us to choose a *finite* subset within these Lie groups and, in particular, offers a rationale for realistic "elementary building blocks" in nature? This is a very puzzling question, though many empirical and algebro-geometric arguments suggest the following particular models:

-the Standard model U(1) x SU(2) x SU(3) of particles/forces [12, 14];
-the E_8 x E_8 model of string theory [12] ;
-the **C** x **Q** x **O** model by Geoffrey Dixon [13];
-the A-D-E model by Frank "Tony" Smith [14].

We propose a *new principle* for the choice of the "correct" inner symmetries: *The inner symmetries of physical laws must be identical with inner symmetries of Probability Theory itself!* Thus, we suggest that only those inner symmetries which are inherent to Probability Theory (and Statistical Physics) represent correctly those topologies which we observe in our (probabilistic, statistico-physical) experiments.

In all particle physics experiments, we evaluate *probabilities* after all. It appears as if we just observe *different inner symmetries of the Probability Theory* itself. This line of thought brings a particularly strong support for the unique and ubiquitous *A-D-E sequence of symmetries* [14, 3]. These fundamental A-D-E symmetries might be those "information atoms" which we percept as the basic building blocks of nature. The A-D-E symmetries guarantee logical reproducibility of our probabilistic assignments. Personally I consider this "probabilistic" clue very remarkable and worth future investigations.

6.3 That Number 4 ...

The number 4 often stands at the crossroads between the realms of topological and euclido-geometric phenomena: We conjecture that the phenomena below are not just "random coincidencies":

-algebraic equations and their solvability in radicals (--> Galois theory);
-4-color theorem for planar graphs;
-4 dimensions of space-time continuum;
-4 fundamental dimensions of Bayesian Probability Theory [3];
-there are exactly 4 normed division algebras [7, 8, 13, 14];
-there are exactly 4 parallelizible spheres (S_0, S_1, S_3, S_7) [13, 14];
-and much more number-4 phenomena for night thoughts ...

7. CONCLUSIONS

Exponential Family is a natural "depository" for both topological (global, discrete) and euclidean (local, continuous) types of information. It stands as *the* link between a continuous and discrete world, and a natural rationale for field-particle duality. The *topological structures* are in-a-sense absolute, coordinate-free, and imposed by intrinsically global quantities: differential forms, simple Lie groups, and exponential "sampling factors" of Bayes' Theorem. The *euclidean structures* are relative, rather "parametrization artifacts", imposed by local coordinates, abelian Lie groups, and prior factors of Bayes' Theorem.

Inverse Problems involve *topological changes*. Many fundamental *physical laws* are intrinsically *topological statements*. The law of *topological evolution* is analogical to a generalized *First Law of Statistical Thermodynamics*.

Given an *Exponential Family/Map*, we can ask questions in terms of *symmetries*, *conservation laws*, *thermodynamic laws*, *sufficient statistics*, *quantization*, etc. This brings us close to questions of modern physics. Particularly promising is a novel amalgamation of Probability Theory with Lie groups, especially those *exceptional* ones, which has much to say to the origin of our basic physical laws. It implies a new *probabilistic* rationale for the ubiquitous A-D-E sequence of symmetries. Work continues.

ACKNOWLEDGEMENT

Supported in part by the projects ÚPT AV ČR #3151 and MŠMT-KONTAKT ME-526. Editorial cooperation of Prof. Marek Rudnicki is greatly appreciated.

REFERENCES

1. R. M. Kiehn: "Cartan's Corner", a bonanza of remarkable unpublished manuscripts, available on-line: http://www.cartan.pair.com; see e.g., "Continuous Topological Evolution", version of March 3, 2002
2. H. Flanders, "Differential Forms with Applications to the Physical Sciences," Dover Publ. Inc., New York, 1989 (first published 1963)
3 A. Gottvald: "Exponential Family and Inverse Problems", see an accompanying paper in the present proceedings
4. A. Gottvald, "Laplace's Demon and Maxwell's Demon: From Historical Myths to Topological Evolution," in Proc. of 8th MENDEL Conf. on Soft Computing, Brno (CR), 2002, pp. 139-146
5. E. M. Chirka, P. Dolbeault, G M. Khenkin and A. G. Vitushkin, "Introduction to Complex Analysis," Springer, Berlin, 1997; esp. A. G. Vitushkin, "Remarkable Facts of Complex Analysis,", pp. 1-17
6. M. A. Akivis and B. A. Rosenfeld, "Élie Cartan (1869-1951)," American Mathematical Society, Providence, 1993
7. J. Baez, "The Octonions," Bull. Amer. Math. Soc. 39 (2002), pp. 145-205; on-line: http://math.ucr.edu/home/baez/Octonions/octonions.htm
8. L. Frappat, A. Sciarrino and P. Sorba, "Dictionary on Lie Algebras and Superalgebras," Academic Press, San Diego, 2000
9. N. Bourbaki, "Lie Groups and Lie Algebras," Springer, Berlin, 1989 (first published 1971, 1972)
10. J. F. Adams, "Lectures on Exceptional Lie Groups," The University of Chicago Press, Chicago, 1996
11. J. H. Conway and N. J. A. Sloane, "Sphere Packings, Lattices and Groups," Springer, New York, 1999
12. M. Kaku, "Introduction to Superstrings and M-Theory," Springer, New York, 1999
13. G. Dixon, "Division Algebras: Octonions, Quaternions, Complex Numbers, and the Algebraic Design of Physics," Kluwer, Dordrecht, 1994; a related on-line source: http://www.7stones.com/Homepage/AlgebraSite/algebra0.html
14. F. "Tony" Smith, a bonanza of information on algebro-geometric and particle-physical aspects of the A-D-E theory available on-line: http://www.innerx.net/personal/tsmith
15. A. Gottvald, "Aharonov-Bohm Effect and Vector Potential: How Weird for Bayesians?" in Non-linear Electromagnetic Systems, P. Di Barba and A. Savini (eds.), IOS Press, Amsterdam, pp. 447-450
16. A. Gottvald, "Planck's Law and the Rise of Information Physics," in Energy and Information in Non-linear Systems, A. Gottvald (ed.), CSAEM, Brno, 2001, pp. 44-47

IDENTIFICATION OF INDUSTRIAL ELECTROMAGNETIC FIELD SOURCES

A. Canova, M. Repetto, M. Tartaglia
Dipartimento di Ingegneria Elettrica Industriale, Politecnico di Torino (Italy)

Abstract: The evaluation of the electromagnetic field emitted by an industrial source under quasi-stationary conditions is usually performed by means of magnetic measurements which gives a spatial distribution of the magnetic flux density amplitude. In fact usual industrial applications produce electric field levels which are far from any known health effects, moreover a quasi-static electric field is quite easy to screen. On the contrary, the design of a magnetic field shielding apparatus requires to simulate both the source and the shield system and to evaluate the main shield parameters like geometry, materials, position with respect to sources. Thus it seems powerful to simulate actual source fields by means of suitable simplified sources which can be handled by analytical relations. In this work, an identification process of industrial magnetic fields with a negligible harmonic content is performed using a genetic algorithm applied to a simplified source. The proposed source is a suitable combination of three-phase ones and it seems able to create a realistic electromagnetic field distribution both in space and in time. Some results related to possible industrial applications are discussed.

Key words: ELF magnetic field, genetic algorithm, identification, power systems.

1. INTRODUCTION

In the design phase of shielding systems for industrial electromagnetic fields under quasi-stationary conditions, for instance electrical sub-stations, it is often necessary to perform electromagnetic analysis in order to predict shield performances. This process requires to know, with a reasonable degree of accuracy, the spatial distribution of magnetic field emitted by the installation. Unfortunately, a model of the source is seldom available

M. Rudnicki and S. Wiak (eds.), Optimization and Inverse Problems in Electomagnetism, 193–203.

because source area is frequently inaccessible, thus this field is usually got from measurements. The identification of sources which gives an approximation of the expected field seems a powerful tool to arrange shields. Some identification procedures have already been proposed in the literature and in particular Olsen in [1] makes use of a "best-fit" procedure where current values of known sources are evaluated by means of a normal equations method under the assumption to know both amplitudes and phases of measured field. However it must be remarked that standard magnetic field measurements, see for instance [2], allow to evaluate only the rms values of each magnetic flux density spatial component and under the hypothesis of a slow time variation of magnetic fields it is possible to collect the results of many measurements points to obtain a spatial map of the field. But no information about phase shifts between different components and between different point field values can be obtained. Therefore it seems important to find a way to handle the above field possible measured values inside an electromagnetic analysis code for screening purposes that means an identification procedure suitable to represent the known field quantities.

A genetic algorithm seems the most powerful tool to identify the source parameters able to represent the distribution of magnetic field amplitude. The genetic algorithm used is based on a continuous variables approach and is following the basic lines presented in [3].

2. METHODOLOGY

The field simulation is based on a system of simple known sources which are able to give, as much as possible, the same time evolution of actual field sources. Due to this consideration a time harmonic behavior of current and field values is adopted by means of complex quantities. Since industrial sources usually include three-phase systems also the elementary source is made up of three circular turns with three symmetrical currents. For stated values of the geometric parameters r_s and a (see Fig. 1), each elementary source is characterized by one degree of freedom that is the rms value of the currents flowing within the coils.

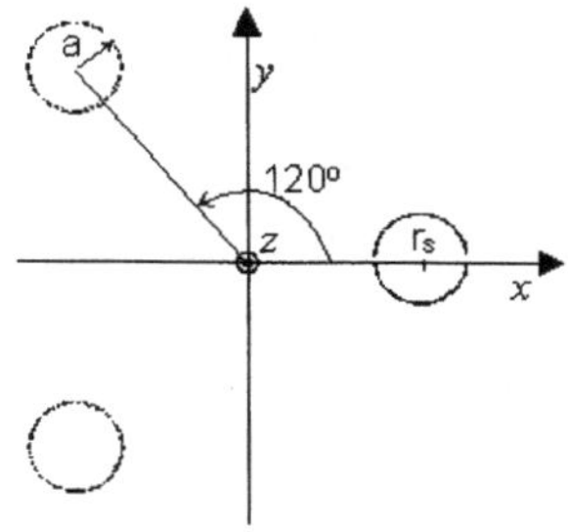

Figure 1. Elementary three-phase source in the x-y plane (z axis) with the centre on the coordinate system origin.

The currents i_1, i_2 and i_3 of each elementary source are:

$$i_1 = \sqrt{2}I \sin(\omega t)$$
$$i_2 = \sqrt{2}I \sin\left(\omega t - \frac{2\pi}{3}\right)$$
$$i_3 = \sqrt{2}I \sin\left(\omega t + \frac{2\pi}{3}\right)$$

A "cluster" of sources includes three elementary sources arranged along the coordinate axis *x*, *y*, and *z*. Each source has its own current value thus each cluster is characterized by 3 degrees of freedom assuming that a negligible phase shift between different sources.

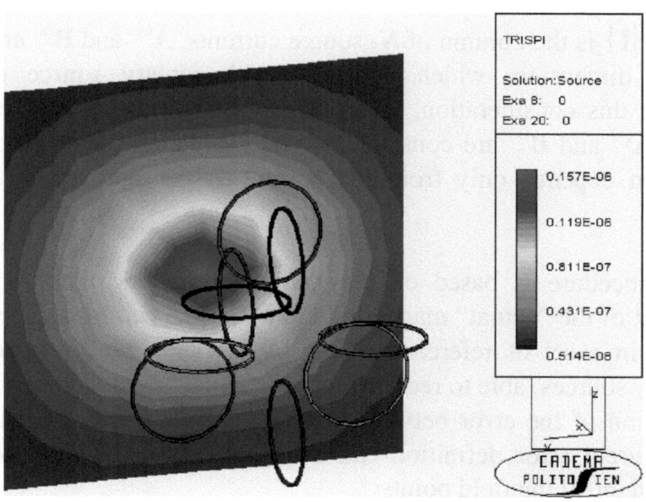

Figure 2. Cluster of three-phase elementary sources and map of magnetic field amplitude produced by the elementary source on the same source plane.

These clusters can be placed in space on the boundary of a region including actual sources or in proximity of sources. In Fig. 2 a "cluster" of coils is shown together with the magnetic induction map generated on a reference plane with a unit rms current I.

The field produced by each elementary source in a point k can be represented by its components in the x,y and z coordinate system. For example, the x-component can be written as:

$$\overline{H}_{kl}^{(x)} = \left(a_{kl}^{(x)} + jb_{kl}^{(x)}\right) \cdot I_l$$

where I_l is the rms amplitude of the current of the l-th elementary source and the coefficient $a_{kl}^{(x)}$ and $b_{kl}^{(x)}$ depend from the source geometry (centre position, distance of a loop centre from source centre r_s, loop radius a) and from the coordinates of the point k.

Having N_s system of sources, which can be placed in different planes (xy, xz or yz), and applying the superposition effect principle one can write:

$$\overline{H}_{k}^{(x)} = \sum_{l=1}^{Ns} \left(a_{kl}^{(x)} + jb_{kl}^{(x)}\right) I_l$$

If the magnetic field is evaluated in N_F field points, the previous relation can be written N_F-times so that in matrix form one has:

$$\vec{\mathbf{H}}^{(x)} = \mathbf{A}^{(x)} \cdot \{\mathbf{I}\} + j\mathbf{B}^{(x)} \cdot \{\mathbf{I}\}$$

where: $\{\mathbf{I}\}$ is the column of N_S source currents, $\mathbf{A}^{(x)}$ and $\mathbf{B}^{(x)}$ are matrix of N_F x N_S dimension, which depend on elementary source geometries. According this consideration, when the cluster geometry is fixed, the two matrixes $\mathbf{A}^{(x)}$ and $\mathbf{B}^{(x)}$ are constant and the evaluation of the magnetic field distribution depends only from the current values and it can be quickly evaluated.

The procedure is based on the knowledge of the amplitude of each component of the "actual" magnetic field H_{act} produced by a generic electric appliance in a set of reference points (N_F). Then, the N_S currents in the elementary sources, able to reconstruct the "actual" field, are provided by the minimization of the error between the "actual" field and the reconstructed field. Different error definition can be adopted like the root mean square error based on all the field points:

$$e_{rms} = \frac{1}{3N_F}\sqrt{\sum_{k=1}^{N_F}\left(\left(H_{actk}^{(x)} - \sqrt{\left(H_{\mathrm{Re}k}^{(x)}\right)^2 + \left(H_{\mathrm{Im}k}^{(x)}\right)^2}\right)^2 + \left(H_{actk}^{(y)} - \sqrt{\left(H_{\mathrm{Re}k}^{(y)}\right)^2 + \left(H_{\mathrm{Im}k}^{(y)}\right)^2}\right)^2 + \left(H_{actk}^{(z)} - \sqrt{\left(H_{\mathrm{Re}k}^{(z)}\right)^2 + \left(H_{\mathrm{Im}k}^{(z)}\right)^2}\right)^2\right)}$$

or the maximum error evaluated on the known field values:

$$e_{\max} = \max_{k=1,N_F}\sqrt{\left(H_{actk}^{(x)} - \sqrt{\left(H_{\mathrm{Re}k}^{(x)}\right)^2 + \left(H_{\mathrm{Im}k}^{(x)}\right)^2}\right)^2 + \left(H_{actk}^{(y)} - \sqrt{\left(H_{\mathrm{Re}k}^{(y)}\right)^2 + \left(H_{\mathrm{Im}k}^{(y)}\right)^2}\right)^2 + \left(H_{actk}^{(z)} - \sqrt{\left(H_{\mathrm{Re}k}^{(z)}\right)^2 + \left(H_{\mathrm{Im}k}^{(z)}\right)^2}\right)^2}$$

3. APPLICATION

3.1 Test case

The proposed identification procedure has been preliminary analyzed on a test case represented in Fig. 3, which can represents a simplify model of a three-phase air core transformer. Three symmetrical currents flow in three cylindrical windings having the same geometric configuration and parallel winding axis. Total ampere-turns rms amplitude is 40 kA. Each cylinder is 1 m long along its axis. The distances between the axes of two consecutive cylinders is 0.5 m and a cylinder diameter equal to 0.2 m is assumed.

The main interest is devoted to identify the magnetic field distribution on a rectangle laying on a plane (identification plane) parallel to the winding axes. Four clusters of elementary sources are placed on the vertices of a rectangle parallel to the identification plane which is assumed as the plane of winding axes. The magnetic field distribution to be identified is predicted according to the procedure proposed in [3] by means of simple numerical computations.

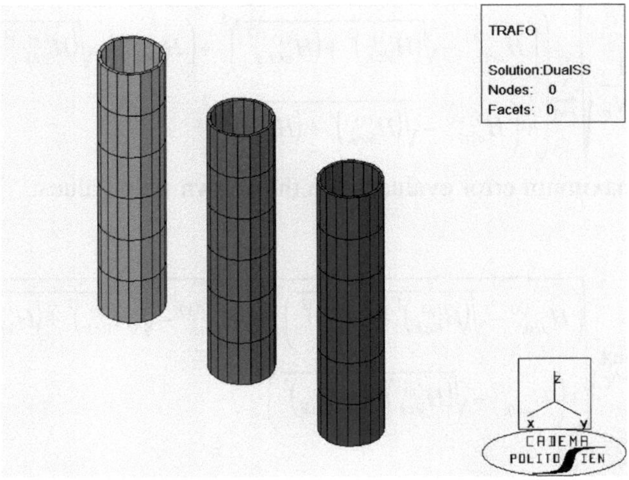

Figure 3. Test case: three-phase air core transformer.

Fig. 4 shows the flux density map on the identification plane and Figs 5, 6 and 7 give the identified field distributions when three different values of the geometric parameter r_s (50 cm, 30 cm and 25 cm) are considered. The parameter a is always taken equal to 7 cm.

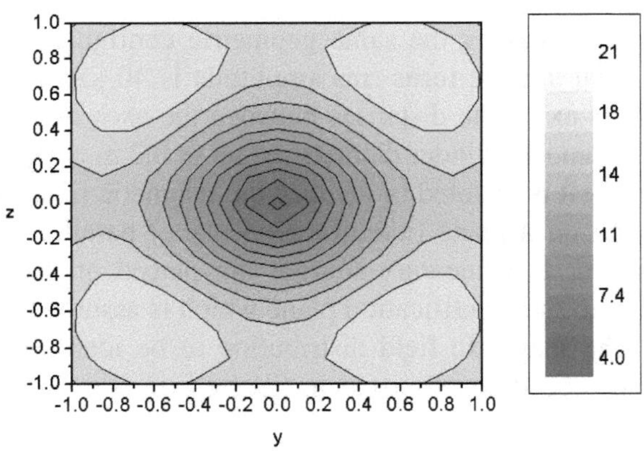

Figure 4. Magnetic flux density values [μT] for the test case of Fig. 3

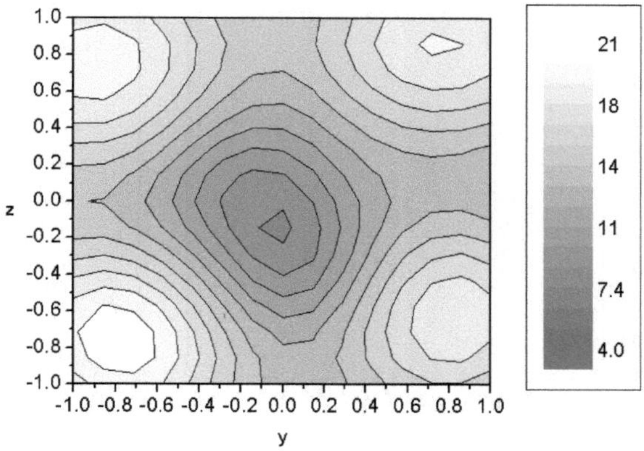

Figure 5. Reconstructed values of magnetic flux density [μT] ($r_s = 0.5$ m)

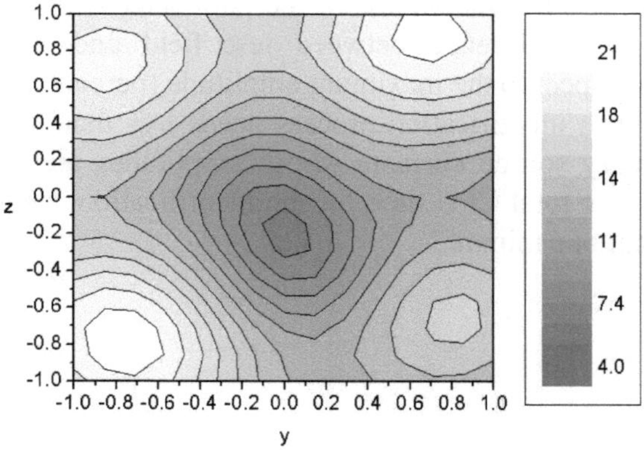

Figure 6. Reconstructed values of magnetic flux density [μT] ($r_s = 0.3$ m)

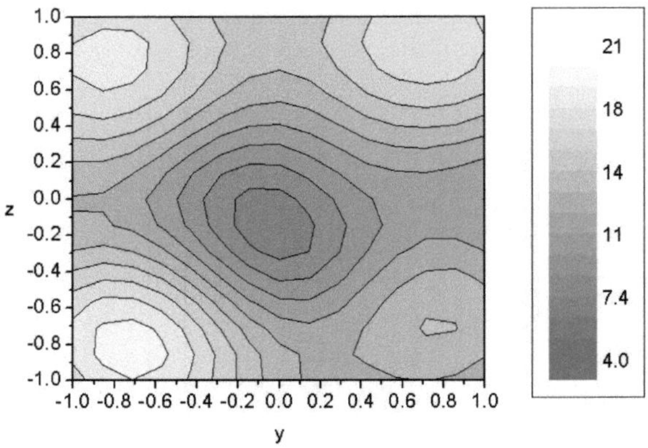

Figure 7. Reconstructed values of magnetic flux density [μT] ($r_s = 0.25$ m)

Figs 8 gives for the same values of the parameter r_s the absolute value distribution of the difference between test field and identified fields normalized with respect to the maximum amplitude (percentage error).

In any case each identification process needs less than 10^5 iterations to evaluate elementary source currents whose amplitudes are detailed in the following table. The total CPU time for the identification, running on a PIII 450 MHz, is about one minute.

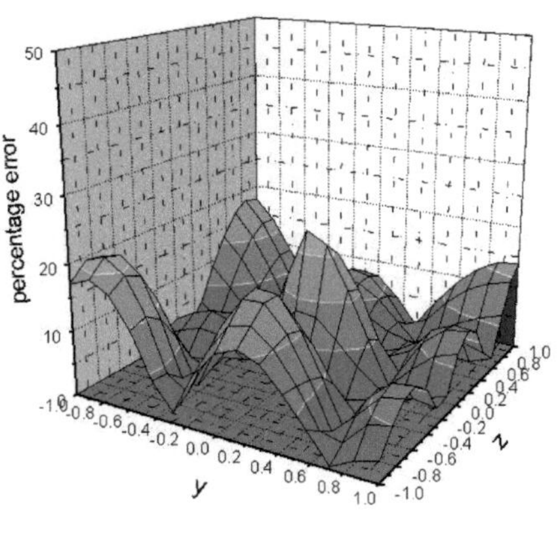

Figure 8: absolute value distribution of the difference between test field and identified fields normalized with respect to the maximum amplitude ($r_s = 0.5$ m)

Table 1. Ampere-turns amplitude of elementary sources

		$r_s = 0.5$ m (Fig. 5)	$r_s = 0.3$ m (Fig. 6)	$r_s = 0.25$ m (Fig. 7)
Cluster 1	Ix	2.0436E+04	2.4506E+04	2.9714E+04
	Iy	2.8937E+04	2.5590E+04	2.6162E+04
	Iz	4.6917E+02	2.6021E+03	1.0376E+03
Cluster 2	Ix	2.1486E+03	4.2953E+01	3.5435E+03
	Iy	2.6159E+04	2.3496E+04	2.5979E+04
	Iz	1.4176E+04	9.3778E+03	1.7449E+04
Cluster 3	Ix	2.7093E+04	2.8057E+04	3.5866E+04
	Iy	4.1156E+02	2.6344E+03	2.1077E+03
	Iz	3.8725E+04	3.5066E+04	3.7772E+04
Cluster 4	Ix	3.7037E+04	4.1219E+04	4.1287E+04
	Iy	1.5543E+03	2.4427E+03	3.5298E+02
	Iz	3.1230E+04	3.1501E+04	3.1355E+04

3.2 Practical case

The proposed computational scheme has been applied to the study of a medium/low voltage substation situated inside an hospital, where requirements about magnetic field emission are quite strict. In this case, the magnetic field measurements performed around the substation necessary to shield one of the two transformers and its low voltage high current busbars.

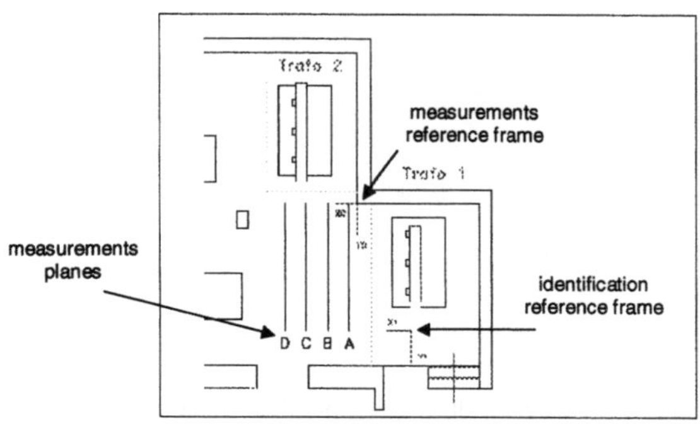

Figure 8. Outline of the substation used for the identification

The map of the site under consideration is shown in Fig. 8, where the transformer to be shielded is indicated as "trafo 1". The figure contains also the traces of the planes *A*, *B*, *C* and *D* where magnetic induction has been measured on regular grids of 6x4 points along *y* and *z* axis respectively. The plane *A* closest to the source was considered suitable for setting an Aluminum shield and thus an identification of the source was there performed.

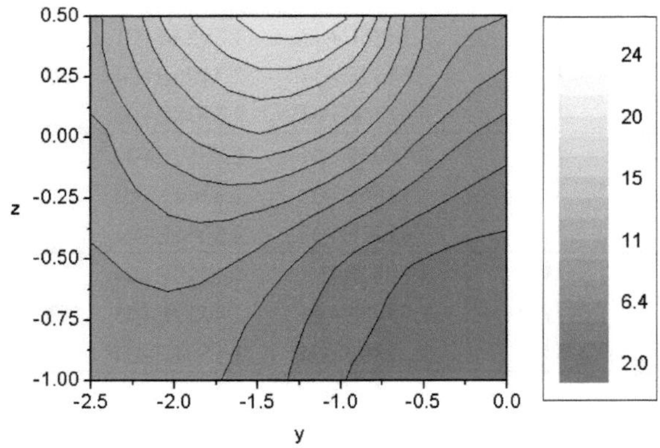

a)

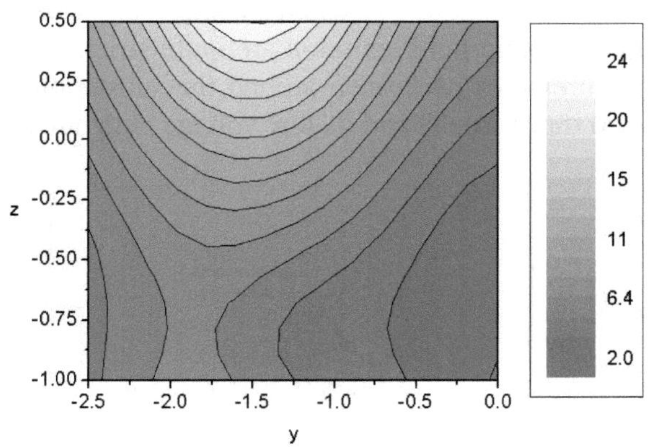

b)

Figure 9. Magnetic flux density values [μT], on plane A of Fig. 2: a) measured values, b) reconstructed values

Four clusters of elementary sources were used for a total of 12 degrees of freedom for the optimization. Measured and reconstructed field amplitudes are reported in Fig. 9. The above figure shows that the pattern of the field is satisfactorily reconstructed by the identification which gives an error on the maximum value of the field lower than 20 %.

These preliminary results are encouraging about the possibility of using this approach in a standard industrial procedure for electromagnetic field sources identification.

ACKNOWLEDGEMENTS

The authors wish to thank Gradenigo Hospital for the permission to measure and to analyse the magnetic field distribution in its main substation, dr. Stefano Ficara and dr. Fabio Freschi who performed some experimental measures and preliminary computations while developing their master degree thesis.

REFERENCES

[1] R. G. Olsen, C.E. Lyon, "Modelling of extremely low frequency magnetic field sources using multipole techniques", IEEE Trans. on Power Delivery, Vol. 11, No. 3, July 1996, pp. 1563-1570.

[2] European Standard – EN50366 – " Measuring methods for determining the magnetic flux density in the space around household and similar appliance with particular regard to human exposure"

[3] P.Alotto, C. Eranda, B. Brandstaetter, G. Furntratt, C. Magele, G. Molinari, M. Nervi, K. Preis, M. Repetto, and K. Richter, "Stochastic algorithms in electromagnetic optimization", IEEE, Trans. on Magn., vol.34, no.5, Sept.98, pp 3674-3684.

CAPITALISATION AND TREATMENT OF MODELS FOR THE OPTIMIZATION OF ELECTRIC DRIVES

L. Allain (*), L. Gerbaud (*), C. Van Der Schaeghe (**)
(*)Laboratoire d'Electrotechnique de Grenoble, CNRS UMR 5529 INPG/UJF, ENSIEG BP 46, 38402 St Martin d'Hères cedex, France, e-mail: Laurent.Gerbaud@inpg.fr, Loig.Allain@leg.ensieg.inpg.fr,

(**) Somfy Industrie, 50 Avenue du Nouveau Monde, BP 152, 74307 CLUSES Cédex France, christian.vanderschaeghe@somfy.com

Abstract: For the optimization of electric drives, both simulation and analytical models have to be developed. As many tools and environment exist for simulation and optimization, the paper proposes to capitalise models for the two aspects, in a structure description based on XML. Then, it proposes a process and tools to transform them to obtain calculation component. In this way, both dedicated simulation components and dedicated calculation components are obtained. This is applied to the sizing of electric drive using optimization algorithms [1].

Key words: Model capitalization, optimization process, electric drives

1. INTRODUCTION

An optimization with numerous criteria (until some hundreds) on the global electric drive is often required. In this way, models fast to compute have to be prioritised for optimization process [1].

An electric drive is made of several parts: digital or analogue control, power electronic and electric machine. In such a system, the sizing criteria may be represented in several manners. In this way, the sizing model may be made of symbolic equations, simple numerical models using Newton Raphson methods, series, etc... However, some sizing criteria need the

M. Rudnicki and S. Wiak (eds.), Optimization and Inverse Problems in Electomagnetism, 205–212.

solving of differential equations, linear or not, and possible hybrid state equations. In order to use simulation, the paper proposes to build specific simulation components according to a dedicated simulation scenario. Such a component allows then to extract the needed criteria from simulation block (see Figure 1).

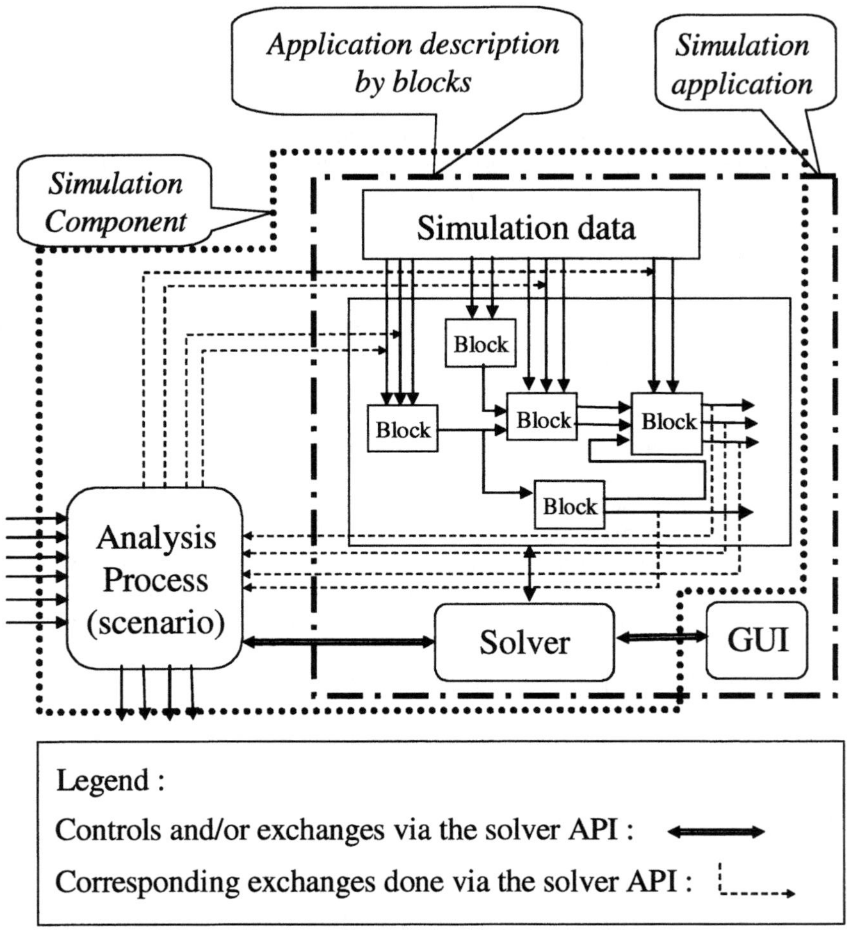

Figure 1 : *Simulation components*

In the optimization model, these components are considered as black boxes with input parameters (e.g. inductance, capacitance, etc.), output criteria (e.g. average values, maximum values, harmonic, etc.) and possibly numerical coefficients required by the numerical methods used in the simulation component. They are combined thanks to their API with the rest of the model, generally an analytical equations in the paper (see Figure 2). The global model appears then under a homogeneous format to the user: equations.

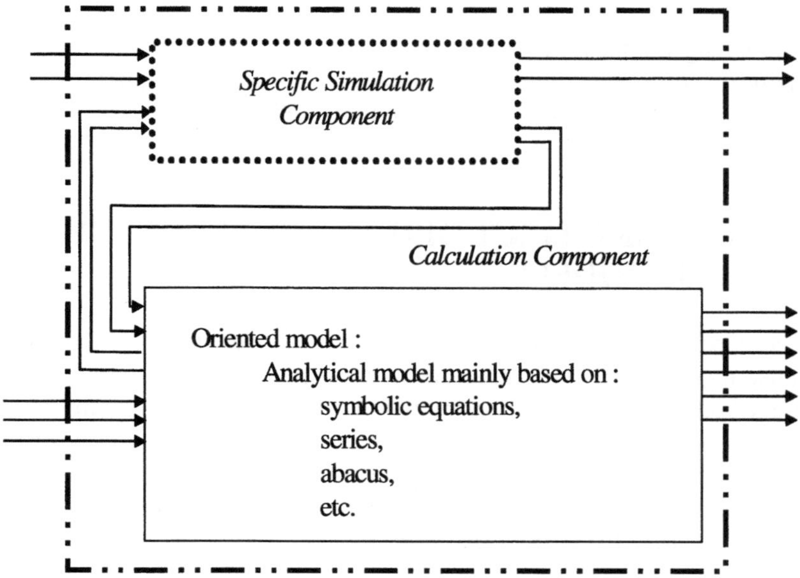

Figure 2 : Calculation component that can be used in an optimization process

2. COMPARISON OF SIMULATION AND SIZING MODELS.

Block description is a convenient way to describe models in simulation tools : see Matlab/Simulink, Saber, etc.. So, the designer has to use blocks from library or to develop his own blocks. By connecting blocks and using a solver, the user gains a simulation model that will be enhanced into an analysis model by adding analysis criteria.

So, two model kinds are considered: the sizing model mainly based on analytical modelling, and the analysis model with simulation.

Both models have to be built with similar treatment: initially non oriented equations and solving sequence and then a treatment to give an orientation to the model. Differentials or derivatives may be defined if the models are non linear, implicit, and in the case of some optimization algorithms (gradient). However, the two kinds of model differ on the following aspects.

The simulation model may be based on state equations and discrete events [2, 3, and 4]. Sometimes, hybrid state equations are used, e.g. to model static converters, or any system with a changeable topology i.e. mixing by continuous equations and event aspects.

The sizing model may use several kinds of calculation process to give the sizing criteria, e.g.: numerical valuation of symbolic equations, series, use of calculation components (thanks to their API).

3. USE OF OBJECT ORIENTED STRUCTURE FOR THE MODELS

The paper proposes to use an object oriented structure for the models to capitalise model [5, 6]. This structure is implemented in XML and allows to represent both simulation and optimization, including the specificity of each one. For simulation aspect, this structure is object oriented and is inspired from Modelica. Then, the paper proposes translators from these structured models into specific representation or programming language, according to the used tools (here for simulation or optimization).

4. MODELLING PROCESS OF AN ELECTRIC DRIVE FOR OPTIMIZATION

The paper firstly presents the process to obtain specific simulation components and then calculation components including the first ones, from a model structured in databases representation (see figure 3).

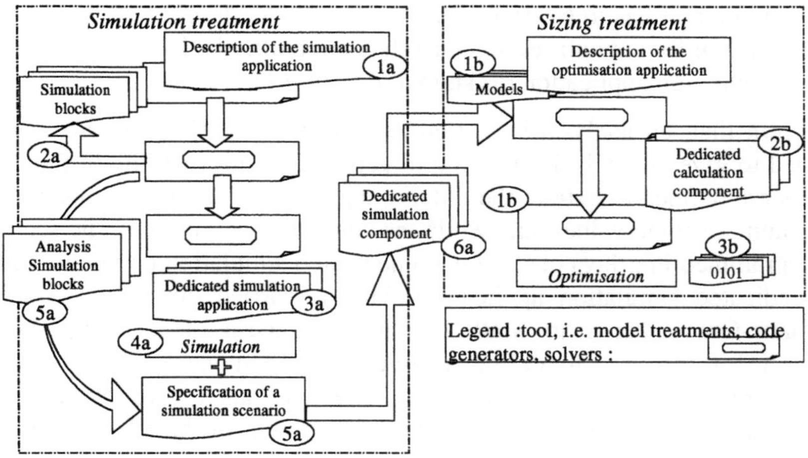

Figure 3: Modelling process for both simulation and optimization

In the first step, the user defines his model according to a structure explained in [7] and derived from Modelica [5]. The description of each component is symbolic and made for each component of the application.

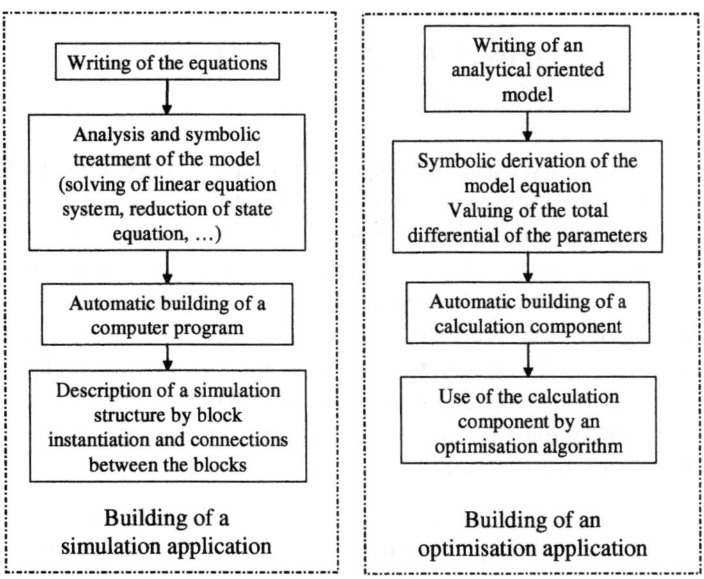

Figure 4 : Building of simulation and optimization application

Due to causality, some of these components are gathered together in one, to be transformed in blocks for a simulation tool : e.g. : S-functions for

Matlab/Simulink and blocks for Gentiane-Oro [4]. A translator per simulation target has to be developed, but the model description is the same, what is useful for capitalisation. Once they are built, blocks are connected to give a simulation application (as shown of Figure 1).

From that, an analysis is carried out and a simulation scenario can be defined to give an analysis allowing to get some sizing criteria. In this way, blocks are defined to extract the criteria, e.g. to compute FFT, r.m.s. values, maximum, average values, etc. while the simulation is running. The scenario is programmed in a Java class to get a dedicated simulation component from the dedicated simulation application. This is carried out with the solver Gentiane-Oro rather than Matlab/Simulink because of its API, and its easiness for the user to define a simulation component. For a gradient optimization algorithm, the simulation component has to give the derivative of its outputs according to the derivative of its inputs, or the differential of its outputs [1]. This is carried out with finite differences.

A similar process is proposed for the building of the sizing application, as shown on figure 4. Here, the models described by the designer are oriented contrary to the previous process where the models are automatically oriented by specifications given by the user.

5. STRUCTURE OF THE MODELS FOR CAPITALIZATION

Now, the paper discusses the way to structure models. For optimization, a model mainly includes symbolic equations, specific simulation components, etc. For simulation, a model includes state equations, algorithm and events (mainly to model control parts). But the information described above may be insufficient. Indeed a model is comprehensive with a large part of knowledge that can not always be transcript into equations or algorithms. The model is useful to designer if he can understand the hypothesis under which the model can be used. Then, it is proposed to build a modelling component that embeds equations and physical representations and all information for the model.

Such a representation has a great interest with emerging languages as XML (eXtensible Markup Language) that is particularly attractive to represents knowledge in a database (see figure 5).

This modelling allows to characterise every variable used in a model and to define relation between them. These relations are non oriented equations (e.g. : power part of an electric drive) or oriented algorithm (e.g. : control part of any application controlled by numerical processor).

This object oriented representation is defined for the simulation models of the components, but also to connect components.

```
<MODEL name= RLC>
<DIMENSION name='R'> </DIMENSION>
<DIMENSION name='E'> </DIMENSION>
<DIMENSION name='L'> </DIMENSION>
<DIMENSION name='C'> </DIMENSION>
<DIMENSION name='Ir'> </DIMENSION>
<DIMENSION name='Ur'> </DIMENSION>
<DIMENSION name='Il'> </DIMENSION>
<DIMENSION name='Ul'> </DIMENSION>
<DIMENSION name='Uc'> </DIMENSION>
<DIMENSION name='Ic'> </DIMENSION>
<DIMENSION name='Ie'> </DIMENSION>
<EQUATION expression='Ur=R*Ir'></EQUATION>
<EQUATION expression='Ul=L*der(Il)'></EQUATION>
<EQUATION expression='Ic=der(Uc)/C'></EQUATION>
<EQUATION expression='Ir=Ie'></EQUATION>
<EQUATION expression='Il=Ir'></EQUATION>
<EQUATION expression='Ic=Il'></EQUATION>
<EQUATION expression='E=Ur+Uc+Ul'></EQUATION>
</MODEL>
```

Figure 5 : Structure representation of a RLC circuit fed by a voltage source

6. TRANSLATOR OF MODEL INTO CALCULATION COMPONENTS.

Methods have been developed to treat these model and to translate them into calculation tools. After a primary treatment carried out by MAEL specific treatments are carried out. MAEL is a tool developed in our laboratory and it is not detailed in the paper. From this treatment, a solving sequence is deduced and structured by using an object oriented representation. This structuration has been implemented in XML and is also a way to capitalise model out of any solver.

Translators are developed for several simulation tools, among them; a Matlab/Simulink translator has been developed. The calculation component for optimization can be automatically obtained by Pro@Design [1] except for simulation component. However, MAEL will carry out this in future works.

7. CONCLUSION

The paper has compared simulation and optimization approaches, by comparing their similarities and their complementarities. Mainly, both are using models made with equations and algorithms. The difference is mainly the way to treat them. So, the paper has proposed to capitalise models under a specific object oriented language and then to combine and translate them into specific solver. In future works, translators will be developed, and sizing model of electric drive, including simulation process, will be developed.

REFERENCES

[1] E. Atienza, J. Bigeon, F. Wurtz and B. Belhabib, "Steps to an Electrical Design Environment", IEEE-IECON'99, San Jose, CA, USA, November 29th – December 3rd, 1999, pp° 815-820

[2] H. Kragh, F. Blaabjerg, J.K. Pedersen, "An advanced tool for optimised design of power electronic circuits", proceeding of IEEE-IAS'98, Saint Louis, Missouri, USA, october 12-15, 1998, pp 991-998

[3] R.B. Ridley, F.C. Lee, "Practical Nonlinear Design Optimization Tool for Power Converter Components", IEEE Power Electronics Specialists Conference and Exposition, Blacksburg, Virginia, June 1987.

[4] L. Gerbaud, A. Bolopion, J. Bigeon, "Gentiane : a simulation package for electromechanical systems", ELECTRIMACS'96, Saint Nazaire, France, Sept. 17-18-19, 1996, pp 689-694

[5] Mattson, Otter, Elmqvist « Modelica Hybrid Modelling and efficient simulation », 38 th IEEE 1999 0-7803-3544-9/96 pp891 à 897,1999

[6] D.A. van Beek 1, J.E. Rooda "Languages and applications in hybrid modelling and simulation: Positionning of Chi" Control Engineering Practice vol. 8, nr. 1, 2000, 81-91.

[7] L.Allain, L. Gerbaud, C. VanDerSchaeghe, Object Oriented Modelling for model capitalisation and simulation of electromechanical systems. IECON'02, Sevilla, Spain, November 5 to 8, 2002

APPLICATION OF SCREENING ANALYSIS TO THE OPTIMIZATION OF AN ELECTROMAGNETIC INDUCTION HEATING DEVICE

S. Alfonzetti, E. Dilettoso and N. Salerno
Dipartimento Elettrico, Elettronico e Sistemistico, Università di Catania, Viale A. Doria, 6 - Catania I-95125, Italy

e-mail: alfo@dees.unict.it, edilettoso@dees.unict.it, nsalerno@dees.unict.it

Abstract: In this paper an induction heating system is optimized using the hybrid FEM-DBCI (Finite Element Method - Dirichlet Boundary Condition Iteration) to evaluate the objective function. In order to reduce the computing time for the whole procedure, Design of Experiments (DOE) techniques are used to screen significant geometric parameters. Once the number of problem variables is decreased, the optimization is performed by means of the Hooke-Jeeves algorithm. The results are compared with those obtained by stochastic optimization of the device.

Key words:

1. INTRODUCTION

The aim of the optimal design of an induction heating system is the realization of a temperature profile in the workpiece, matching the specifications of the required technological process. In general, given the workpiece characteristics (geometry and materials) and having selected the best frequency according to efficiency and economy criteria, the inductor geometry (e.g. position and shape of the inductor turns) is varied with a cut-and-try procedure in order to obtain a power density distribution in the

M. Rudnicki and S. Wiak (eds.), Optimization and Inverse Problems in Electomagnetism, 213–221.

workpiece which assures a temperature distribution as close as possible to the required one.

Alternatively, the design of the electromagnetic device may be expressed as an optimization problem:

$$\begin{aligned} &\min \quad y = g(x_1, x_2, \cdots, x_N) \\ &L_i \le x_i \le U_i \quad i = 1, \cdots, N \end{aligned} \tag{1}$$

where x_1, x_2, ..., x_N are the design variables (e.g. position and/or shape of the inductor turns) and g is a suitable objective function to be minimized. Quite often, the estimate of g may involve an FEM analysis.

DOE techniques [1] are used in optimization of electromagnetic devices when the number of design variables is too large and/or the evaluation of the objective function has a high cost, as, for example, when a long computing time is required by field analysis with FEM [2, 3]: the aim is to save calculation time and to find a global optimum.

In this paper, DOE is used in the first step of optimization of an induction heating system: screening analysis of optimization variables gives useful information about objective function behavior and optimum location. So, in the second step, the reduced geometric variable set can be optimized by means of the Hooke-Jeeves algorithm [4], a deterministic optimization method.

The results are compared with those obtained by stochastic optimization of the same device [5]: the optimum is slightly worse but the computing time for the whole procedure is dramatically reduced.

2. THE INDUCTION HEATING SYSTEM

The inductor-workpiece system is shown in Figure 1. It consists of a solid graphite disk plate (R_1=20 mm, R_2=290 mm, t=14 mm, $\sigma=151.5 \cdot 10^3$ S/m, $\mu_0=4\pi \cdot 10^{-7}$ H/m) faced by a pancake inductor coil with 12 concentric turns (height: h=10 mm; widths: $w_1=w_2=w_3=w_4$=10 mm, w_5=12 mm, w_6=14 mm, w_7=19 mm, w_8=20 mm, w_9=21 mm, w_{10}=11 mm, $w_{11}=w_{12}$=10 mm), in which a time-harmonic current flows (I = 712 A, frequency f = 4 kHz). Owing to mechanical constraints, minimum distances must be kept between the inductor and the disk (a = 16 mm) and the internal inductor turn and the z-axis (b=49.5 mm); they define the so-called forbidden regions, indicated by dashed lines in Figure 1.

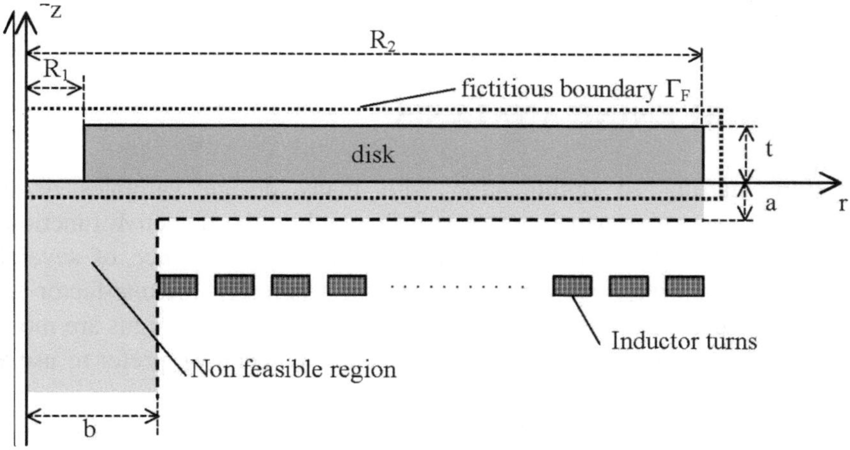

Fig. 1 – Axial section of the induction heating system to be optimized

The goal of the optimization is to find the axial co-ordinates of the turns that allow the power distribution in the graphite disk to approach the target power distribution as closely as possible. The first two turns are fixed at a distance of 16 mm from the disk, while the other ten may be vertically moved from 16 mm to 41 mm from the disk. The objective function y has been selected as:

(2)
$$y = \sqrt{\sum_{i=1}^{55} \left(\frac{p_i - p_{0i}}{54} \right)^2}$$

where p_i and p_{0i} are the calculated and desired power densities, respectively, at 55 equidistant points on the horizontal axis of the disk. The ten (N = 10) design variables are the axial co-ordinates of the centers of the turns: $x_1 = z_3$, $x_2 = z_4$, …, $x_{10} = z_{12}$; $L_i = -46$ mm and $U_i = -21$ mm, i = 1, …, 10.

The actual power distribution due to a particular configuration of the axial positions of the turns is estimated by means of the hybrid FEM-DBCI method for the solution of static and quasistatic field problems in infinite domains [6]. The domain D discretized by FEM is that enclosed by the fictitious truncation boundary Γ_F, the dotted line shown in Figure 1, placed at a distance d = 7 mm from the disk. All the turns are left out as external sources: the mesh remains the same for every position of the turns. The domain was discretized with 344 second-order triangular finite elements and 783 nodes.

3. SCREENING ANALYSIS

Before starting an optimization with many design variables, it is advisable to determine which are most influential on the objective function. Traditionally, an experimenter wanting to know the influence of several factors on a given response varies every factor separately: the one-factor-at-a-time approach. It is, however, demonstrated that factorial designs are more efficient than one-factor-at-a-time experiments [1], hence we prefer to use a factorial design.

In our case the problem lacks "measurement errors" in the evaluation of the objective function, so we may consider a single replicate of a factorial design. Since there are ten factors, a simple two-level full factorial design leads to 2^{10} designs, each of which requires an estimate of the objective function by means of an FE simulation. A way to reduce the number of experiments is to neglect the high-order interactions, applying the sparsity of effects principle: that is, most systems are dominated by some of the main effects and low-order interactions. Information on these effects may be obtained by running a fraction of the complete factorial experiment: in effect, these fractional factorial designs are widely used in screening experiments. A two-level fractional factorial design of resolution IV is considered: it consists of 32 experiments and allows the main effects (and some 2-factor interactions) to be estimated without "confounding". The design matrix is shown in Table 1: the design variables are coded and the signs + and – correspond, respectively, to the high and low positions of each turn. One row of the design matrix is an experiment, so 32 evaluations of the objective function (2) are performed for the screening: the calculated values are listed in the last column in Table 1.

The results of Yates analysis [7] are summarized in Table 2. The main effects and some 2-factor interactions are ranked by means of their T value from the most to the least significant: hence the positions of turns x_7, x_8, x_9, and x_{10} notably affect the objective function; the positions of turns x_1, x_2, x_3 and x_6 have some influence, while turns x_4 and x_5 are unimportant.

Table 1 – Design matrix for the 2^{10-5}_{IV} design.

x_1	x_2	x_3	x_4	x_5	x_6	x_7	x_8	x_9	x_{10}	y (W/m^3)
–	–	–	–	–	+	+	+	+	+	9.12176E+6
+	–	–	–	–	–	–	–	–	+	3.29531E+6
–	+	–	–	–	–	–	–	+	–	3.62328E+6
+	+	–	–	–	+	+	+	–	–	5.26632E+6
–	–	+	–	–	–	–	+	–	–	3.86802E+6
+	–	+	–	–	+	+	–	+	–	5.08465E+6
–	+	+	–	–	+	+	–	–	+	4.80536E+6
+	+	+	–	–	–	–	+	+	+	6.32616E+6
–	–	–	+	–	–	+	–	–	–	3.91176E+6
+	–	–	+	–	+	–	+	+	–	5.42401E+6
–	+	–	+	–	+	–	+	–	+	5.12314E+6
+	+	–	+	–	–	+	–	+	+	5.71102E+6
–	–	+	+	–	+	–	–	+	+	5.19983E+6
+	–	+	+	–	–	+	+	–	+	5.87197E+6
–	+	+	+	–	–	+	+	+	–	6.32051E+6
+	+	+	+	–	+	–	–	–	–	2.99633E+6
–	–	–	–	+	+	–	–	–	–	3.96912E+6
+	–	–	–	+	–	+	+	+	–	6.64383E+6
–	+	–	–	+	–	+	+	–	+	6.26559E+6
+	+	–	–	+	+	–	–	+	+	4.86103E+6
–	–	+	–	+	–	+	–	+	+	6.24413E+6
+	–	+	–	+	+	–	+	–	+	5.03066E+6
–	+	+	–	+	+	–	+	+	–	5.42471E+6
+	+	+	–	+	–	+	–	–	–	3.10998E+6
–	–	–	+	+	–	–	+	+	+	6.85250E+6
+	–	–	+	+	+	+	–	–	+	5.06019E+6
–	+	–	+	+	+	+	–	+	–	5.41458E+6
+	+	–	+	+	–	–	+	–	–	3.25334E+6
–	–	+	+	+	+	+	+	–	–	5.78767E+6
+	–	+	+	+	–	–	–	+	–	3.42166E+6
–	+	+	+	+	–	–	–	–	+	3.51253E+6
+	+	+	+	+	+	+	+	+	+	8.03295E+6

Table 2 – Yates analysis results.

Identifier	Effect estimate	T value
mean	5151059.50	
x_8	1524523.75	2.2
x_9	1411207.50	2.1
x_7	1279415.00	1.9
x_{10}	1112147.50	1.6
x_6	523170.00	0.8
x_1	-378442.50	-0.6
x_2x_3	298757.50	0.4
x_2	-296265.00	-0.4
x_1x_2	261871.25	0.4
x_3x_4	221592.50	0.3
x_1x_3	217392.50	0.3
x_3	-172478.75	-0.3
x_2x_4	150616.25	0.2
x_1x_5	-128706.25	-0.2
x_2x_5	-95616.25	-0.1
x_3x_7	-94750.00	-0.1
x_1x_4	84561.25	0.1
x_4	-65370.00	-0.1
x_5x_8	-62270.00	-0.1
x_5	58440.00	0.1
x_2x_7	-53691.25	-0.1
x_2x_6	-47918.75	-0.1
x_3x_5	-47007.50	-0.1
x_1x_9	41443.75	0.1
x_4x_5	38666.50	0.1
x_2x_8	-27197.50	0.0
x_1x_8	14110.00	0.0
x_5x_6	11498.75	0.0
x_1x_7	-7863.75	0.0
x_1x_6	-7811.25	0.0
x_4x_6	-243.75	0.0

A graphical way to show the same information is the half-normal probability plot (see Figure 3) in which the absolute values of the effect estimates are plotted against their cumulative normal probabilities. The effects that are negligible are normally distributed and will tend to fall along a straight line on this plot, whereas significant effects will not lie along the straight line.
The mean plot of the main effects is analyzed in Figure 4. In this plot, each point represents the mean of the objective function values when the relative design variable is at a low or high level. It shows that when one of the last four turns is at its low position, the mean of the objective function values drops. So, the optimum is likely to be in the region of the search space where the four most external turns are at their lowest position.

4. OPTIMIZATION

Making the most of this information, the optimization is carried out in order to determine the axial co-ordinates of turns x_1, x_2, x_3 and x_6, keeping the remaining turns fixed: x_7, x_8, x_9 and x_{10} in a low position (–46 mm), and x_4 and x_5 in a middle position (–33.5 mm). Because of the reduced number of parameters and the localization of the optimum region, a deterministic approach may be used for optimization. A direct method, the Hooke-Jeeves algorithm [4], is used: starting from this good initial

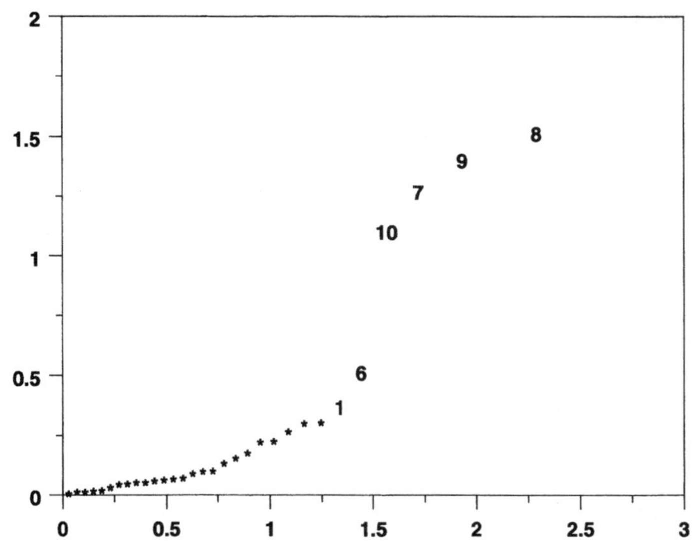

Fig. 3. Half-normal plot of the effects (MW/m^3).

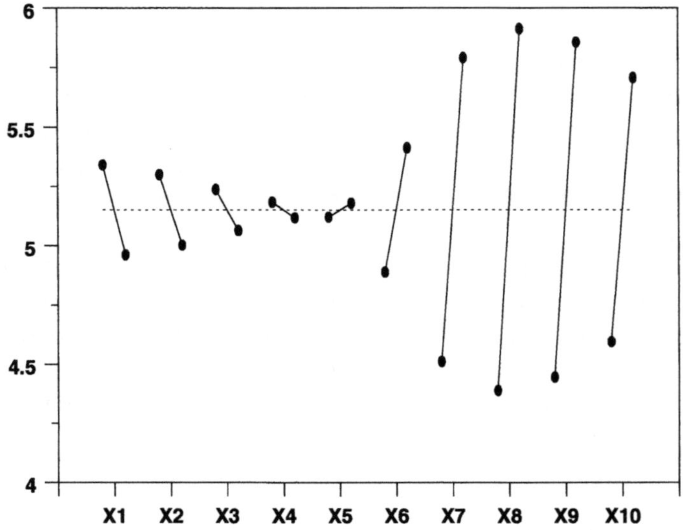

Fig. 4. DEX mean plot of the response (MW/m^3).

point, it achieves the optimum with 200 objective function evaluations. The optimal position of the turns is $x_1 = -21$ mm, $x_2 = -24.28$ mm, $x_3 = -46$ mm, $x_6 = -46$ mm, and the objective function is $y_{opt} = 1.66088$ MW/m^3. The same device was optimized by means of stochastic algorithms [5] without any information on the influence of the design variables on the objective function: an optimum less than 1% better than that of this paper was found, but it was obtained with 3000 objective function evaluations, instead of the 232 runs employed here.

5. CONCLUSIONS

In this paper the results of screening analysis, by means of DOE techniques, were used in order to reduce the number of significant design variables and to choose the starting point for the Hooke-Jeeves algorithm. In this way the optimum found is slightly worse than that obtained by stochastic optimization but the computing time for the whole procedure is dramatically reduced.

The FEM computations were performed by means of ELFIN, a large finite element code developed at the University of Catania for electromagnetic CAD research [8].

The screening analysis was performed by means of a free software tool available on the Web [9].

REFERENCES

[1] D.C. Montgomery, "Design and Analysis of Experiments," Wiley, 1997.

[2] P. G. Alotto, M. Gaggero, G. Molinari and M. Nervi, "A Design of Experiment and Statistical Approach to Enhance the Generalized Responce Surface Method in the Optimisation of Multiminima Problems," IEEE Transactions on Magnetics, vol. 33, 1997, pp. 1896-1899.

[3] S.Brisset, F.Gillon, S.Vivier and P.Brochet, "Optimization with Experimental Design: an Approach Using Taguchi's Methodology and Finite Element Simulations," IEEE Tran. Mag., vol.37, 2001, pp.3530-3533.

[4] R. Hooke and T.A. Jeeves, "'Direct search' solution of numerical and statistical problems," Journal of the Association for Computing Machinery, 8, 1961, pp. 212-229.

[5] S. Alfonzetti, E. Dilettoso, F. Dughiero and N. Salerno, "Stochastic Optimization of an Induction Heating System by means of DBCI," COMPEL, vol. 19, n. 2-3-4, 2000, pp. 569-575.

[6] G. Aiello, and S. Alfonzetti, "Finite Element Computation of Axisymmetric Eddy Currents in an Infinite Domain," COMPEL, vol. 19, n. 2-3-4, 2000, pp. 167-172.

[7] F. Yates, "The design and analysis of factorial experiments," Imperial Bureau of Soil Science Bulletin, 35, Hafner Publishing Company, 1937.

[8] G. Aiello, S. Alfonzetti, G. Borzì, and N. Salerno, "An overview of the ELFIN code for finite element research in electrical engineering," Proceedings Electrosoft'99, Seville, Spain, May 17-19, 1999.

[9] J.J. Filliben, and A. Heckert, "DATAPLOT Interactive Graphics & Data Analysis Language," Comp. & Appl. Math. Lab., Nat. Inst. of Standards & Technology, http://www.itl.nist.gov/div898/software/dataplot/.

REFERENCES

IDENTIFICATION OF CONDUCTIVITY DISTRIBUTION IN A AU-AL JUNCTION FOR APPLICATION TO INTEGRATED CIRCUITS

Alessandro Bramanti*, Paolo Di Barba*, Antonio Savini*, Umberto Anselmi Tamburini**, Filippo Maglia**

** Department of Electrical Engineering, University of Pavia, I-27100 Pavia, Italy*

e-mail: di_barba@etabeta.unipv.it

*** Department of Physical Chemistry, University of Pavia, I-27100 Pavia, Italy*

Abstract: The junction between Au and Al in a current-carrying contact for application to integrated circuits is considered. The conductivity of the compound is identified by means of a field-inversion technique.
The simulation of the growth process of the Au-Al compound relies on the transient field analysis of the current flow through the junction. Results of the identification are compared with measurements.

Key words: integrated circuits, two-metal junction, conductivity identification.

1. INTRODUCTION

The paper is oriented towards the application of inversion techniques to the solution of reconstruction problems [1], in which the configuration of conductive materials must be retrieved basing upon global characteristics like the voltage-current curve at the terminals of a device.

M. Rudnicki and S. Wiak (eds.), Optimization and Inverse Problems in Electomagnetism, 223–231.

A class of power integrated-devices for automotive applications were considered, in the frame of a scientific cooperation with ST Microelectronics (Agrate, Italy). For these systems a few requirements about mean lifetime and operating conditions still represent an upper limit to modern technologies. In particular, current flowing through the Au-Al contacts between Al pads upon the Si die and the external Cu leads stimulates the growth by diffusion of various inter-metallic layers. Moreover, depending on the value of specific current involved, mass transportation and consequent formation of interface voids of Au are observed too. Both diffusion and electromigration cooperate increasing the electric resistance of the contact and decreasing its mechanical strength. Inside such solder contacts, the time evolution of the degenerative processes was focused on, in order to build a predictive model based on non-destructive external measurements. The non-uniqueness of the solution is the main result of ill-conditioning of the inverse problem, because several conductivity profiles give rise to the same contact resistance. More, the associated direct problem is quite complex too, because it involves the coupling of electric and thermal fields under dynamic conditions on a microscopic space-scale.

On one hand, such phenomena are still quite unexplored from the chemical and physical point of view; only elementary yet unsatisfactory analytical or circuit models are used, despite the remarkable deal of experimental data available in the literature [2][3]; the only field-based model of diffusion was found in [4] where the effect of 2D geometry on compound formation was investigated. On the other hand, maximizing the mean lifetime of the device, - i.e. the mean time to the failure of the contact - represents an urgent target from the industrial viewpoint.

2. TEST PATTERN

In the frame of a previous study [5], a test pattern has been used for measurement, consisting of a cylindrical furnace and Copper electrodes. The experiments were performed under high vacuum, controlling the furnace temperature by means of a feedback thermocouple. A typical sample consisted of one 127 μm thick Au foil, pressed between two 100 μm thick Al foils; the diameter was 5 mm. Tests were performed under different current densities (0-1019 Acm^{-2}) and temperatures (400-500 °C), and for times up to 90 h. At the end of each experiment, the thickness profile and composition of each compound layer were acquired. Experimental results are fully reported and discussed in [6].

3. SIMULATION

The simulation of the growth process implies space and time discretization. To reduce cost and storage, a number of simplifying assumptions have been made.

The influence of the plastic case encapsulating the junction has been neglected.

Owing to the relatively low value of specific current, electromigration effects have been disregarded. In other words, a thermally-activated diffusion process has been assumed.

The growth of an equivalent single-layer has been considered, with the growth speed of the multi-layer compound.

Referring to a Cartesian frame, the growth of the compound has been assumed to be symmetric with respect to both x- and y-axis; moreover, the distribution of temperature has been assumed to be dependent on x-direction only, while the boundary temperature is constant due to external control.

The model parameters (electrical and thermal conductivities σ ($\Omega^{-1}m^{-1}$) and k ($Wm^{-1}K^{-1}$) of each material, respectively and pre-exponential factor D_0 (m^2s^{-1}) of diffusivity for the Au-Al pair) are considered to be time-independent. k in particular is assumed constantly equal to 50 $Wm^{-1}K^{-1}$. Having fixed the parameter values, the solution of the analysis problem is based on the following iteration:

knowing the material distribution at a given instant, a 2D finite–element analysis [7] of the current flow in the junction determines the current density field (computational domain is shown in Fig. 1);

the temperature distribution is then found by integrating twice the source term $\frac{J^{2}}{\sigma k}$ where J(x) is the current density profile in the junction;

diffusivity D is updated by means of Arrhenius law $D = D_0 e^{-\frac{E_0}{kT}}$ (with $D_0 = 1.35 \cdot 10^{-2} cm^2 s^{-1}$, $E_0 = 1.54 \cdot 10^{-19} eV$) at each node of the interface, where E_0 and k are activation energy and Boltzmann constant, respectively, and T is the absolute temperature;

the y-coordinates of moving nodes along the interface are updated according to the parabolic growth law $y^2=Dt$, which is the well-known solution of diffusion equation in the 1D case under isothermal conditions.

A new material configuration is so obtained and the finite-element mesh is regenerated. Time evolution is discretized by means of a finite-difference implicit scheme with time step Δt=60 s in the first six iterations and Δt=360 s subsequently; the iterative process stops at t=7200 s.

Table 1. Material data

Material	σ [S/m]
Gold	$4.46\ 10^7$
Aluminum	$3.87\ 10^7$
Copper	$5.77\ 10^7$

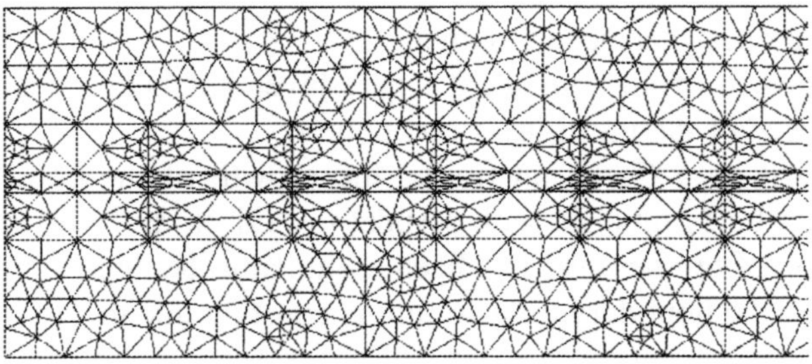

Figure 1. Detail of the finite-element mesh.

4. IDENTIFICATION PROBLEM

The physical quantity of major interest to be identified is the electrical conductivity of the compound. To this end, let initial conditions be defined: electrodes are supplied by a step of current with density 1019 Acm^{-2} at t=0, when the compound thickness is supposed to be null and temperature is uniform and equal to 450°C. The identification problem reads as follows: *find the electrical conductivity σ of the compound such that the error between simulated and measured value of thickness at t=7200 s is minimum starting from initial conditions.*

To start the optimisation process, a tentative value of conductivity equal to $3.33\ 10^6\ \Omega^{-1}m^{-1}$ is considered, while the measured value of thickness was 75 μm. A deterministic minimizer of lowest-order (simplex method) is used; whenever a tentative value of conductivity is chosen by the minimizer, a full transient simulation of the growth process is run [1].

5. RESULTS

The identified σ is about 5.5 10^5 $\Omega^{-1}m^{-1}$, a value remarkably lower than that used in the literature [6].

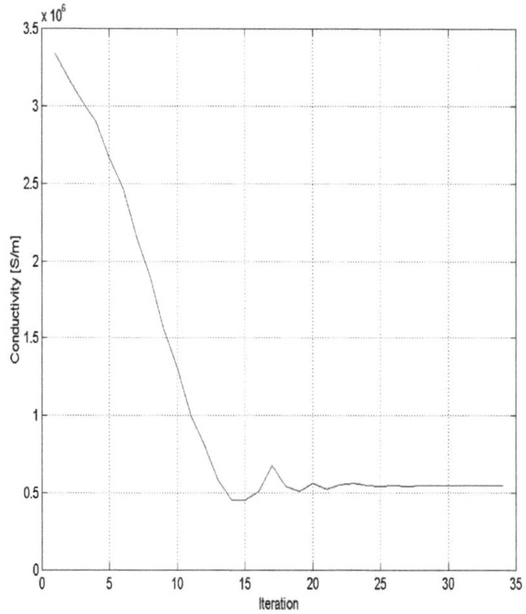

Figure 2. Variations of electrical conductivity during optimization.

With such a value, the maximum thickness of the compound along the profile grows, to a first order proportionally to the square root of time, up to the measured value of 75 μm after 2 hours.

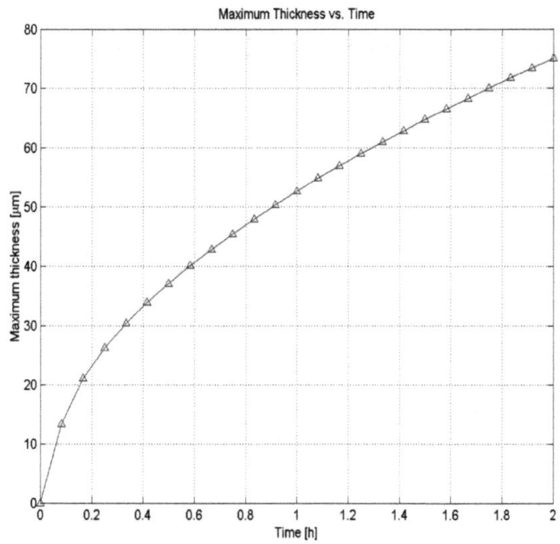

Figure 3. Maximum thickness of compound *vs* time.

The position of such maximum is always on the symmetry axis of the system, and the profile decreases, in a grossly parabolic fashion, to the outer minimum, with a gap less than 8 μm.

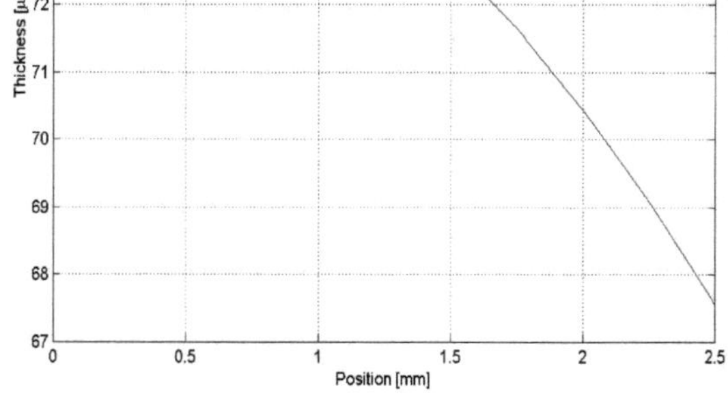

Figure 4. Compound profileafter 7200 s.

The current density, after the initial uniformity, tends to lump towards the peripheral zone (right part of the domain), corresponding to the smallest compound thickness.

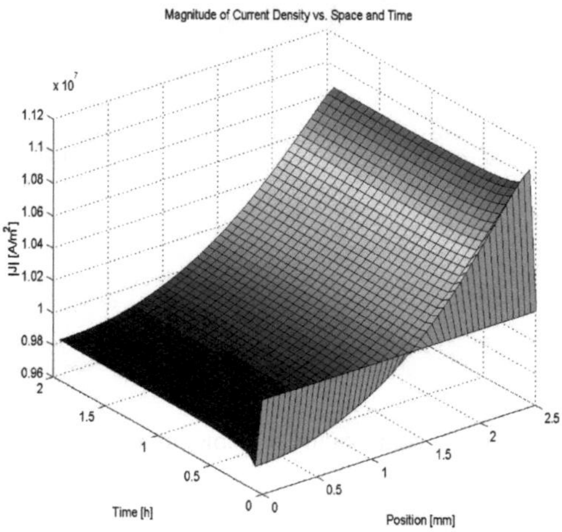

Figure 5. Current density in the compound *vs* space and time.

The current density variation vs. space is quite independent of time, for the total current is kept constant throughout the simulation. Temperature obeys the same qualitative behavior as the profile, with a maximum gap of about 10°C from the 'hot' point in the middle of the device and the thermostatically controlled outer interface.

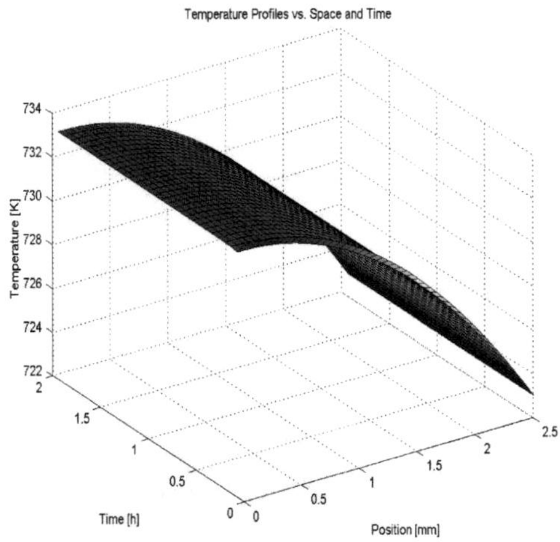

Figure 6. Temperature of compound *vs* space and time.

6. CONCLUSION

The model adopted follows quite closely the known experimental results; the optimisation procedure allows identifying a useful global parameter, which, at the time being, still cannot ba carefully measured.

REFERENCES

1. P. Neittaanmaki, M. Rudnicki and A. Savini: Inverse Problems and Optimal Design in Electricity and Magnetism. Oxford Science Publications, 1996.
2. E. Philofsky: "Compound formation in Gold-Aluminum systems", Solid-State Electronics, 13 (1970) 1391-1399.
3. M. Braunovic: "Compound compounds at Aluminum-to-Copper electrical interfaces: effect of temperature and electric currrent", IEEE Trans. Components, Packaging and Manufacturing Technology A, 17 (1994) 78-85.
4. G.G. Harman and C.L. Wilson: "Materials problems affecting reliability and yield of wire bonding in VLSI design", Materials Research Society Symposium Proceedings, 154 (1989) 401-413.
5. N. Bertolino, J. Garay, U. Anselmi-Tamburini and Z.A. Munir: "Electromigration Effects in Al-Au Multilayers", Scripta Materialia, 44 (2001) 737-742.

6. N. Bertolino, J. Garay, U. Anselmi-Tamburini and Z.A. Munir: "High-flux current effects in interfacial reactions in Au-Al multilayers", Philosophical Magazine B, 88 (2002) 969-985.
7. /www.infolytica.com/ (visited in July 2002).

INVERSE PROBLEM OF VOLTAGE OVER ELECTROSTATIC SHIELD GAP IN HELICAL TRANSFORMER WINDINGS

Petar Bodlovic, Kemal Hot and Tomislav Kelemen
Electrical Engineering Department, Polytechnic of Zagreb, 10000 Zagreb, Croatia, e-mail: pbodlovi@tesla.vtszg.hr; Koncar R&D, Bastijanova bb, 10000 Zagreb, Croatia

Abstract: Multi-conductor transmission line model of a helical winding for an analysis of the very fast electromagnetic transients will include the lumped voltage between the edges of the open electrostatic screen, previously found to be responsible for experimentally found anomalous distortion of the travelling waves. Contrary to the former model with distributed voltage and axial symmetry, the new model requires parameters beyond those of classical theories of wave propagation in transformer windings. The additional parameters are found by matching the simulated voltage transients with the experimental results.

Key words: Multiconductor transmission-line theory, anomalous distortion, travelling waves, inverse problem

1. INTRODUCTION

With multi-conductor transmission-line theory in the previous work [1] it was found that experimentally established anomalous distortion of penetrating voltage in a single helical transformer winding [2] comes from reflections of the travelling waves on the insulating gap of the electrostatic shield. There is a voltage difference over the gap, induced by changing magnetic field penetrating inside the shield. Following the axially symmetric

M. Rudnicki and S. Wiak (eds.), Optimization and Inverse Problems in Electomagnetism, 233–244.

field-theoretic model in ref. [3], the induced lumped voltage across the gap has been replaced with distributed voltage around the shield. Model parameters were set and calculated according to classical theories of [4].

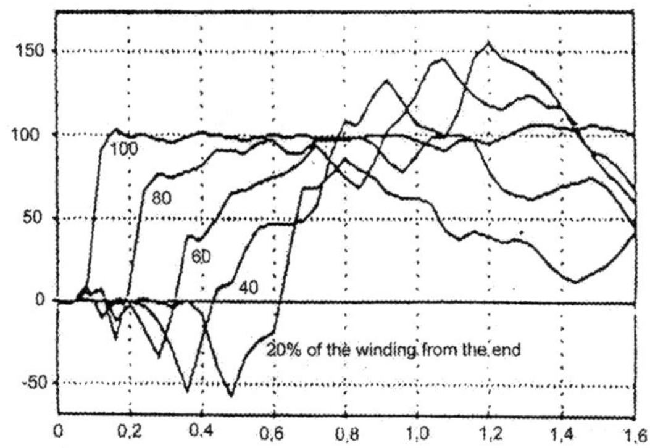

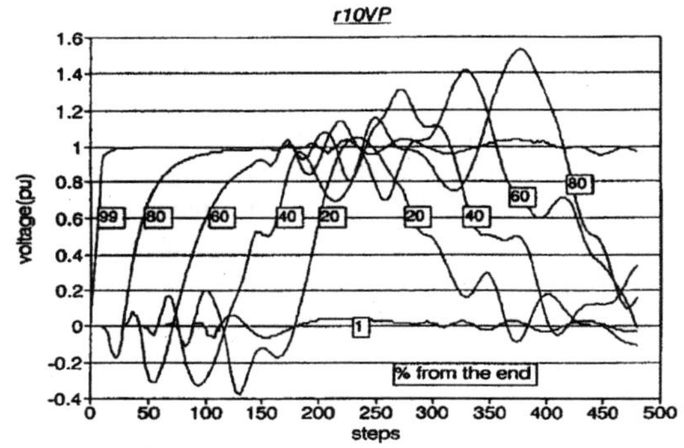

Figure 1. Voltage transients for 42 ns surge rise time: Top measured [2] and bottom calculated [1]

This work returns to the real lumped gap voltage and the anomalous distortion due to the selective reflections of traveling waves, instead of modal velocities dispersion. Besides checking the validity of the conjecture that there exists the gap voltage explaining the anomalous distortion, the new model does not require modal analysis, being significantly simpler then previous one. On the other hand, it is necessary to introduce additional model constants for the magnetic field inside shield cylinder, modified by remaining eddy currents in the shield. Instead of modeling eddy currents, the additional model constants are determined, as inverse problem, from agreement with experimental results. The solution gives additional physical insight into the propagation of surge waves in transformer windings.

Especially it becomes possible to investigate the influence of the insulating gap in electrostatic shield to the resonant frequencies, observed with frequency analysis of very fast transient over-voltages [5] occurring in gas insulating switchgear (GIS) due to the operation of the disconnector switch.

2. MULTI-CONDUCTOR TRANSMISSION LINE MODEL OF THE WINDING WITH THE INSULATING SHIELD GAP

In the multi-conductor transmission line model of a helical winding, depicted in Fig.2, electromagnetic transients are represented by line currents $i_n(x,t)$ and voltages $e_n(x,t)$ to the grounded screen as functions of longitudinal coordinate x and time t. Subscript "n" is at the same time the discrete transversal coordinate n = 1,2....N for a winding consisting of N turns. Longitudinal coordinate "x" is continuous in the interval [0,w) where w is the length of each turn. The insulating shield gap g-h is positioned in the middle between the terminals.

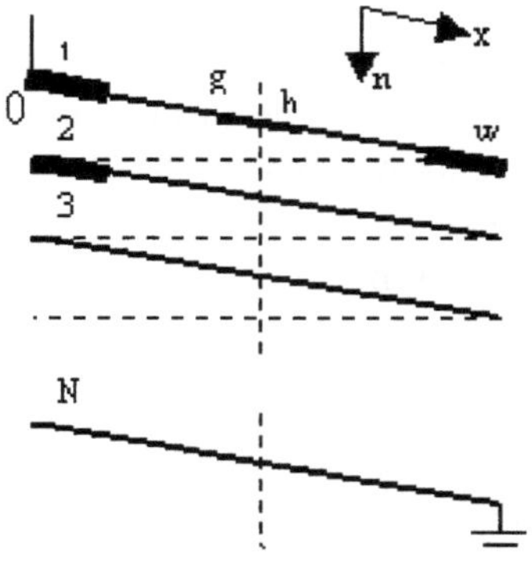

Figure 2. Multi-conductor transmission line model of a helical winding with insulating gap g-h in the middle

The N-column vectors [i(x,t) and [e(x,t) are supposed to satisfy the vector telegraphist's equations for lossless lines

$$- \partial[e(x,t)/\partial x = [L_1] [\partial i(x,t)/ \partial t \tag{1a}$$

$$- \partial[i(x,t)/\partial x = [C] [\partial e(x,t)/ \partial t \tag{1b}$$

Eqn. (1a) interprets the induction law by N×N matrix $[L_1]$ of self- and mutual -inductances per unit length (p.u.l.) due to field between the winding and the shield cylinder. Empirical exponential approximation of mutual inductances

$$[L_1]_{n,k} = L_1 \zeta_1^{|n-k|} \tag{2}$$

is used according to [4]. The inverse $[L_1]^{-1}$, called the reluctance matrix [R], of the Toeplitz's matrix $[L_1]$ is tri-diagonal. The tri-diagonal capacitance matrix [C] from ladder capacitances network is, in accordance with [5,6], equal to v^{-2}[R] for guided TEM waves, where v is the velocity constant. Thus, computing [C] and v from winding geometry and material constants, gives also L_1 and ζ_1 constants.

The electromagnetic transients are solutions of system (1ab) satisfying mixed boundary conditions: initial, external and internal. Zero initial conditions

$$[e(x,t) = [0, \quad [i(x,t) = [0, \quad (t < 0, \ 0 \leq x < w) \tag{3a}$$

and external

$$e_1(0,t) = u(t), \ e_N(w,t) = 0, \ (t \geq 0) \tag{3b}$$

correspond to the experimental arrangement with the imposed surge voltage u(t) at front end and grounded far end of the winding. Terminal input $i_1(0,t)$ and output $i_N(w,t)$ currents belong to the solution of the problem. Remaining 4*N - 2 internal conditions supplement the Eqs. (1) at artificial turn sticking points w-0 and g-h:

$$e_n(0,t) = e_{n-1}(w\text{-}0,t), \quad i_n(0,t) = i_{n-1}(w\text{-}0,t), \ (n = 2,...N) \tag{4a}$$

$$e_n(h,t) = e_n(g,t) + egh_n(t), \quad i_n(h,t) = i_n(g,t), \ (n = 1,...N) \tag{4b}$$

where $egh_n(t)$ is the voltage across the gap in the electrostatic screen. Later $egh_n(t)$ will also include the voltage drop due to n-th turn distributed resistance. Modeling the reactive part of $egh_n(t)$ is the main concern of this paper.

According to [1] $egh_n(t)$ comes from time derivative of the magnetic field flow inside the screen cylinder. It is supposed that

$$[egh(t) = [B]\,[Z_{gap}]\,([i(w,t) - [i(0,t)\,), \quad [B]_{n,k} = M_2\, \beta^{|n-k|} \tag{5}$$

where

$$[Z_{gap}] = v\ ([L_2] - [L_1]), \quad [L_2]_{n,k} = L_2\, \zeta_2^{|n-k|} \tag{6}$$

and ($[L_2]$ is the matrix of mutual inductances for a winding without electrostatic shield. For simple use of ref. [1] the matrix [B] should be the unity matrix, obtained from (5) with $M_2 = 1$ and $\beta = 0$ (with $0^0 = 1$), meaning that no eddy currents exist so that the screen is transparent to the magnetic field. Plots in Fig.3 demonstrate overwhelming distortion of transients. Significant reduction of the constant M_2 brings the results towards to the expectations. It appears that the electrostatic screen, with only one axial gap, does not prevent all eddy currents, leaving shield to be semi-transparent to the magnetic field. The first order exponential smoothing with the constant $0<\beta\leq 1$ provides for difference between the distributions of the penetrating magnetic field and the field in the traveling wave between turns and the screen. Growing β flattens the magnetic flux through the turns inside cylinder as well as the voltages over the gap approaching the case of the closed electrostatic shield.

There are equally simple and successful alternative expressions for [egh(t). Almost the same voltage and current transients are obtained from the first order temporal exponential smoothing with the same smoothing constant β and a somewhat different weight M_1. Explanation lies in strong space-time correlation created by traveling waves.

For better match with experiment analysis, the model includes resistance R_{gnd} at ground terminal entering in the voltage equation at the end of the last turn. Analogously the internal resistance R_{surge} of the surge generator is introduced only for reasons of completeness. Distributed resistance is lumped in the middles of turns giving contribution

$$R_{turn}\,([i(w,t) + [i(0,t)/2 \qquad (5a)$$

to the voltage [egh(t). The distortion-less losses are respected with the "*atten*" constant by scaling the incoming waves with *(1–atten)*.

Two model constants M_2 and β originate from the eddy currents in open electrostatic shield and should be determined as the inverse problem from the experimentally established electromagnetic transients.

3. DISCRETE SPACE- TIME SOLUTION METHOD

The per unit surge impedance matrix [Z] from the method of characteristics is given by

$$[Z]_{n,k} = \zeta_1^{|n-k|} \qquad (7)$$

while per unit surge admittance $[Y] = [Z]^{-1}$ is tri-diagonal. The Cholesky's factorization of [Y] is used for transformation between current and voltage traveling waves in both directions. Discrete time variable t counts half-turn passages of waves with velocity v. The sent current wave $[i^f(0,t)$ at $x = 0$

and the reflected $[i^b(w,t)$ at $x = w$ are explicitly expressed through the history data $[i^b(0, t) = [i^b(g, t-1)$, $[i^f(w, t) = [i^f(h, t-1)$, $[e^b(g, t-1)$ and $[e^f(h, t-1)$. Then, using (5, 6) the gap voltage $[egh(t)$ is calculated which drives current

$$[igh(t) = [Y] [egh(t)/2 \tag{8}$$

so that

$$[i^b(g, t) = [i^b(h, t) - [igh(t), \quad [i^f(w, t) = [i^f(g, t) - [igh(t), \tag{8a}$$

where

$$[i^b(h, t) = [i^b(w, t -1), \quad [i^f(g, t) = [i^f(0, t -1). \tag{8a}$$

4. RESULTS

Voltage transients in Fig. 3, calculated according to the presented model, for the transformer winding in ref. [2] with magnetically transparent electrostatic screen, implied in refs. [3] and [1], are very far from experimental transients in ref. [2] and calculated in ref. [3], both shown in Fig. 1.Besides substantial differences there are some organizational changes in the following graphs: Turns are numerated from beginning, plot of the 60-th turn transient is omitted for clearness and discrete time steps correspond to traveling time for half of the turn, approximately equal to 2 ns. Surge rise time of 20 time steps is approximately the same as 42 ns as in ref. [2]. Having the discrete time solution method for given model constants, the solution of the inverse problem consists of finding model constant which give agreement with the experimental transients in Fig. 1

The most relevant part of the introduced model constants is shown as the column headings in Table 1. The values of the two coupling constants ζ_1 ζ_2 and L_2/L_1 ratio are considered as well known. The 7-dimensional search space for the values of the remaining decomposes into subspaces of one or two dimensions. The fitting is accomplished by the visual inspection of the resulting voltage transient plots. Table 1 contains characteristic cases of model constants that are used to obtain the charts in Figs. 4-13.

While fitting the model constants it was interesting to get insight into the traveling waves as elementary components of the solution. As an alternative completion of the results, the charts of the current transients are shown besides the voltage transients in Figs. 4-6. The graphs in the Fig. 6 show transients for the case (d) in the Table 1, obtained from the case (e): firstly by fitting the constants M_2, secondly introducing R_{gund} for the evident rise of the voltage above zero when reflected wave arrives and thirdly with minor improvements with loss constants R_{turn} and *atten*. Transients in the Fig 5 for case (b) in the Table 1 follow from near optimal constants M_2 and β with further improvements as for the case (d). The results for near optimal fitting with the constant M_1 = 0,05 (temporal filtering) are almost the same as for spatial filtering with constants in row (b) of the Table 1. In order to understand the complexity of the distorted transients of the case (b), they are compared with transients in the Fig. 4 for the winding with a closed electrostatic shield having constants of the case (a) in the Table 1.

Table 1. The model constants for: (a) Transients without shield gap, (b) Near optimal match with experiment, , (d) Match without filtering (b=0), (e) Magnetically transparent screen, ζ_1 = 0,642

Case	ζ_2	L_2/L_1	R_{grnd}	M_2	R_{turn}	β
(a)	0,642	1	0,4	0	0,004	0
(b)	0,86	2,7	0,4	0,1	0,001	0,725
(d)	0,86	2,7	0,4	0,05	0,001	0
(e)	0,86	2,7	0,4	1	0,001	0

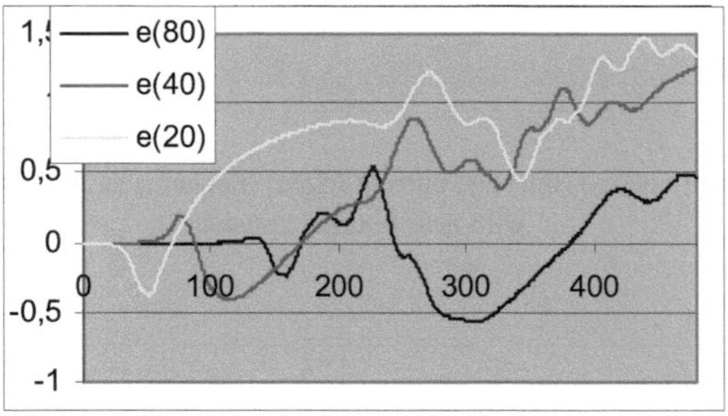

Figure 3. . Calculated voltages for the helical winding with magnetically transparent screen (case (e) in Table 1)

The voltage transients in Fig. 6, based on the simple scaling of the magnetic flux inside the cylinder, are distorted quite like the experimental [2] and calculated [1]. The flux filtering according Eq. (5) with the constants for the case (b) in the Table 1 eliminates the only one significant discrepancy for the peek of the 80-th turn voltage *e(80)*. The current transients in Figs. 5,6 are, besides distortion, significantly reduced.

Further results of spatial distributions in Figs. 7-11 are aimed to explain the under- and over-voltages. The terminal currents in Figs. 12-13 are relevant to modeling the electromagnetic transients in circuits containing transformer windings.

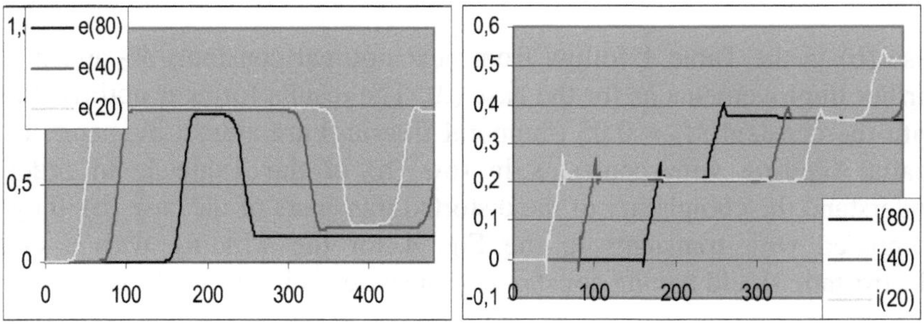

Figure 4. Calculated distortion-less voltage (left) and current (right) transients at 20-th, 40-th and 80-th turn

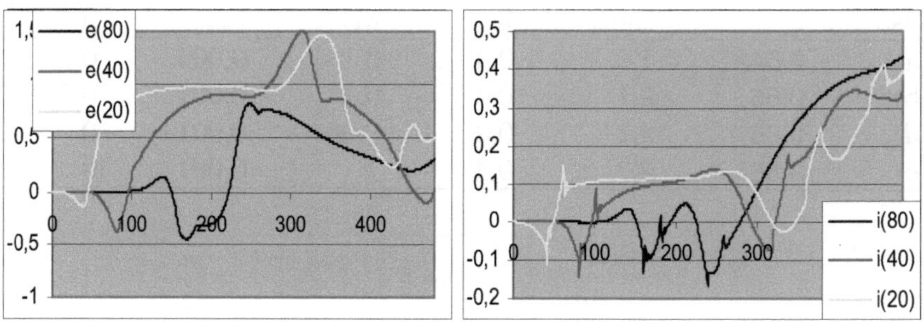

Figure 5. Calculated voltage (left) and current (right) transients at 20-th, 40-th and 80-th turn with near optimal constants

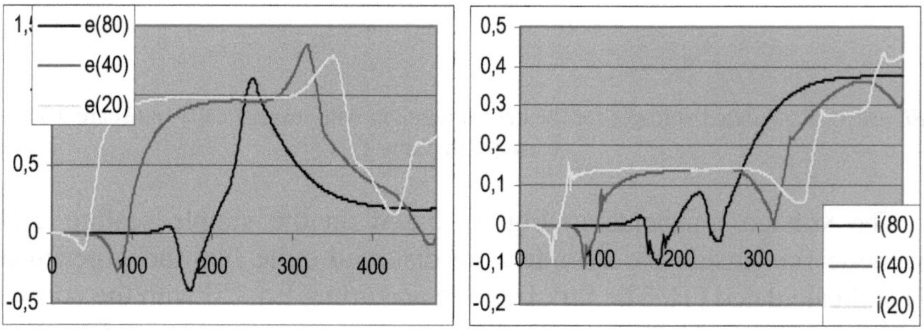

Figure 6. Calculated voltage (left) and current (right) transients at 20-th, 40-th and 80-th turn without filtering

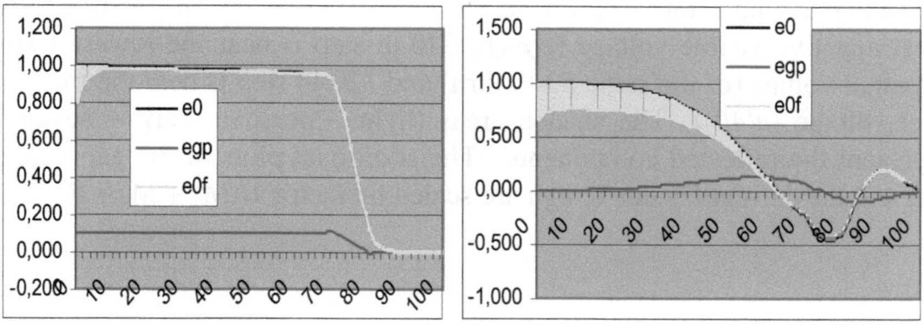

Figure 7. Distortion-less (left) and near optimally fitted (right) spatial distribution of the voltage transients at 160-th step

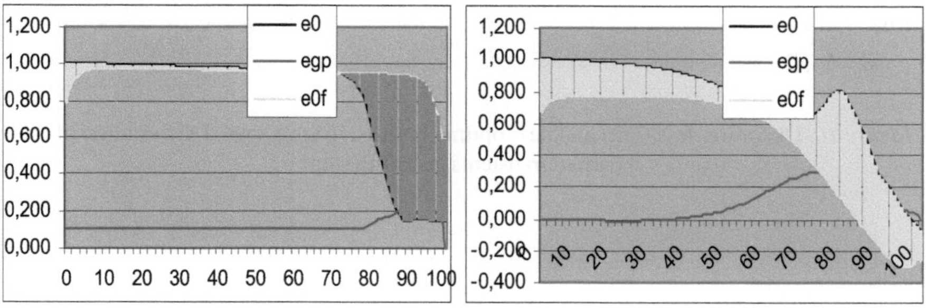

Figure 8. Distortion-less (left) and near optimally fitted (right) spatial distribution of the voltage transients at 240-th step

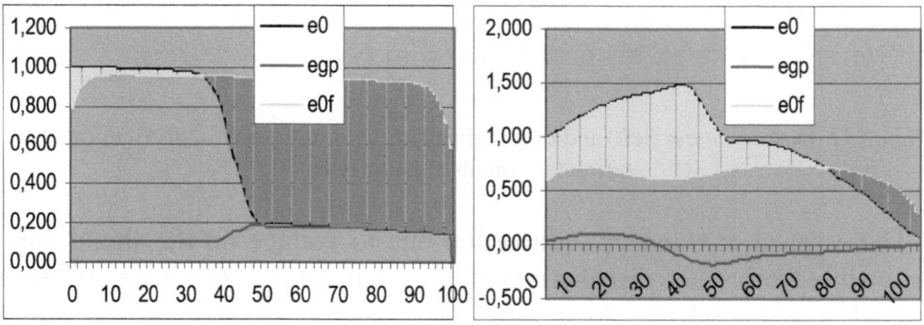

Figure 9. Distortion-less (left) and near optimally fitted (right) spatial distribution of the voltage transients at 320-th step

The space distribution of voltages in Figs. 7-9 and currents in Figs.10-11 are shown at equidistant intervals of 80 steps such that 160-th step is near the most negative visible voltage (*e(80)*), 240-th step is near the lowest visible maximal voltage(of the same 80-th turn) and 320-th step is near the maximal over-voltage (*e(40)*) The shaded areas (light – positive, dark – negative) represent the reflected components. The additional plots of the gap voltage *egh* and the current *irg* due to *egh* are scaled by factor $1/(M_1 + M_2 + R_{turn})$.

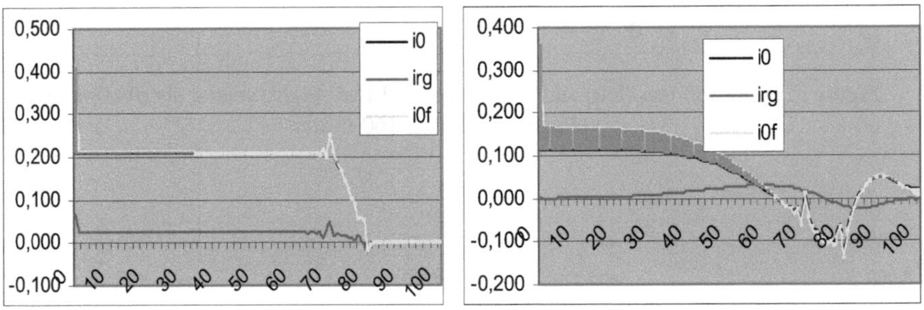

Figure 10. Distortion-less (left) and near optimally fitted (right) spatial distribution of the current transients at 160-th step

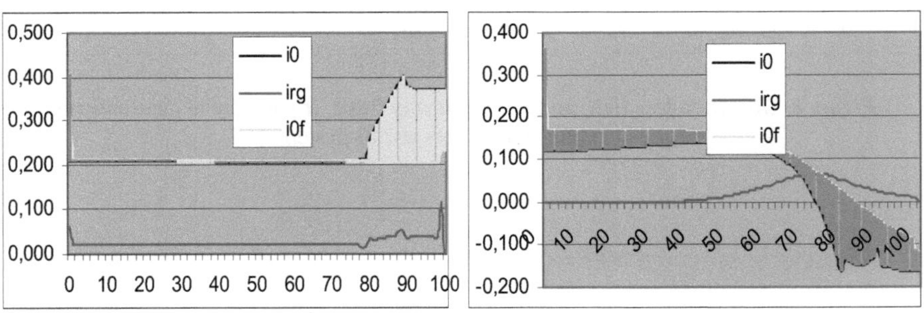

Figure 11. Distortion-less (left) and near optimally fitted (right) spatial distribution of the current transients at 240-th step

The essence of transients' distortion due to voltage on the shield gap is best visible from space distributions in Figs 7 and 9 for 160-th step when the forward waves reach the 80-th turn. The chart on the right of the Fig. 10 shows that the reflected, due to voltage on the gap, negative current drains the electric charge from the wave front to the background creating under-voltage at the front, visible in the right chart of the Fig. 7. Reflections on the low resistance grounded terminal transform the under-voltages of the forward to over-voltages of the backward wave.

The graphs in the Figs. 12-13 show currents at both terminals where there are boundary layers due to discrete irregularities of the terminal conductors configuration. Besides total currents *i0* = *i(1)* at the surge terminal and *in* = *i(N)* at the grounded terminal there are traveling components, slightly moved into interior of the winding to avoid contributions of the boundary layer. So *i1b* = $i^b(2)$ is the exciting return current arriving at the end of the first turn, *i1f* = $i^f(2)$is the forward current leaving the end of the first turn, *in1f* = $i^f(N-1)$ is the exciting forward current arriving at the beginning of the last turn and *in1b* = $i^b(N-1)$ is the return current leaving the beginning of the last turn.

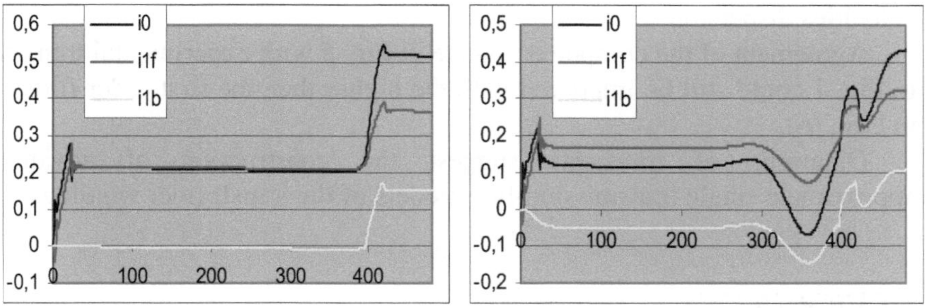

Figure 12. Surge terminal currents for distortion-less (left) and near optimally fitted (right) transients

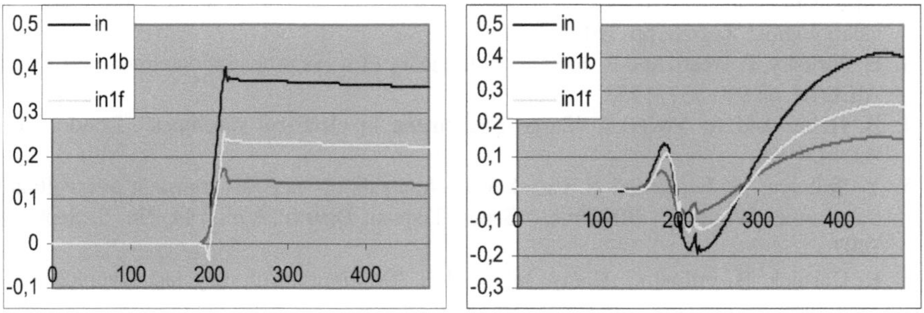

Figure 13. Ground terminal currents for distortion-less (left) and near optimally fitted (right) transients

Exhibited traveling current components at the winding terminals in Figs. 12-13 lead to workable definition of the characteristic admittance of the winding, related to the characteristic admittance of the single turn and magnetic flux linkages. Also, they support the idea of the approximate single transmission-line model in Ref. [5] as well as give suggestions how to improve such models.

5. CONCLUSION

As supposed in [1], the anomalous distortion of very fast electromagnetic transients in helical transformer winding with open electrostatic shield can be quantitatively explained with reflections on the insulating gap. Model constants M_2 and β are efficiently determined by inverse method of matching the calculation with the experimental results. There are some coupling between fitting constants M_2, β and the ratio L_2/L_1 but it was found that M_2 has dominant influence on the minimal values of the voltage transients, β on the maximal values and L_2/L_1 on the shape near extremes. The fitted ratio L_2/L_1 agrees with the calculated in [1], found responsible for velocities dispersion.

Agreement of the computed voltage in Fig. 5 with experimental transient in Fig. 1 could still be improved with the higher then the first order filtering in Eqn. (6).

Diagrams of transients suggest the construction of advanced approximate single transmission-line models of the transformer windings.

REFERENCES

1. P. Bodlovic, T. Kelemen and K. Hot: "Anomalous distortion of travelling waves in helical transformer windings", 10th Int. Symp. on Electromagnetic Fields in El. Eng., Cracow (Poland), pp. 319-322 , 2001.
2. T. Kelemen, L.Brkic, R. Gardijan, Z. Gojceta and A. Mikulecky: " Steep front wave propagation in transformer winding", Proc 3rd SONT Conference on the new technologies, Zagreb, pp. 161-164, 1991
3. H. Poritsky, P. Abetti and R. Jerrard, Field theory of wave propagation along coils", Trans. AIEE/III, pp 930-937, 1953
4. B. Heller and A. Veverka, "Surge phenomena in electrical machines", London: Iliffe Books, 1968
5. Y. Shibuya, S. Fujita and N. Hosokawa: „Analysis of very fast transient overvoltage in transformer Winding", IEE Proc. –Gener. Transm. Distrib., Vol. 144, No. 5, September 1997
6. K. Cornick, B. Filliat, C. Kieny, W. Muller: "Distribution of very fast Overvoltages in Transformer windings", CIGRE report 12-204, Paris 1992

MIXING OF FEM AND ANALYTICAL MODELING: PRELIMINARY DESIGN OF A TRANSFORMER

Benoit Delinchant, Frédéric Wurtz and Jaime Fandino

Laboratoire d'Electrotechnique de Grenoble (INPG/UJF - CNRS - UMR 5529), ENSIEG, B.P. 46,38402 Saint Martin d'Hères Cedex, FRANCE

Abstract: We are interested in this paper in methodologies of preliminary design. We applied our study on the optimization of a three-phase transformer. Preliminary design requires low time cost. Although analytical models answer this issue, they brought some loss of quality, as they are subject to many hypotheses. Analytical model can be mixed with numerical ones to improve the modeling quality. The originality of this paper is that we used different modeling methods and software to perform a better design. In this paper, we compared transformer optimizations with 5 different specifications sheets, using two models. On the one hand, a pure analytical model was used, with an automatic symbolic differentiation. On the other hand, we have introduced a FEM evaluation of the leakage reactance in the analytical model. This second model has been implemented using software links and numerical differentiation, which are also discussed.

Key words: preliminary design, analytical modeling, numerical modeling, sizing.

1. INTRODUCTION

1.1 Preliminary design methodology: mix modeling

Nowadays industrial design corresponds to find solutions to different specification sheets on a product as fast as possible. By mixing time

M. Rudnicki and S. Wiak (eds.), Optimization and Inverse Problems in Electomagnetism, 245–252.

efficiency of analytical models and modeling quality of numerical ones, the optimization process has to be more efficient. That is why it is important to develop methodologies and tools that allow to build models and processes, mixing analytical and numerical models as well as to experiment them using software.

Despite this conceptual design methodology, the implementation may bring some troubles. Mixing analytical and numerical models, corresponding to very independent and heterogeneous software, is quite difficult. Moreover, sensitivity computation aspects must be considered, when integrating such a mixed model in a design problematic.

Classical algorithms used in optimal shape design are gradient based. Indeed, even with numerical differentiation, gradient sensitivity helps to converge faster than using a stochastic algorithm or an artificial neural network approach [4]. Thus, to be more effective in preliminary design, these techniques have to be implemented.

1.2 The application: a three-phase transformer

Our study describes the sizing of the magnetic core and windings of a three-phase transformer. An entire analytical model for sizing this device is given by [1]. It calculates magnetic parameters (induction, current density ...), dimensions, weight, and cost. Among the calculated parameters, an important one is the leakage reactance X_1, seen from the primary side. To express this leakage reactance, the energy is calculated into windings and between them during a short circuit (*equation 1*). This corresponds to a linear increase of the leakage field inside of the windings and a constant max value between them. From this field distribution, the flux can be easily expressed (*equation 2*) *(see Table -1. for variables description).*

$$\frac{1}{2} L \cdot i^2 = \frac{1}{2} \mu_0 \iiint H^2 \cdot dv \quad (1) \qquad X_1 = \mu_0 . \pi . Dm . N_1^2 . \varpi . \left(D_2 + \frac{A+G}{3} \right) . \frac{1}{h} \quad (2)$$

Table - 1. Variables descriptions

L :	*leakage inductance*	*X1 :*	*leakage reactance*	*A :*	*primary coil thickness*
i :	*coil current*	*N1 :*	*coil turn number*	*G :*	*secondary coil thickness*
μ_0 :	*vacuum permeability*	*Dm :*	*central leg average diameter*	*h :*	*coils height*
dv :	*annular space of both coils and inter-space.*	*D2 :*	*distance between primary and secondary coil*	*H :*	*magnetic field*

This analytical modeling of leakage reactance can be substituted with a numerical one. A finite element model has been developed to compute this reactance, and has been included in the rest of the analytical equations, which contains electromagnetic expressions linked to economical equations.

1-a. transformer CAD view *1-b. parametric geometry* *1-c. head coil effects*

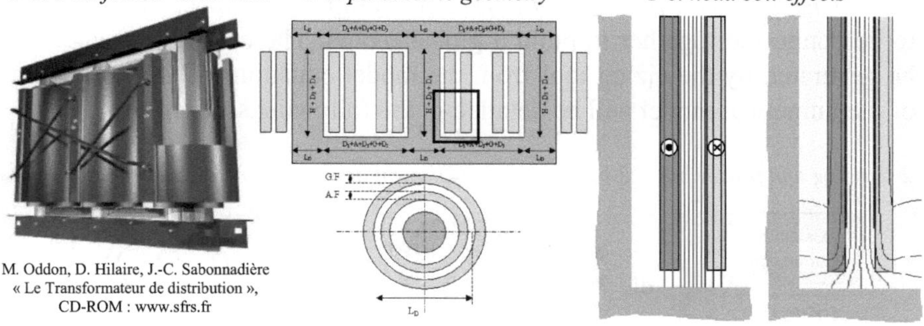

Figure - 1. Modeling the transformer with and without coil head effects.

This second model, named mixed-model for its mix of analytical and numerical modeling, has been devised to be as close as possible to the analytical one. Indeed, this work aims at studying the impact of mixing numerical modeling with analytical modeling. The main model difference is that numerical modeling takes into account the coil head effect, where the magnetic field is not supposed to be rectilinear, as in the analytical one (*figure 1.c.*). Analyzes of both models have shown that they gave quite similar results on X_1 computation. Comparisons have been made on a wide range of values for several critical input variables, leading to an average difference of 5%. The introduction of the leakage reactance FEM computation, which is quite similar to the analytical one, may be studied to evaluate this influence for future mixed-modeling.

2. MIXED MODELING FOR SIZING

2.1 Optimal sizing with an SQP algorithm

In our study, the preliminary design consists on the optimal sizing of the transformer. To do this, we may use dedicated tools, which perform optimization with SQP *(sequential quadratic programming)* algorithm. There can be different implementations of these tools, depending on the sensitivity propagation method being used. The one used in our study (Pro@Design™ [1] [3]) is based on a methodology described by [2] and is dedicated to analytical models. Indeed, it can generate, from an equation file, a component able to compute output values from input ones, as well as output differential values, thanks to symbolic derivation. To use this tool, we must define two components, in which analytical and numerical models are defined (*figure 2.c.*). Such a model composition is generally used in MDO (*MultiDisciplinary Optimization [6]*). Both components (inner models) have to be connected together to build a global model. The analytical model can be generated by the sizing tool from the model equations. The introduction of the numerical model will be detailed in the following section.

2-a. sizing tool framework *2-b. component implementation* *2-c. mixed model component*

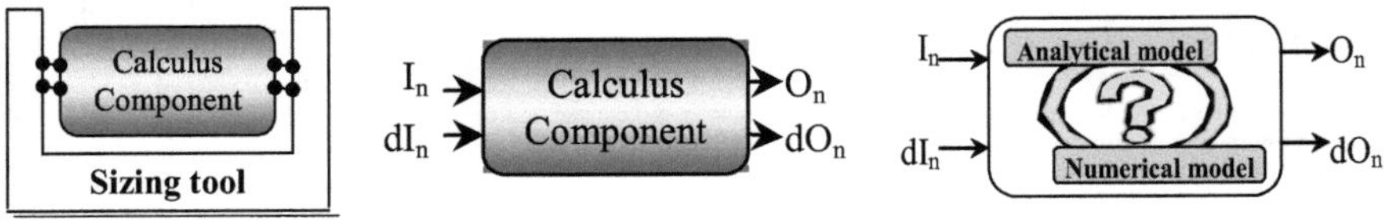

Figure - 2. Sizing tool architecture in which the mixed model has been integrated.

2.2 Numerical model integration

Two stages are needed during this integration: the automatic control of numeric simulator, and the projection into an optimizable component.

Indeed, to introduce the numerical model of the leakage reactance into the mixed-model component, some particular treatments have to be done. First of all, the FEM tool (Flux2D™ [2]) must be controlled dynamically with different input parameters and may return the computed inductance value. This can be done thanks to a tool integration methodology [5] applied to software controlled by files. A *Template Editor*[3] allows generating wrapper to control the FEM calculation on a specific device. The transformer was then modeled into the FEM software and the *Template Editor* was applied to encapsulate the leakage reactance computation into a *black box*[4]. Details of this particular stage can be found in [5].

The second step was the projection of this *black box* to the sizing tool component specifications. Some troubles may appear with this *semantic wrapping*, which corresponds to the mapping of meaning between the source and the target of information flow. In our case, we may have Flux2D™ configuration parameters, which do not aim at being optimized; but plugged into pro@DESIGN™, all parameters are considered as optimizable ones. Other issues appear during the implementation of sensitivity computation, as it will be described in the following section.

2.3 Derivation implementation issues

The FEM software sensitivity analysis must be implemented with regards to the sizing tool requirements. For instance, it may impose its own propagation mechanisms like differential or partial derivative mechanisms.

Moreover, some strategy has to be implemented to optimize time cost of numerical differentiation, due to model coupling. Indeed, model coupling leads to propagation issues, regarding with the global sensitivity analysis. GSE (Global Sensitivity Equation [6]) allows solving this issue, when both composition information and necessary global sensitivities are known. In our optimization structure, sensitivity analysis is implemented at the level of each inner model, without knowing how composition was made of. Thus, different methodologies have been implemented, leading to the reduction of optimization time by a factor 4.

This work was only interested with results, which have been computed with a method using systematic partial derivatives, based on forward finite difference. The partial derivation implementation shows stable results during the sizing process, thanks to the configuration of the differentiation step for each parameter. Despite of it, numerical differentiation may bring bad sensitivity analysis, if the finite difference step is too fine. Numerical step was set to 10^{-2} for the sizing process, but as it can be seen in the next section, such an imprecise step will extend the algorithm research time.

3-a. plot of $X_1=f(h)$ *3-b. plot of $\dfrac{\partial X_1}{\partial h}$ with several steps for the finite difference*

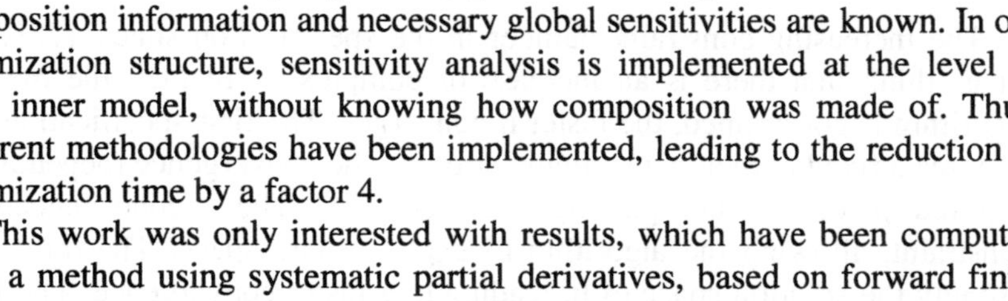

Figure -3. Numerical differentiation precision analysis.

3. ANALYSIS FROM 5 SPECIFICATIONS SHEETS

Table - 2. Model parameters constraining the sizing process.

Inputs		**Outputs**	
H :	Coil height (m)	X_1 :	Primary leakage reactance (%)
J :	Current density (A/m²)	A :	Leg width (m)
N_1 :	Coil turn number	Ltt :	Magnetic circuit total length (m)
Bt :	Flux density (T)	cost :	Transformer cost for 30 years of use (French francs).

Optimizations on both models have been realized for 5 different specification sheets. Due to our two very similar models, the sizing converges to the same final point (less than 1% for main input parameters, cost is different by 0.5%). Thanks to this result, we are now able to compare the behavior of the mixed-model with the analytical one.

3.1 Specification sheets characteristics

The increasing constraint number in the specification sheets (*Table -3.*) let us think that there is an increase of complexity. In fact, the more the algorithm is constrained, the faster it converges. The first specification sheet imposes a fixed output, constraining strongly the convergence (depending on required precision) but guiding the algorithm. The second relaxes this constraint, allowing the algorithm to explore more space. The third limits strongly the solution space of h, leading to a faster convergence. The fourth imposes a new output constraint (A), giving the algorithm a search direction instead of all space. The last one does it again with Ltt.

Table -3. 5 specification sheets applied on the transformer model for sizing.

		Spec1		Spec2		Spec3		Spec4		Spec5	
		min	max	min	max	min	max	min	max	min	max
Optimizable Input parameters	H	**1**	**10**	1	10	**1**	**1.4**	1	1.4	1	1.4
	J	**5e5**	**4.5e6**	5e5	4.5e6	5e5	4.5e6	5e5	4.5e6	5e5	4.5e6
	N_1	**100**	**1000**	100	1000	100	1000	100	1000	100	1000
	Bt	**0.5**	**1.9**	0.5	1.9	0.5	1.9	0.5	1.9	0.5	1.9
Constrained outputs	cost	**Objective function to minimize**									
	X_1	**Fixed: 7.2**		**5.76**	**8.64**	5.76	8.64	5.76	8.64	5.76	8.64
	A							**0**	**0.04**	0	0.04
	Ltt									**0**	**3**

3.2 Sizing time cost

Iteration number decreases along the five sizing process (*Table -4.*). The longest iteration number for mixed model, compared with the analytical one, for specifications 1, is due to the fixed constraint on X_1, which must converge exactly to 7.2% (more or less the required precision). In fact, a 10^{-2} step was too imprecise to reach easily a fixed value (10^{-3} makes converging the algorithm faster in 12 iterations). We do not explain the differences for specifications 2, where the mixed-model allows converging faster to a quite

Table -4. Sizing duration and iteration number for the five specification sheets.

		Analytical model (symbolic differentiation)	Mixed model	Mixed model Microscope Metaphor
Spec1	time (sec)	0,21s	8183(2h16)	2200(37min)
Spec2	time (sec)	0,28	5300(1h28)	2200(37min)
Spec3	time (sec)	0,21	5000(1h23)	1300(22min)
Spec4	time (sec)	0,16	3100(51min)	840(14min)
Spec5	time (sec)	0,16	3100(51min)	2150(36min)

similar result. A more precise step (10^{-3}) makes the optimization converging faster (10 iterations). Expected for this specification sheet, the analytical model converges more quickly than the mixed one. This is probably due to a differentiation imprecision, compared with the end precision required for the sizing.

Finally, huge computation time differences between analytical model and mixed-model can be observed (more than one hundred times). The absolute time cost is very dependent of the computer configuration[5]. Solving the FEM problem is not the main time consumer. Actually, most of the time was spent during the automatic problem setting, modules launching and GUI displaying. Indeed, for FEM computing (including geometry modification, physical parameters modification, resolution and result extraction), only 10% are enough to solve the FEM problem. A bigger problem would probably modify the solving time only and would then increase this 10% ratio to a more realistic one.

3.3 The microscope metaphor

Last column of table 4 describes iteration number and time cost for a microscope metaphor of the solution space exploration. In a design process, we consider the computation time cost as being as important as the modeling quality. This specific process tries to optimize both, by exploring a huge solution space as fast as possible (analytical model), followed by a refined model (numerical or mixed), which starts at the previous result *(see an illustration in figure 4).*

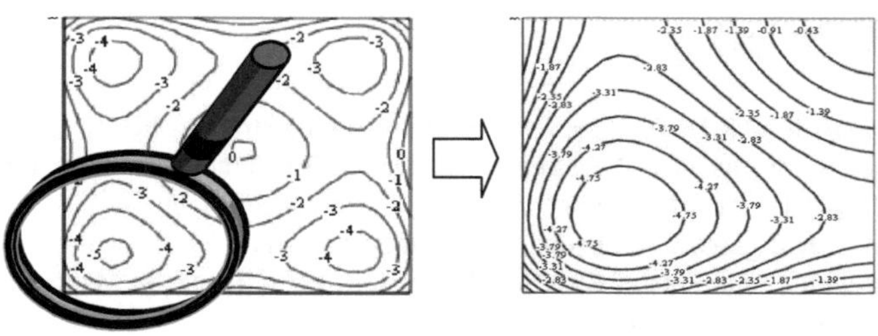

Figure -4. Microscope metaphor: a first model is used to ensure fast converge and large space search, a second model use previous result ensuring finer convergence.

In our study, both models are quite similar, but the mixed model takes into account head coil effects, affecting X_1 values. The experiment shows that the mixed model, which starts at the final point of the optimized analytical model, has just to adjust the solution to fit with the new computation of X_1. The consequence is a high decrease of iteration number during the sizing of the mixed-model, leading to better optimization times.

4. CONCLUSION

In this study, we have successfully encapsulated an available commercial FEM software thanks to a *Template Editor* tool. The numerical differentiation has been implemented with regards to the sizing tool specifications. From this mixed-model, which has been made as close as possible to the analytical one, we studied its behavior on 5 different specifications sheets. The main result is the feasibility of mixed modeling, using different simulation tools to achieve a design goal. Another result confirms an obviousness which is a heavy time cost, when introducing a numerical computation. But a preliminary design methodology using analytical modeling, followed by mixed or pure numerical modeling would be a good way to improve convergence behavior.

NOTES

1. **Pro@DESIGN™** optimizes solution of an approximate analytical model, representing nonlinear heavily constrained problems. www.designprocessing.com
2. **Flux2D™** is an electromagnetic and thermal analysis software, based on finite element method. www.cedrat.com
3. **Template Editor** is a software enabling file controlled software to be driven by a program, thanks to the automatic generation of command files from a template one.
4. **Black Box** is a model exhibiting a behavior that approximates what is observable from the outside of the modeled object, without knowledge about its internal make-up.
5. **Computer specifications**: Pentium III 450MHz, RAM 256Mo, but time has been reduced by 5 on another computer (PIII 1GHz, RAM 256Mo, and a better graphic card).

REFERENCES

1. M. Poloujadoff, R.D. Findlay, "A Procedure for illustrating the Effect of Variation of Parameters on Optimal Transformer Design", IEEE Transactions on Power Systems; Vol. PWRS-1, No. 4, November 1986
2. F. Wurtz, J. Bigeon, C. Poirson, "A Methodology and a Tool for the Computer Aided Design with Constraints of Electrical Machines", IEEE Trans. on Magnetics, vol. 32, no. 3, May 1996, pp. 1429-1432
3. E. Atienza, J. Bigeon, F. Wurtz, V. Leconte, "EDEN : A generic integrated component software for constraint optimisation", COMPUMAG'01, 13th Conference on the Computation of Electromagnetic Fields, July 2-5, 2001, Evian, France, pp.232-233
4. E. Ratajewicz-Mikolajczak, J. Sikora, J. Starzynski "Comparison of optimal shape design algorithms", Published in Journal of Technical Physics, 41 , pp. 311-320, 2000
5. B. Delinchant, F. Wurtz, J. Bigeon "Automating tool integration to preliminary design request", CEFC'02, 10th Biennial IEEE Conference on Electromagnetic Field Computation, Perugia, Italy, June 16-19 2002
6. Sobieszczanski-Sobieski, J., "Sensitivity of Complex, Internally Coupled Systems," AIAA Journal, Vol. 28, No. 1, January 1990, pp. 153-160

AN EVOLUTIONARY ALGORITHM FOR THE OPTIMAL DESIGN OF A RARE EARTH PERMANENT MAGNET MACHINE

Özdemir Göl, Bijan Sobhi-Najafabadi
Electrical Machines and Drives Research Group, University of South Australia

Abstract: This paper deliberates on the suitability of evolutionary algorithms for design optimisation of electromagnetic devices. Development and application of such an algorithm for a rare earth permanent magnet generator is discussed. Test results for a proof of concept generator are given for verification.

Key words: permanent magnet generator, optimization, genetic algorithm, evolutionary computation

1. INTRODUCTION

Electrical machine design optimisation has been traditionally undertaken using deterministic techniques. However stochastic techniques are increasingly making an impact on the design practice because of the attractive features they are able to offer [1]. Advantages include the ability to locate a near optimal solution with far greater certainty independently of the initial values of design variables, and less reliance on designer experience than is the case with deterministic alternatives.

Of the various stochastic techniques, Evolutionary Computation, inspired by the notion of biological evolution, appears particularly suitable for adoption in electrical machine design for a variety of good reasons. For instance, unlike their deterministic counterparts, they do not require the use

M. Rudnicki and S. Wiak (eds.), Optimization and Inverse Problems in Electomagnetism, 253–260.

of derivatives in executing the search; thus, they can deal with both continuous and discrete functions. They are also capable of dealing with the problem of mixed integer programming (MIP), intrinsic in electrical machine design [2].

In the following, the development of an evolutionary algorithm will be discussed. Its usefulness will be demonstrated by applying it to the design optimisation of a rare earth permanent magnet generator. Although the developed suites of algorithms [3] have included those for multi-objective optimisation, only single-objective optimisation algorithms are to be discussed here, mainly because an existing stator core was used to effect constrained optimisation, as will be clear from the following discussion. The emphasis will be on the credibility of design outcomes from the algorithm execution as verified experimentally.

2. DEVELOPMENT OF A STOCHASTIC SINGLE-OBJECTIVE OPTIMISATION ALGORITHM

A stochastic single-objective optimisation algorithm for a PMG can be devised by combining a generator design algorithm with a single-objective Genetic Algorithm (GA) [3,4]. A non-stationary penalty function can be employed to enable the GA to handle constraints. For algorithm execution, the user needs to be prompted to enter such information as the design specifications as designated by the end-user, data on magnetic properties of core and magnet materials, and electrical properties of conducting materials. It is also required to nominate the objective function, constraints and a non-stationary penalty function. Information will be also required on benchmark values and the upper and lower values of independent design variables. In configuring the GA, decisions have to be made concerning the representation of variables, population size, probability of selection, selection method, probability of crossover, crossover method, probability of mutation and the mutation method. These configurations can be set as default values for the algorithm. Alternatively, they can be saved for future use when modifications and further executions are required, avoiding the need to re-enter data. The algorithm provides feedback on the computational status during execution. Number of generations is the termination criterion. The execution stops upon satisfying the termination criterion.

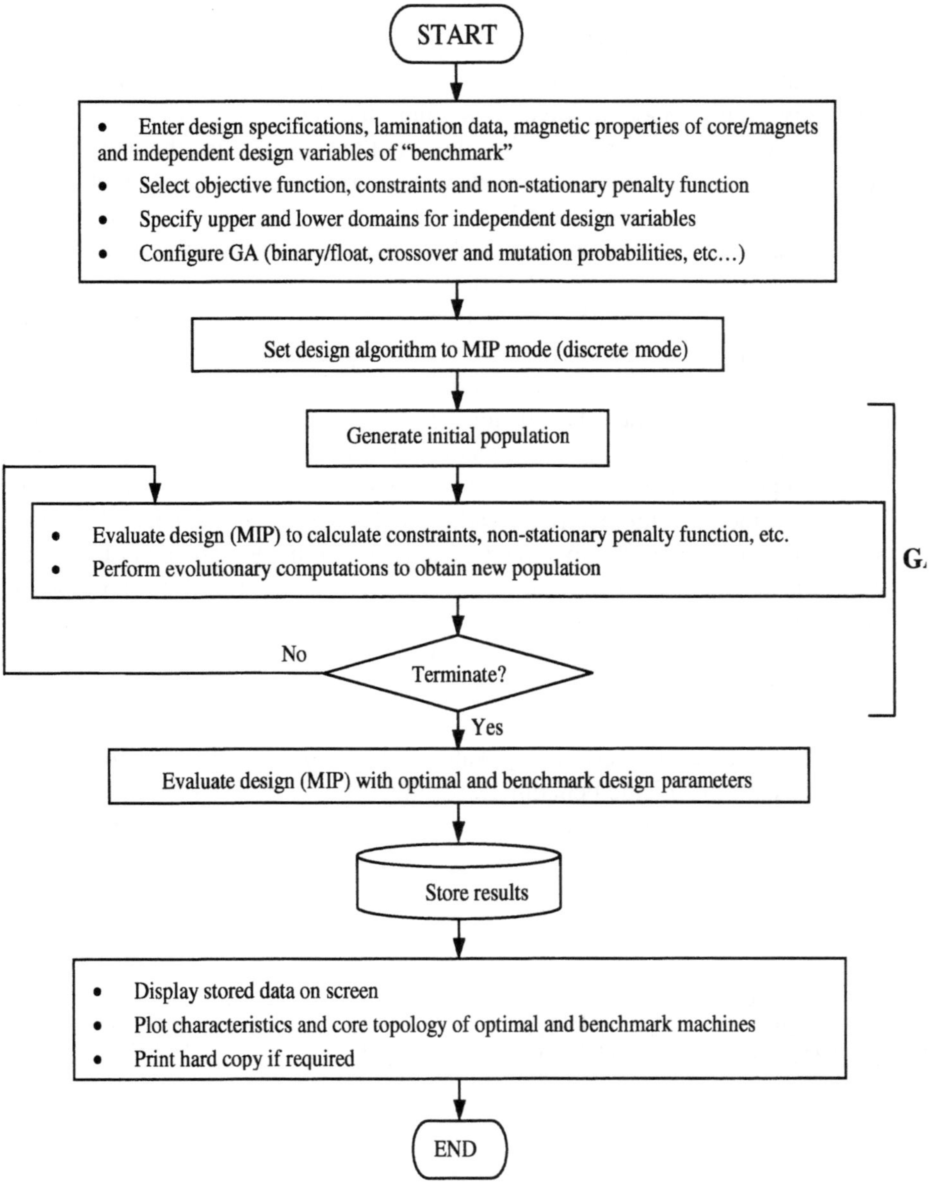

Figure 1 Stochastic single-objective optimisation algorithm: flowchart

APPLICATION OF THE ALGORITHM: DERIVATIVE DESIGN BASED ON AN EXISTING LAMINATION

It is an accepted practice in industry to base the development of improved electrical machine series on existing core geometries for reasons of scales of economy. This may involve changing the core length and winding configurations, using an existing lamination design. In the case to be discussed below a similar approach was adopted, in that an existing stator core of a 2-phase squirrel cage induction motor was used. The machine was to be refurbished to operate as a 3-phase permanent generator. The existing 2-phase winding was replaced with a balanced 3-phase winding. The cage rotor was substituted with a solid iron rotor housing surface-mounted rare earth permanent magnets. Both the winding configuration and the rotor topology were to be the object of optimisation. Single objective optimisation was to be adopted since the full load efficiency was deemed to be the most significant objective. The consequence of this for the algorithm used was that independent design variables were reduced to three. These were: the field form factor, total electrical loading and the air gap clearance. All others were fixed by virtue of using an existing lamination. Design constraints were normalised for use. The design algorithm was modified to incorporate constraints into the objective function. In doing so, a non-stationary penalty function was employed [2,5,6]. The fixed core length is incorporated in the design optimisation as an equality constraint.

Table 1 gives the efficiency values obtained from Run 1 and Run 2 together with that of the benchmark machine. Fig. 2 enables the performance of the optimised generators to be compared with that of the benchmark generator for resistive load. Fig. 3 exemplifies the convergence of the independent design variables to optimal.

Table 1. Comparison of efficiency

Objective function	Benchmark	Run 1	Measured
Efficiency at full load	0.82802	0.85378	0.85559

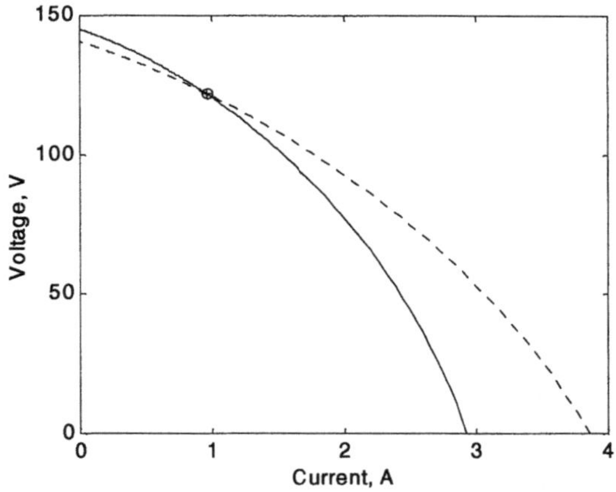

(a)

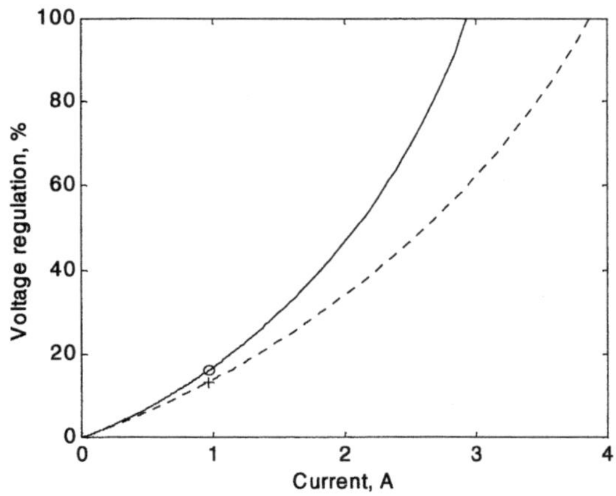

(b)

Figure 2. Performance comparison for resistive load
(a) External characteristic (b) Voltage regulation

----- Optimal performance ——— Benchmark performance
\+ Optimal at full-load O Benchmark at full-load

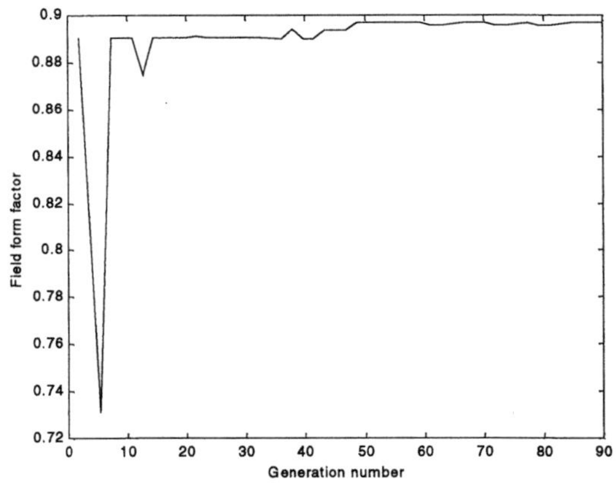

(a)

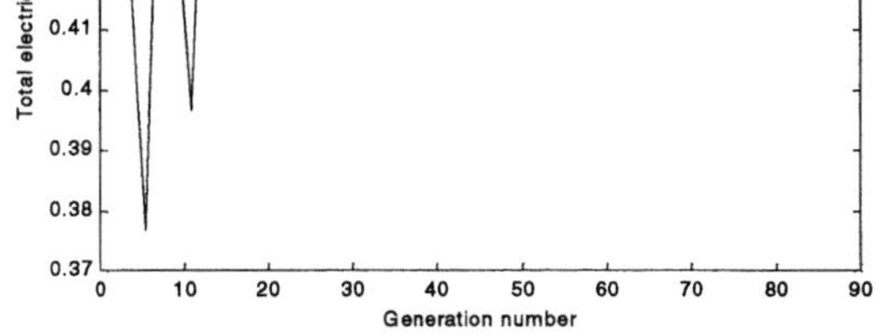

(b)

Figure 3. Convergence to optimal:
(a) field form factor (b) total electrical loading

3. EXPERIMENTAL VERIFICATION

The optimised generator was manufactured and laboratory tested. Test results are presented in Figure 4 below. Predicted performance characteristics are superimposed to facilitate a direct comparison. As can be seen, the correlation is excellent throughout.

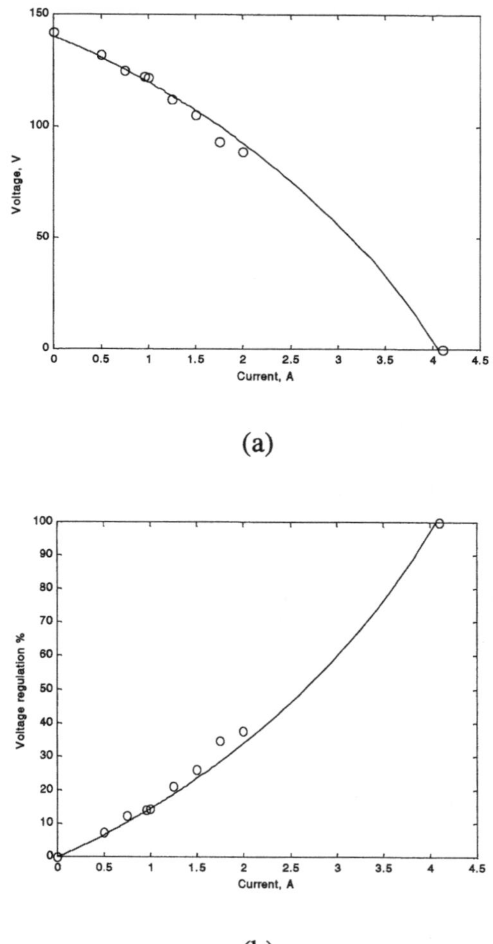

Figure 4. Operating characteristics for unity power factor
(a) External characteristic
(b) Percentage voltage regulation for unity power factor

—— Predicted o o Measured

4. SUMMARY AND CONCLUSIONS

Stochastic design optimisation techniques offer a great deal of scope for adoption in electrical machine design. They converge reliably, deliver credible results and require far less decision making from the user. This paper has endeavoured to demonstrate their efficacy using the case of a generator. An existing machine was retrofitted as a rare earth permanent magnet generator, optimising its efficiency, using a user friendly stochastic design optimisation algorithm developed by embedding a design algorithm in a GA shell. The results presented convincingly demonstrate the validity of the approach.

REFERENCES

[1] Mohammed O. A., Russenschuck S., Kasper M.: Design Optimization of a Superferric Octupole Using Various Evolutionary and Deterministic Techniques, IEEE Transactions on Magnetics, Vol. 33, No.2, pp. 1816-1821, 1997.

[2] Göl Ö., Wieczorek, J. P.: Design Optimisation of Electromagnetic Devices using Genetic Algorithms, Proceedings of EMAC98 3rd Biennial Engineeering Mathematics and Applications Conference, Adelaide, Australia, pp. 217-220, 1998.

[3] Sobhi-Najafabadi B.: Optimal Design of Permanent Magnet Generators, PhD Thesis, University of South Australia, Adelaide, Australia, 2002.

[4] Göl Ö., Sobhi-Najafabadi B.: Development of Optimisation Algorithms Applicable to Design of Permanent Magnet Generators, Proceedings of the Second IASTED International Conference, Crete, Greece, pp. 484-489, 2002.

[5] Pahner U., Hameyer K.& Belmans R., A Parallel Implementation of a Parametric Optimization – Numerical Optimization of an Inductor for Traction Drive Systems, IEEE Transactions on Energy Conversion, Vol. 14, No. 4, 1999, pp. 1329-1334.

[6] Michalewicz Z., Attia N.: Evolutionary Optimization of Constrained Problems, Proceedings of the 3rd Annual Conference on Evolutionary Programming, World Scientific, pp. 98-108, 1994.

OPTIMIZATION OF THE FORCE CHARACTERISTIC OF A SOLENOID ELECTROMAGNET WITH FERROMAGNETIC DISC IN THE COIL

Vultchan Gueorgiev, Alexander Alexandrov and Ivan Yatchev
Technical University of Sofia, Dept. of Electrical Apparatus, 1156 Sofia, Bulgaria
E-mail: vulchy@vmei.acad.bg, alexandr@elap.vmei.acad.bg, yatchev@vmei.acad.bg

Abstract: The force characteristic of a solenoid electromagnet can be significantly changed by inserting ferromagnetic disc in the coil. The optimization of the force characteristic is carried out by varying the position and the dimensions of the disc and seeking maximal initial force. Finite element method has been used for field analysis and force computation. Results have been obtained for solenoid electromagnets with one and two ferromagnetic discs in the coil.

Key words: Solenoid electromagnets, force characteristics, optimization, response surface methodology, finite element method.

1. INTRODUCTION

Electromagnetic actuators are widely used devices in a variety of applications. Their design and optimization is subject of permanent interest in recent years [1-12], employing different optimization strategies and design approaches.

Obtaining desired force characteristic is of major importance in the design of solenoid actuators. Usually the change of the characteristics is obtained either by varying geometry, e.g. the shape of the pole pieces, or by external control. Another way to change the characteristic is inserting a ferromagnetic disc inside the coil. The paper presents optimization of the force characteristic of solenoid electromagnet with one and two ferromagnetic discs.

M. Rudnicki and S. Wiak (eds.), Optimization and Inverse Problems in Electomagnetism, 261–268.

2. SOLENOID ELECTROMAGNET WITH FERROMAGNETIC DISC IN THE COIL

The electromagnet under consideration features a ferromagnetic disc placed in the coil in such a way that the coil is divided in two parts – Fig. 1. It is subject of Bulgarian patent [13]. The location of the disc is somewhere in the zone of the air gap of the electromagnet.

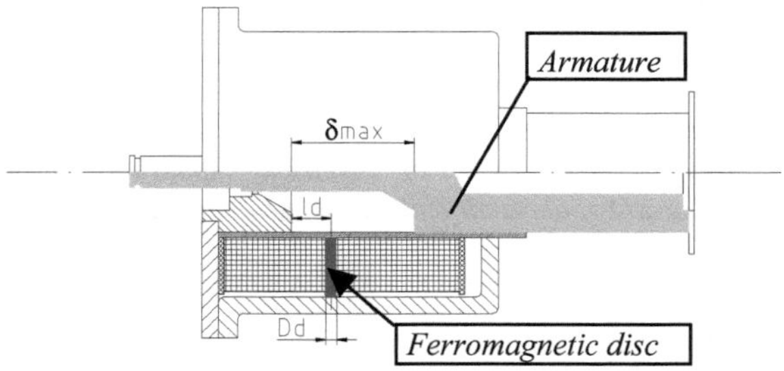

Figure 1. Solenoid electromagnet with a ferromagnetic disc in the coil

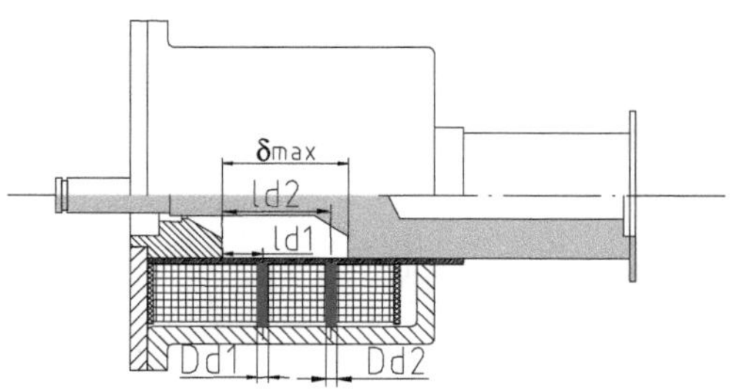

Figure 2. Solenoid electromagnet with two ferromagnetic discs in the coil

The ferromagnetic disc is cut in radial direction in order to avoid its behaviour as a secondary winding in dynamic conditions. Only the magnetic field redistribution due to the presence of the disc is thus obtained and its influence on the force has been studied.

Varying the position of the armature, the disc can have different relative positions with respect to the air gap. When the disc is inside the air gap, i.e. at large gaps, the disc strengthens the working magnetic flux through the armature. This results in increasing the electromagnetic force compared to the case without disc. The flux density in the disc is relatively low, the disc is not saturated and its influence on the force is significant. When the disc is outside the air gap zone, i.e. at smaller gaps, the disc weakens the working magnetic flux, which leads to lower electromagnetic force than in the case without disc. The ferromagnetic disc is more saturated and its influence on the force is relatively low.

Another similar construction is the one with two ferromagnetic discs in the coil, shown in Fig. 2. Both discs are located in the air gap zone.

Typical force-stroke characteristic is shown in Fig. 3. The influence of the disc is clearly seen. Its position is close to the crossing point of the two curves. More details are given in [14].

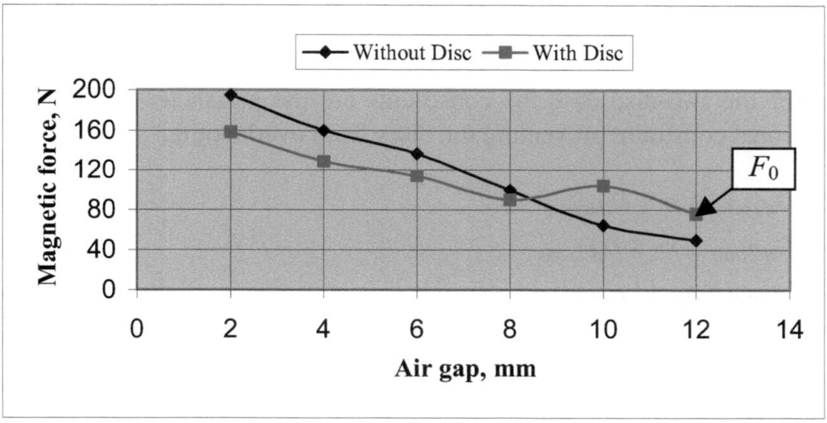

Figure 3. Typical force-stroke characteristic without and with ferromagnetic disc in the coil

The change of the characteristic due to the disc is in favourable direction for actuators for car starters, as the initial force there is very important. That is why the optimization is carried out for obtaining higher initial force.

3. OPTIMIZATION OF THE FORCE CHARACTERISTIC

The optimization problem is formulated on the basis of fixed geometry of the electromagnet (an existing electromagnet for car starter was taken as a basis), unchanging material properties and constant apparent current density in the coil. Optimization parameters are the position and the dimensions of the disc(s) with the corresponding geometry constraints. The objective function is taken to be the initial force F_0 - Fig. 3. Thus the optimization problem is:

$$F_0(\mathbf{x}) \rightarrow \max \tag{1}$$

where $\mathbf{x} = (x_1, x_2)$ for the case with one disc; $x_1 = D_d$; $x_2 = l_d$ - Fig. 1, and $\mathbf{x} = (x_1, x_2, x_3, x_4)$ for the case with two discs; $x_1 = D_{d1}; x_2 = D_{d2}; x_3 = l_{d1}; x_4 = l_{d2}$ - Fig. 2.

For the one-disc case the constraints are only the lower and upper bounds of the parameters:

$$\begin{aligned} D_{d\min} &\le x_1 \le D_{d\max} \\ l_{d\min} &\le x_2 \le l_{d\max} \end{aligned} \tag{2}$$

For the two-disc case the constraints are the parameter bounds and an additional constraint preventing the discs from overlapping:

$$\begin{aligned} D_{d1\min} &\le x_1 \le D_{d1\max} \\ D_{d2\min} &\le x_2 \le D_{d2\max} \\ l_{d1\min} &\le x_3 \le l_{d1\max} \\ l_{d2\min} &\le x_4 \le l_{d2\max} \\ x_4 - x_3 &\ge \frac{x_1 + x_2}{2} + l_{\min} \end{aligned} \tag{3}$$

where $l_{\min}$ is the minimal distance between the discs.

The above stated optimization problem is solved with the help of response surface methodology [10-12]. For the force evaluation the finite element method is employed using the program FEMM (finite element method magnetics - by David Meeker [15]) and the Lua® scripting language. The obtaining of the response surface model and the optimization are performed with the help of Matlab® package [16,17]. Thus minimal amount of code writing has also been achieved.

For obtaining the response surface model, first a series of numerical experiments is carried out. Full factorial design of experiments has been applied with 4 or 5 levels of each factor. Each experiment consists of FEA of the magnetic field of the electromagnet and postprocessing calculation of the force. After having the data from the designed experiment, the response surface model is obtained and optimised. The response models are polynomials of different degree. For ensuring that the global optimum is obtained, a grid of points covering the parameter domain has been generated and the optimization has been performed for each point of this grid as an initial estimate.

Generalised flowchart of the optimization process is shown in Fig. 4. The interaction between the FEM program and Matlab® is realised through an intermediate file containing the data from the numerical experiment.

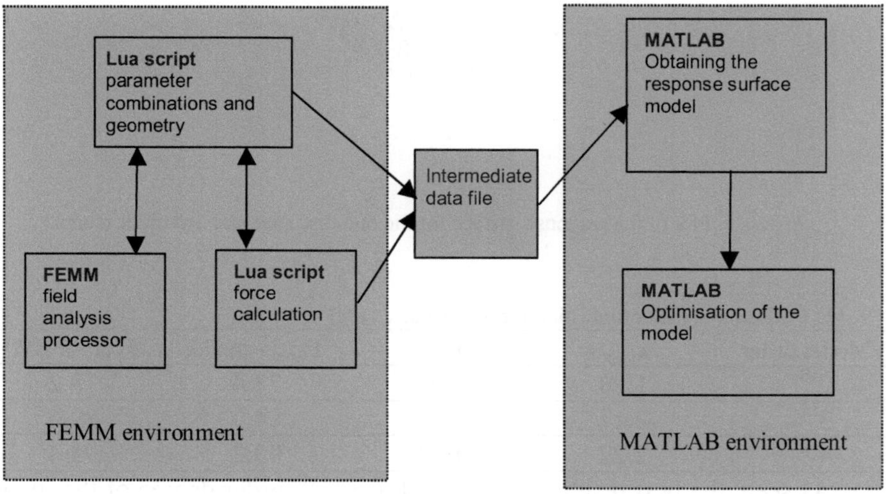

Figure 4. The essence of the optimization procedure.

4. RESULTS

Results are obtained for the one- and the two-disc case and for the following values of the dimensions: $\delta_{max} = 12$mm; $D_{d\ min} = D_{d1\ min} = D_{d2\ min} = 0.3$ mm; $D_{d\ max} = D_{d1\ max} = D_{d2\ max} = 6$ mm; $l_{d\ min} = l_{d1\ min} = l_{d2\ min} = 0$ mm; $l_{d\ max} = l_{d1\ max} = l_{d2\ max} = 12$ mm; $l_{min} = 0.3$ mm. For the one-disc case the number of numerical experiments was 20 and polynomials of 2nd, 3rd and 4th degree full polynomials have been employed as response surface models. A plot of the 3rd order model is shown in Fig. 5.

After obtaining the optimization parameters, finite element analysis and force calculation at the optimum point have been carried out. The results are given in Table 1. The parameters are in millimetres, the force - in Newtons. The results for the finite element analysis at the optimal points are placed in the last column..

The obtained optimal values of the initial force for the different polynomial degrees are given in graphics form in Fig. 6. The last bar corresponds to the case without ferromagnetic disc in the coil.

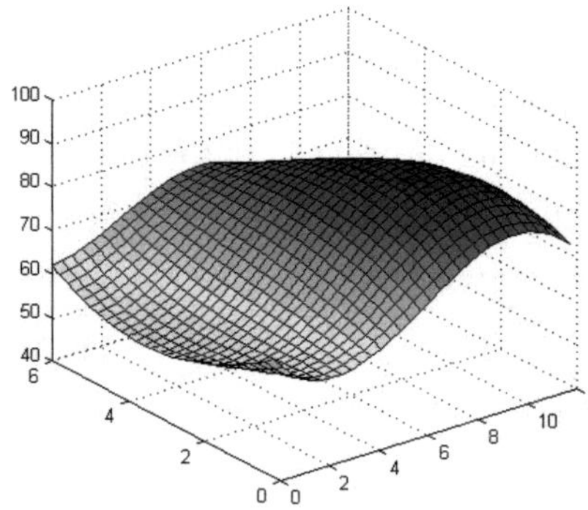

Figure 5. Plot of the response surface for the one-disc case and 3rd order model.

Table 1. Results of the optimization for the one disc case]

Model order	$x_{1\ opt}$	$x_{2\ opt}$	$F_{0\ opt}$ - model	$F_{0\ opt}$ - FEM
2nd	1.28	8.23	77.4	78.2
3rd	2.17	9.21	84.9	89.6
4th	2.22	10.1	93.2	95.4

As seen, the best results are obtained using 4th order degree of the model polynomial.

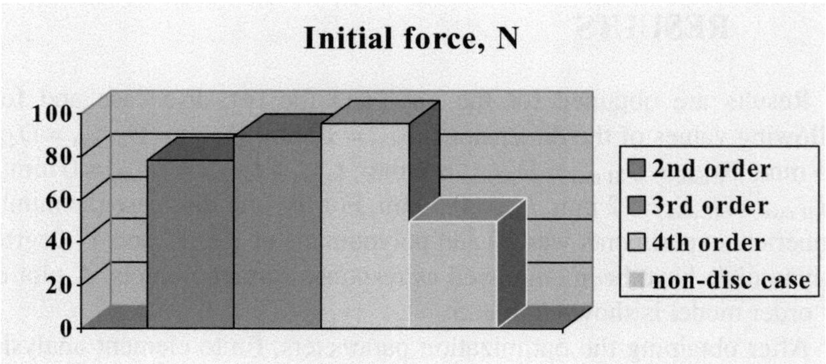

Figure 6. Optimal solution for the one-disc case.

For the case of two discs, the number of numerical experiments was 139 and 2nd, incomplete 3rd (without the terms of the type x^3), and full 3rd order polynomials have been used. The results are shown in table 2, where also the results from the finite element analysis at optimal points are given in the last column. In Fig. 7, a bar chart with the results is shown..

Table 2. Results of the optimization for the two-disc case.

Model order	$x_{1\ opt}$	$x_{2\ opt}$	$x_{3\ opt}$	$x_{4\ opt}$	$F_{0\ opt}$ - model	$F_{0\ opt}$ - FEM
2nd	0.3	0.3	11.0	11.6	94.4	84.4
Incomplete 3rd	1.1	0.3	11.0	12.0	98.1	95.4
3rd	0.3	0.74	11.0	12.0	94.8	90.1

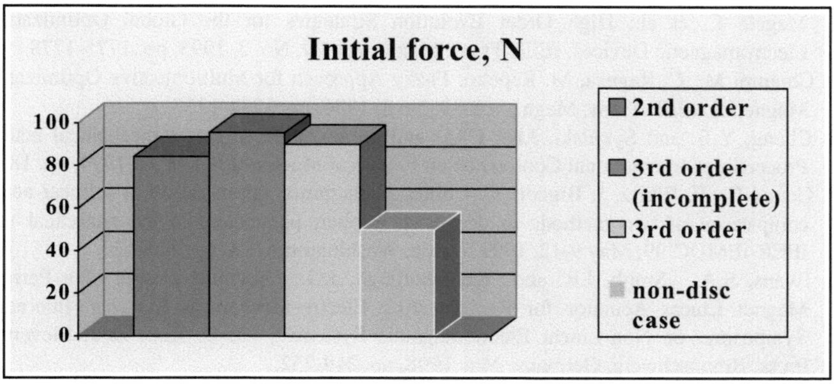

Figure 7. Optimal solution for the two-disc case.

As seen, different response surface models give different optimal solutions. The best among the tested models are 4th order for the one-disc case and incomplete 3rd order for the two-disc case. What is worth noting is that both the one- and the two-disc cases lead to maximal initial force of the same value. This means that for obtaining maximal initial force under the specified conditions, placing of two discs in the coil does not give advantage with respect to one disc. The increase of the initial force with respect to the case without disc is about 90%.

5. CONCLUSION

A partial optimization problem – optimization of the force characteristic of a solenoid electromagnet with one and two ferromagnetic discs in the coil – has been solved. The employed approach was found to give satisfactory results for increasing the initial force - up to 90% with respect to the electromagnet without disc. Inserting both one and two discs leads practically to the same initial force and for its increasing the second disc is not necessary. When considering the whole characteristic, however, insertion of two discs can significantly help in obtaining a characteristic with flat segment, which can be one direction for further research.

REFERENCES

1. Neittaanmaki P., M. Rudnicki and A. Savini: Inverse problems and optimal design in electricity and magnetism. Oxford, Clarendon press 1996 XIII, 367 p.
2. Gottvald A., K. Preis, C. Magele, O. Biro, and A. Savini: Global optimization methods for computational electromagnetics, IEEE Transactions on Magnetics, 1992, vol. 28, No. 2, pp. 1537-1540.
3. Magele C. et al.: High Order Evolution Strategies for the Global Optimization of Electromagnetic Devices, IEEE Trans. Magn., vol. 29, No. 2, 1993, pp. 1775-1778.
4. Chiampi M., C. Ragusa, M. Repetto: Fuzzy Approach for Multiobjective Optimization in Magnetics, IEEE Trans. Magn., vol. 32, No.3, 1996, pp. 1234-1237.
5. Cheng, Y.B. and Sykulski, J.K.: CAD and optimization of electromechanical actuators, Proceedings International Conference on Electrical Machines ICEM'94, 1994, pp. 187-92.
6. Coutel C., F. Wurtz, J. Bigeon, C. Chillet, Constrained optimization of a linear actuator: comparison of two methods to deal with implicit parameters in the analytical model, IEEE-IEMDC'99, May 9-12, 1999 Seattle, Washington, USA, pp. 625-627.
7. Evans, S.A., Smith, I.R. and Kettleborough, J.G.: Optimum Design of a Permanent Magnet Linear Actuator for Reciprocating Electro-Mechanical Systems, International Symposium on Non-Linear Electromagnetic Systems , 13, V. Kose & J. Sievert, IOP Press, Braunschweig, Germany, May 1998, pp. 249-252.
8. P. Di Barba, Farina, M., and A. Savini: Multicriteria Optimization of Air-cored Solenoids with Multiple Windings, Proceedings of the International Symposium on Non-linear Electromagnetic Systems, ISEM'99, IOS Press, Pavia, Italy, May 1999.
9. Wang J., G. Jewell, D. Howe: Design optimization and comparison of tubular permanent magnet machine topologies. IEE Proceedings, Electric Power Applications Vol. 148(5), 2001, pp. 456-463.
10. Papasow S. Optimal design of electromagnetic systems. Tehnika, Sofia, 1975, 279 p. (In Bulgarian)
11. Myers R.H., D.C. Montgomery, Response surface methodology. Process and product optimization using designed experiments, John Wiley&Sons, New York, 1995.
12. Rashid, K., Farina, M., Ramirez, J.A., Sykulski, J.K. and Freeman, E.M.: A comparison of two generalized response surface methods for optimization in electromagnetics COMPEL, v. 20, No. 3, 2001, pp. 740-752.
13. Alexandrov A., A. Pisarev and S. Pavlov, "DC solenoid electromagnet", Bulgarian patent No. 18140, 1972.
14. Georgiev V., I. Yatchev, A. Alexandrov, Force Characteristics of Solenoid Electromagnet with Ferromagnetic Disc in the Coil, Proceedings of 3rd Japanese-Bulgarian-Macedonian Seminar on Applied Electromagnetics JBMSAEM'2K, Ohrid, Macedonia, 2000, pp. 94-99.
15. Meeker D., Finite Element Method Magnetics version 3.1. User's Manual, 2002.
16. MATLAB 5.2. Statistic Toolbox User's Guide v. 2.1, Mathworks, 1998.
17. MATLAB 5. Optimization Toolbox User's Guide, Mathworks, 1997.

OPTIMIZATION OF PERMANENT-MAGNET DC MOTORS USING ORTHOGONAL ARRAYS

Kemal Hot and Petar Bodlović
Electrical Engineering Department, Polytechnic of Zagreb, Konavoska 2, 10000 Zagreb, Croatia, e-mail: khot@public.srce.hr, pbodlovi@vtszg.hr

Abstract: The performance of permanent-magnet DC motor can be improved by optimization of its magnetic circuit taking into account important effects of magnetic density distribution in motor air-gap and neutral zone. The optimization can be achieved by means of finite element method in field and torque calculation and statistical analysis using orthogonal arrays. The electromagnetic torque as an objective function was calculated directly from magnetic vector potentials of moving conductors as an average value through a slot pitch shift. A fast improvement in shape of some motor parts was achieved by a very limited number of field and torque calculations. The optimization results verified this approach in modeling of DC motors with permanent magnets.

Key words: optimization, orthogonal arrays, permanent magnet, DC motor

1. INTRODUCTION

The design and development of various electric machines include the methods of optimization described in [1] and some recent methods of genetic algorithm, experiment planning, quality engineering, etc. The design is governed by the aim of maximum efficiency in electromechanical energy conversion with minimum needs of material and space.

An optimal magnetic field distribution giving the acceptable local magnetic saturation and permanent magnetization is essential prerequisite in order to reach this aim. Such a field distribution can be achieved using appropriate optimization method together with numerical field calculation

M. Rudnicki and S. Wiak (eds.), Optimization and Inverse Problems in Electomagnetism, 269–276.

method and parametric sensitive geometric model. With presented methods and technique it is possible to calculate magnetic field and optimal shapes for arbitrary parts of a permanent-magnet DC motor.

Parametric CAD modeling software can generate the geometry of the motor for determined parameter values. The magnetic field analysis program calculates nonlinear FEM model prepared by CAD program and passes the electromagnetic torque values to an optimization procedure based on a method introduced by G. Taguchi and known from quality engineering [2].

Our motor optimization package contained three major parts: parametric (CAD) pre/post-processor, electromagnetic field (FEM) solver and numerical optimization program (TAM). The mathematical method and software accessories must fulfil the conditions outlined in recent papers [3,4]. The governing equations included also the space with permanent magnetization and nonlinear magnetic properties. Every solution was consisted of a spatial distribution of the magnetic vector potential throughout the considered domain. The electromagnetic torque was calculated according to previously introduced procedure [5]. Basic equations, figures and diagrams for calculated PMDC motor are presented comparing with the recent results [6].

2. ORTHOGONAL ARRAYS AND OBJECTIVE FUNCTION

We choose a set of field calculations where we change the settings (levels) of the parameters we want to study from one calculation to another. The data from all calculations in the set taken together are analysed to determine the effects of the various parameters. The use of special matrices called orthogonal arrays [1] allows the effects of several parameters to be determined very efficiently. As the name suggests, the columns of an othogonal array are mutually orthogonal. Consider the linear form

$$L_i = w_{i1} \cdot \eta_1 + w_{i2} \cdot \eta_2 + w_{i3} \cdot \eta_3 + \ldots + w_{ik} \cdot \eta_k \tag{1}$$

which is a weighted sum of k calculations. Recall that the linear form L_i is called a contrast if the weights add up to zero

$$w_{i1} + w_{i2} + w_{i3} + \ldots + w_{ik} = 0 \tag{2}$$

Two contrasts, L_i and L_j, are said to be orthogonal if the inner product of the vectors corresponding to their weights is zero

$$w_{i1} \cdot w_{j1} + w_{i2} \cdot w_{j2} + w_{i3} \cdot w_{j3} + \ldots + w_{ik} \cdot w_{jk} = 0 \tag{3}$$

The orthogonality is interpreted in the combinatoric sense, i.e. for any pair of columns all combinations of parameter levels occur in an equal number of times. This is a balancing property that implies orthogonality.

The columns of the standard orthogonal arrays are commonly arranged in increasing order of the number of changes. The number of times the level of a parameter has to be changed in running the calculations in the numerical order is less for the columns on the left when compared to the columns on the right hand side. Hence we should assign a parameter whose levels are difficult to change to columns on the left hand side.

We use a quadratic quality loss function and a larger-the-better characteristic (continuous and non-negative)

$$L(y) = A_0 \cdot \Delta_0^2 \cdot \left(1/y^2\right) \tag{4}$$

where A_0 is the loss at the tolerance Δ_0 for variable y. We would like the characteristic to be as large as possible and we do not have any adjustment factor. The objective function to be maximized in this case is given by

$$\eta = -10 \cdot \log_{10}\left(\frac{1}{n} \cdot \sum_{i=1}^{n} \frac{1}{y_i^2}\right) \cong -10 \cdot \log_{10}\left[\frac{1}{\mu^2} \cdot \left(1 + 3 \cdot \frac{\sigma^2}{\mu^2}\right)\right] \tag{5}$$

where μ and σ are the mean and variance of the quality characteristic. As we need to focus our attention on (μ^2/σ^2), it is common practice to call it S/N ratio (because σ^2 is the the effect of noise factors and μ is the desirable part of the calculated data) and take log transform to express it in decibels.

3. MAGNETIC FIELD AND TORQUE CALCULATION

The vector potential is defined by $B = rot\ A$. Using Stokes theorem in two-dimensional domain we obtain

$$\frac{1}{\partial x}\left(\frac{1}{\mu_y}\frac{\partial A}{\partial x}\right) + \frac{1}{\partial y}\left(\frac{1}{\mu_x}\frac{\partial A}{\partial y}\right) = -(J + J_m) \tag{6}$$

where J is real current density and J_m is equivalent permanent magnet excitation through current defined by

$$J_m = \oint \left(M_{0x}/\mu'_x + M_{0y}/\mu'_y\right) dl \tag{7}$$

Quasi-Poisson equation (6) is commonly solved by finite element method (FEM) with prescribed boundary conditions. A very small shift of the rotor $d\alpha$ will cause the change of conservative (magnetic) field energy that must be equal to the difference of input (electrical) and output (mechanical) energy

$$dW_m = u \cdot i \cdot dt - T_e \cdot d\alpha \tag{8}$$

The electromagnetic torque has to be independent on the current change during a small shift of rotor and hence given by

$$T_e = -\frac{\partial W_m}{\partial \alpha} + i \cdot \frac{\partial \Psi}{\partial \alpha} \tag{9}$$

where $\Psi = u{\cdot}dt$ is a linked flux. The magnetostatic treatment of the field can be applied after any discrete shift of rotor and its energy is

$$W_m = \frac{1}{2} \int_V \vec{A} \cdot \vec{J} \cdot dV \tag{10}$$

and the linked flux is

$$\Psi = l \cdot \sum_{j=1}^{N} \sum_{k=1}^{w'} \left(A'_{j,k} - A''_{j,k} \right) \tag{11}$$

where A' and A'' are the average vector potentials of the left and the right hand side of a coil in proper (two electrically nearly diametrical) slots, l is axial length of the slots and $w' = w/N$ where w is total number of conductors and N is number of slots all around the armature. If the winding has $2{\cdot}a$ parallel branches and p poles, the voltage E_i induced during the shift $\Delta\alpha$ at speed n (rpm) is

$$E_i = \frac{p}{a} \cdot \frac{n}{60} \cdot l \cdot \frac{\alpha_p}{\Delta\alpha} \cdot \sum_{i=1}^{w} \left(A'_i - A''_i \right) \tag{12}$$

where α_p is the angle of the pole pitch. As the current I is assumed the only independent variable one can obtain a total electromagnetic torque

$$T_e = \frac{p}{\pi} \cdot \frac{I}{2 \cdot a} \cdot l \cdot \frac{\alpha_p}{\Delta\alpha} \cdot \sum_{j=1}^{N} \sum_{k=1}^{w'} \left(A'_{j,k} - A''_{j,k} \right) . \tag{13}$$

4. OPTIMIZATION PROCEDURE AND RESULTS

As an example we have analyzed a PMDC two-pole motor with 12 armature slots and 192 turns. Its magnetic field at load is shown in Fig.1 with material boundaries in the middle cross-section and geometry parameters we have chosen. Consider the project where we are interested in determining the effect of five shape parameters: A is thickness of magnet holding part, B is housing thickness, C is tooth width, D is inner slope of tooth-tip overhang and E is armature core thickness. Suppose for each parameter four levels are chosen to cover the range of interest.

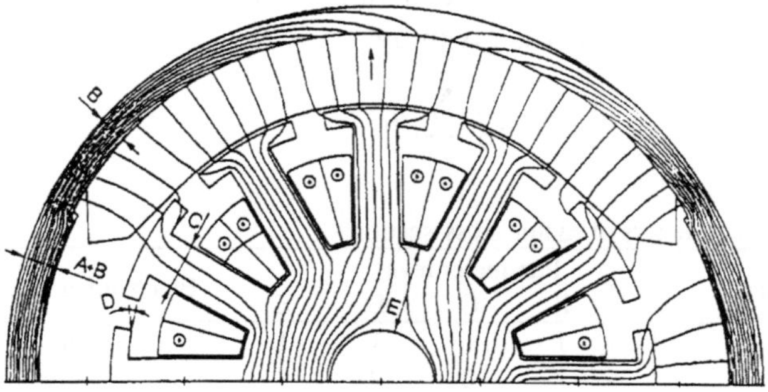

Figure 1. Motor cross-section with shape parameters

The dimensions of ceramic permanent magnets (inner 57.0 mm and outer diameter 71.1 mm), steel shaft (10.5 mm) and air gap (2 mm) were constant for the manufacturing reasons. The magnets were radially excited and already shaped to minimize problems in current commutation and vibrations. The air gap is limited by tolerances of sliding bearings that are self-adjusted to shaft direction.

The parameter and their levels are listed in Table 1 for four levels in steps (mm or deg) that are equal and it is not a limitation by itself. These parameter levels define the experimental region and our goal is to determine the best setting for each parameter so that the electromagnetic torque is maximized.

Table 1. Shape parameters of motor cross-section

Level #	1	2	3	4
A (mm)	0.00	0.50	1.00	1.50
B (mm)	3.15	3.45	3.75	4.05
C (mm)	4.00	4.60	5.20	5.80
D (deg)	0.0	6.0	12.0	18.0
E (mm)	7.00	8.00	9.00	10.00

The matrix experiment selected for this problem is based on standard orthogonal array L_{16} and given in Table 2. It consists of 16 individual calculations corresponding to the 16 rows. The five columns of the matrix represent the five parameters as indicated in the table. The entries in the matrix represent the levels of the parameters. Thus, calculation 1 is to be conducted with each parameter at the first (starting) level. The settings of calculation 1 can also be referred to concisely as $A_1B_1C_1D_1E_1$.

The electromagnetic torque was evaluated under the 16 shape conditions with parameters given in an orthogonal array (Tab. 2). The magnetic field calculations were used according to [5]. Further analysis was performed by the statistical technique based on equation (5) to determine the contribution of each parameter. The overall mean value of η for the region defined by the parameter levels in Table 1 is given by

$$m = \frac{1}{16} \cdot \sum_{i=1}^{16} \eta_i \quad (14)$$

Table 2. Shape parameters of motor cross-section

Calculation #	Parameter level A	B	C	D	E	Electromagnetic torque (Nm)
1	1	1	1	1	1	$y_1 = 0.719$
2	1	2	2	2	2	$y_2 = 0.792$
3	1	3	3	3	3	$y_3 = 0.863$
4	1	4	4	4	4	$y_4 = 0.926$
5	2	1	2	3	4	$y_5 = 0.834$
6	2	2	1	4	3	$y_6 = 0.835$
7	2	3	4	1	2	$y_7 = 0.893$
8	2	4	3	2	1	$y_8 = 0.893$
9	3	1	3	4	2	$y_9 = 0.861$
10	3	2	4	3	1	$y_{10} = 0.888$
11	3	3	1	2	4	$y_{11} = 0.879$
12	3	4	2	1	3	$y_{12} = 0.899$
13	4	1	4	2	3	$y_{13} = 0.873$
14	4	2	3	1	4	$y_{14} = 0.881$
15	4	3	2	4	1	$y_{15} = 0.882$
16	4	4	1	3	2	$y_{16} = 0.887$

By examining columns A to E of the orthogonal array in Table 2, we observe that all four levels of every parameter are equally represented in 16 calculations. Thus, *m* is balanced overall mean over the entire investigated region. The effect of a parameter level is defined as the deviation it causes from the overall mean. Let us examine how the calculated data can be used to evaluate the effect of tooth thickness C at level 3, i.e. case C3. We had this case for calculations 3, 8, 9 and 14. The average S/N ratio for these calculations is given by

$$m_{C3} = (\eta_3 + \eta_8 + \eta_9 + \eta_{14})/4 \quad (15)$$

and the effect of tooth thickness is given by ($m_{C3} - m$). From Table 2, observe that for four mentioned calculations (3, 8, 9 and 14) every of other parameters take levels 1, 2, 3 and 4. So the quantity m_{C3} represents an average η when the tooth thickness is at level C3 where the averaging is done an a balanced manner over all levels of each of other four parameters.

The results are presented as S/N ratio for every parameter at every level (Tab. 3 and Fig. 2). The S/N ratio's overall mean is –1.300 dB. The decrease of S/N ratio for parameter A (thickness of magnet holding part) in level 4 can be explained by strongly increased leakage flux of permanent magnet ends.

Table 3. Signal/noise ratios and parameter effects

Parameter		A (mm)	B (mm)	C (mm)	D (deg)	E (mm)
S/N ratio (dB)	Level 1	-1.709	-1.730	-1.648	-1.485	-1.492
	Level 2	-1.294	-1.430	-1.404	-1.327	-1.354
	Level 3	-1.094	-1.135	-1.165	-1.232	-1.237
	Level 4	-1.103	-0.904	-0.982	-1.156	-1.116
Effect (%)		24.30	37.69	24.50	5.89	7.59

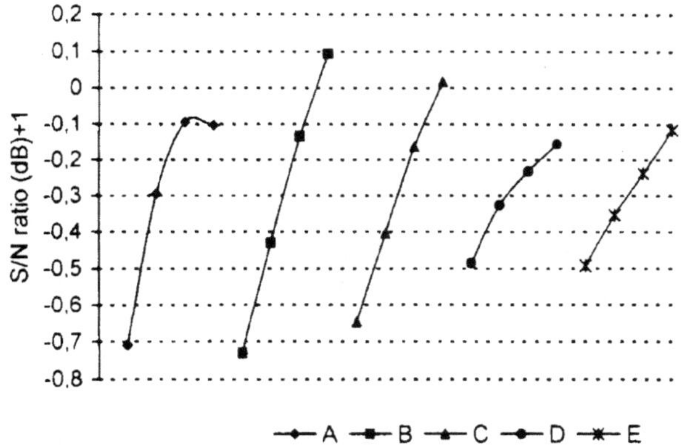

Figure 2. Signal/noise ratio for parameter levels

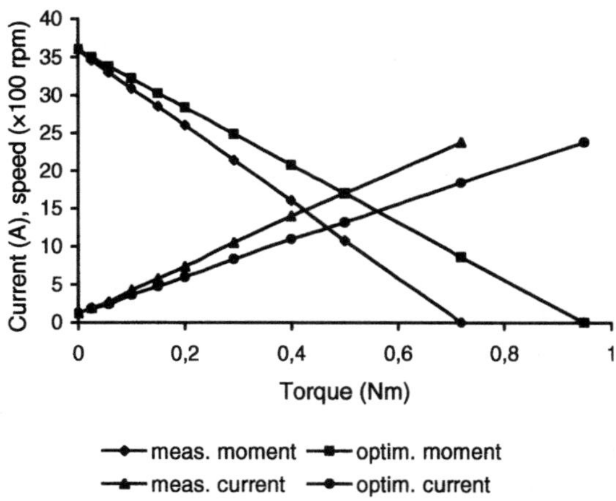

Figure 3. Speed-torque and current-speed characteristics

The total influence of the parameter change was calculated as design effect for every parameter from Table 1. Hence, in the case of additional limitations one is free to chose another combination of parameters than that larger-the-better characteristic (levels A3, B4, C4, D4 and E4) giving maximum torque 0.942 Nm. It is obviously that the inner slope of tooth-tip overhang D and armature core thickness E have smaller influence in electromagnetic torque than the other parameters. The influence of every parameter can be re-calculated per millimeter or degree in design analysis according to Table 1.

5. CONCLUSIONS

The optimal geometry was determined by Taguchi method using the combination of values given in an appropriate table. Instead of a number of possible combinations for five parameters and four discrete values (levels) for each of them (4^5=1024), one had to consider only 16 patterns defined by L_{16} orthogonal array. The optimal shape had to be selected from given patterns in 16-dimensional design space. The effect of each parameter change was estimated by means of FEM calculation and Taguchi method from the point of view of quality engineering. Signal-to-noise (S/N) ratio for each calculated combination of parameters was evaluated by the analysis using FEM to get results of starting electromagnetic torque as an objective function, Taguchi method to reduce the number of calculations and standard statistical analysis. S/N ratio of every level on every parameter had to be calculated. The relative influence effect (in dB) was presented for all parameters. The optimal model from given levels was selected on S/N ratio's maximization and confirmed by an additional calculation (maximum torque increased 31%).

REFERENCES

1. G. N. Vanderplaats, Numerical optimization techniques for engineering design, McGraw-Hill, New York, 1984
2. G. Taguchi and S. Konishi, Orthogonal arrays and linear graphs, ASI Press, Dearborn (MI), USA, 1987.
3. B. J. Reece and T. W. Preston, Finite element methods in electrical power engineering, Oxford: Oxford University Press, pp. 195-200, 2000
4. Benhama, A. C. Williamson and A. B. J. Reece, Force and torque computation from 2-D and 3-D field solutions, IEE Proceedings – Electric Power Application, Vol. 146, pp. 25-31, 1999
5. K. Hot and P. Bodlovic, Direct approach in performance calculation on permanent-magnet DC motor, 10th Int. Symp. on Electromagnetic Fields in El. Eng., Cracow (Poland), pp. 55-58, 2001
6. T. Tsunoda, Y. Ishihara, T. Todaka and K. Hirata, Geometry optimization of DC motor by Taguchi method, 10th Int. Symp. on Electromagnetic Fields in El. Eng., Cracow (Poland), pp.489-492, 2001

IDENTIFICATION OF FIELD MODEL'S PARAMETERS IN THE TURBINE GENERATOR END ZONE USING MEASURED DATA

Ilona Iatcheva, Rumiana Stancheva

Department of Theoretical Electrical Engineering, Technical University of Sofia, 1156 Sofia, Bulgaria, E-mail: ilona@ea.vmei.acad.bg, rds@vmei.acad.bg

Abstract: The present paper deals with an identification problem in the end zone of turbine generator. Parameters of the mathematical model of the coupled - electromagnetic and temperature fields have been determined, using measured temperature data on the surface of the flange and fingers in the end zone. Different modes (rated, at the leading power factor and at the lagging power factor) are considered. Coupled field problem - electromagnetic and temperature field distribution in the end zone of a 200 MW turbogenerator is studied using FEM for forward problem. The secondary model of the temperature values at points with measuring control has been obtained using combination of response surface methodology and design of experiments. The identification problem has been solved as optimisation one. The optimisation criterion used was the quadratic sum of the differences between the computed and measured temperature data.

Key words: Identification, optimisation, response surface methodology, design of experiments, finite element method, coupled problems, turbine generators, end zone.

1. INTRODUCTION

When large turbine generators work at the leading power factor, leakage fluxes in the end zone increase. This leads to increase in the eddy currents and losses in the conductive parts in this zone and can cause overheating and failures. These facts are one of the main restricting factors for optimal use of generators to work at variable regimes. That is way the detailed determination of the electromagnetic and temperature field distribution in the end zone of the turbogenerator working under different operating conditions

M. Rudnicki and S. Wiak (eds.), Optimization and Inverse Problems in Electomagnetism, 277–286.

is very important. The solution of this problem is solution of a coupled field problem – electromagnetic and thermal. Nevertheless, the main part of the works devoted to this task deal with only one of the problem parts - electromagnetic [1,2] or temperature [3,4] distribution analysis.

In this work, investigation has been carried out dealing with coupled field problem using finite element method (FEM). Eddy current losses in conductive parts, obtained in electromagnetic field analysis have been applied as heat sources for the thermal analysis. But in solving the thermal task there are some problems with correct definition of the boundary conditions. The reason is that it is difficult to clearly define convection coefficients along different parts of the boundary. These coefficients can vary in significant way depending on many factors [4] – cooling media properties and its moving direction, temperature, shape and properties of the boundary region, heat flow direction and other constructive features of the machine. Usually the convection coefficients are determined on the basis of laboratory experiment [3,4] or according empirical formulas. But in this way it is not possible to take into account the specific characteristics and changes in particular working machine. The aim of this paper is correct determination of the convection coefficients for the end zone of a 200 MW turbogenerator, by solving an inverse problem using measuring temperature data at the control points shown in Fig.1. This is an identification problem and it has been solved as optimisation one. The optimisation criterion used was the quadratic sum of the differences of the computed and measured temperature data at the control points. Computed temperature values in optimisation procedure have been calculated using mathematical model based on Response Surface Methodology (RSM) and the Design of Experiments (DOE).

2. FORWARD PROBLEM FORMULATION

The axial section of the 200MW turbine generator end region is shown in Fig.1. The aim of forward problem formulation is to constitute mathematical models of the temperature at the control points. As these mathematical models will be applied in solving the inverse problem – identification of convection parameters, there are two main requirements for them:

- they have to be significantly depending on the convection conditions of the different parts of the cooled stator core end surface (Fig.2).
- they have to be easy applicable in optimisation procedure.

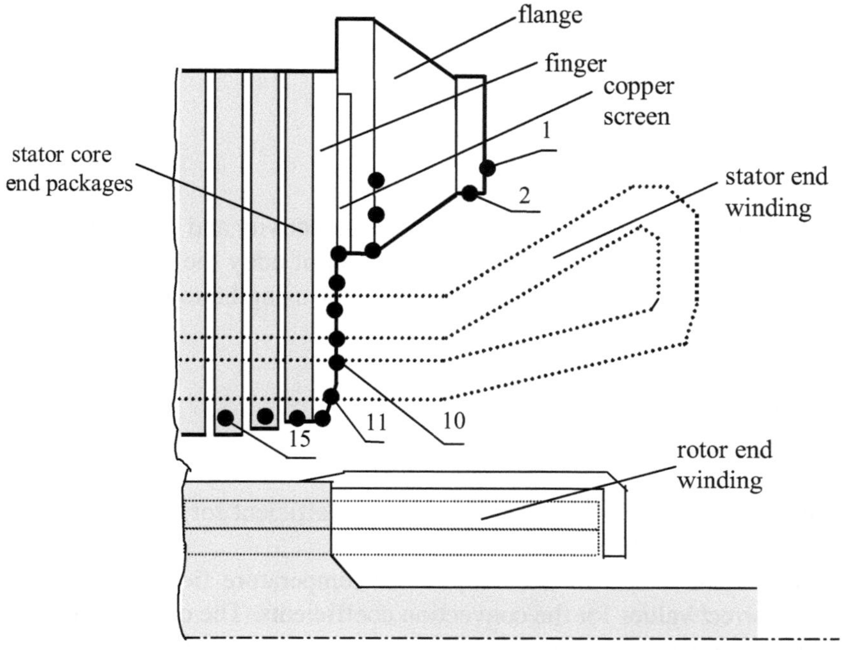

Figure 1. Axial section of the investigated turbogenerator end region with control points

3. THEORETICAL FIELD MODELS

The problem to be analysed is a coupled field problem – electromagnetic and thermal.

The electromagnetic field is analysed as quasi-three-dimensional and time-harmonic taking into account the actual currents of the three-phase stator and rotor windings, anisotropy in the tooth zone, eddy currents in the conductive media and the generator state (load angle and power factor) [5,6]. Thus the describing equations are:

$$\begin{aligned} &\operatorname{rot}\left(\hat{\nu}\operatorname{rot}\vec{A}\right)=\hat{\sigma}\vec{E}+\vec{J}_e \\ &\vec{E}=-\operatorname{grad}V-\frac{\partial\vec{A}}{\partial t} \end{aligned} \tag{1}$$

where $\vec{A}$ is magnetic vector potential, $\vec{E}$ is electric field intensity, V is the scalar electric potential, ν and γ are tensors due to magnetic and electric anisotropy of the medium.

The temperature field is solved as axisymmetric [7] and the governing equations are:

$$div(\lambda grad T) + Q = 0$$

$$Q = \frac{[\sigma]^{-1} JJ^*}{2} \qquad (2)$$

where T is the temperature, λ is the thermal conductivity and Q is the density of the heat sources obtained after determination of eddy the current density J. The cooling conditions are taken into account using boundary conditions of the type:

$$-\lambda \frac{\partial T}{\partial n} = \alpha_k (T - T_0) \qquad (3)$$

In the expression (3) α_k is the convection coefficient for different region boundaries and T_0 is the ambient temperature.

An important factor for accuracy of the temperature field modelling is applying correct values for the convection coefficients. The cooled boundary is divided into parts (Fig.2.) and different convection coefficients for each part are assigned. Along each part the convection coefficients are assumed to be constant.

Both electromagnetic and temperature fields are studied using FEM with a code specially developed for this purpose. A finite element mesh of 11880 elements and 6097 nodes has been applied for solving the problem.

The secondary model of the temperature value at each $j^{\text{-th}}$ control point $\hat{T}^j$ has been obtained using combination of response surface methodology and design of experiments [8]: approximation of the temperature values in considering points depending on the convection coefficients α_k (k=1,4) and on the working mode (defined with active and reactive power P and Q). Thus the secondary models are second orders polynomial expressions of the following type:

$$\hat{T}^j = b_0^j + \sum_{k=1}^{p} b_k^j h_k + \sum_{k=1}^{p} b_{kk}^j h_k^2 + \sum_{\substack{k=1 \\ r=k+1}}^{p} b_{kr}^j h_k h_r \qquad (4)$$

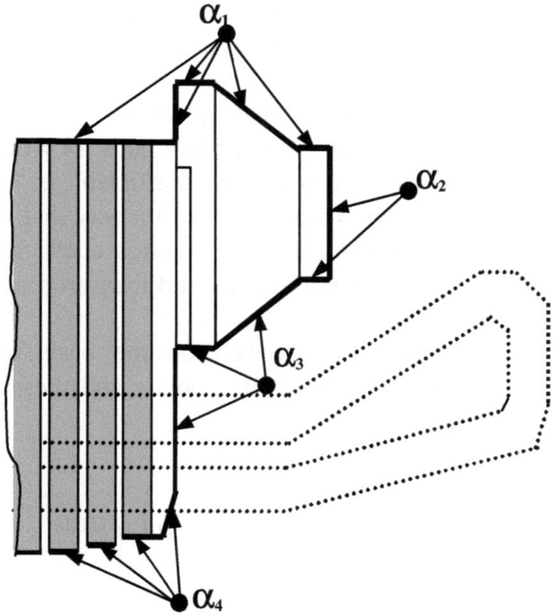

Figure 2. Convection coefficients for different parts of the cooled boundary

In (4), design variables $h_1 \div h_4$ correspond to the convection coefficients α_k used in normalised units and h_5, h_6 are design variables which are relevant with the working mode.

Unknown coefficients b_0, b_k, b_{kk} and b_{kr} are found using the method of least squares. Number of design variables is p (in this case $p = 6$). The experiments used for obtaining these models were numerical FEM analysis of the electromagnetic and temperature fields in the end region and temperature determination in control points. In the experiments different reasonable [3,4] values for convection coefficients have been used. The number of FEM experiments when using Central Composite Design (CCD) was:

$$M = 2^p + 2p + 1 = 77.$$

4. IDENTIFICATION PROBLEM

The identification problem – determination of the exact convection coefficients has been solved as optimisation one by means of non-linear programming.

The goal of optimisation is to find vector $\vec{\alpha}$ - which components are convection coefficients for different boundary parts $\vec{\alpha}=(\alpha_1, \alpha_2, \alpha_3, \alpha_4)$. For each separate part (Fig. 2) convection coefficient is considered to be constant. The optimal vector is sought in such a way that when applying in the thermal analysis its components as convection coefficients the quadratic sum of the differences between computed and measured temperature values at the control points was minimal.

The objective function is described on the basis of the secondary temperature value models at the control points and measured temperature data at the same points:

$$F\min(\vec{\alpha}) = \left\{ \sum_{k=1}^{m} \sum_{j=1,}^{n} \left(\widehat{T}_k^j - T_{k,measuredd}^j \right)^2 \right\} \min \tag{5}$$

In (5) j is the control point number, k is number of the working mode, $\widehat{T}_k^j$ is the calculated temperature value and $\widehat{T}_{k,measured}^j$ measured data for the temperature.

The optimisation was carried out under following constraints for the vector $\vec{\alpha}$ components values:

$$\begin{aligned} 300 &\le \alpha_1 \le 900 \\ 100 &\le \alpha_2 \le 400 \\ 100 &\le \alpha_3 \le 400 \\ 300 &\le \alpha_4 \le 900 \end{aligned} \tag{6}$$

The measured data values for different operating modes used in the task are shown in Table 1.

5. RESULTS

The study has been carried out for 200 MW turbogenerator with additional thermo-resistors posed at control points on the surface of the flange and fingers in the end zone (Fig.1). Different modes: rated, at the leading power factor and at the lagging power factor have been investigated (Table 1). Special attention has been paid to the most dangerous regimes at the leading power factor.

Table 1. Investigated modes

mode N	1	2	3	4	5	6	7	8	9	10
P, MW	200	180	160	200	180	160	200	180	160	140
Q, MVAr	80	80	80	40	40	40	40	0	0	0
mode N	11	12	13	14	15	16	17	18	19	20
P, MW	140	140	140	140	160	200	180	180	160	140
Q, MVAr	80	40	0.0	-20	-42	-20	-22	-40	-40	-57

During the process of creating secondary field model FEM analysis of electromagnetic and temperature field distribution has been done using different combinations of design variables ($h_1 \div h_6$), corresponding to convection coefficients and operating mode. Design of experiment has been used for obtaining of these combinations. The distribution of the eddy current amplitude J_m at rated mode is shown in Fig. 3a in axial direction and in Fig. 3b in radial direction for the first stator core end package. Eddy current losses in conductive parts, obtained in electromagnetic field analysis have been applied as heat sources for the thermal analysis.

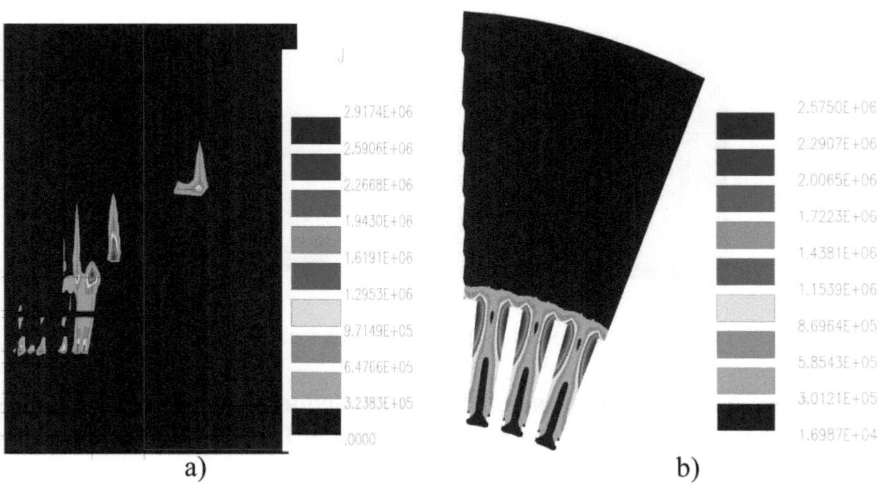

Figure 3. Eddy current distribution in the end zone under rated mode:

a) axial direction b) radial direction

The results obtained from the solution of the identification problem are the values of convection coefficients at the different boundary parts of the stator core end surface of the investigated 200 MW turbine generator:

$\alpha_1 = 620\ W/m^2.K$

$\alpha_2 = 130\ W/m^2.K$

$\alpha_3 = 300\ W/m^2.K$

$\alpha_4 = 680\ W/m^2.K$

Using the obtained convection coefficients temperature distribution in the end zone of the considered turbo generator was obtained for different operating modes. Temperature field distribution in axial section for mode N 20 is shown in Fig.4.

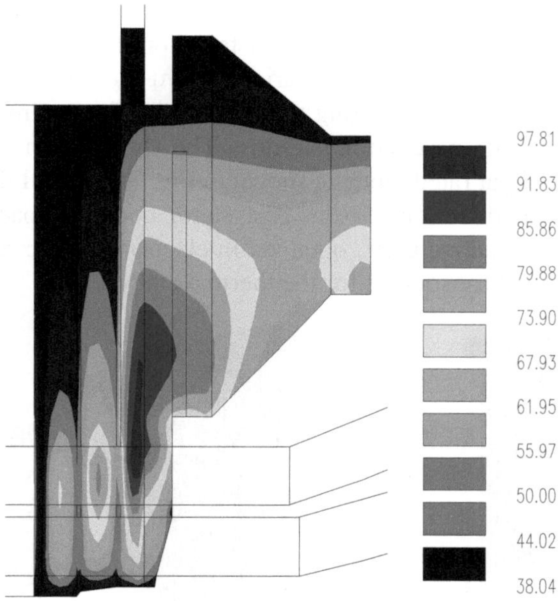

Figure 4. Temperature field distribution for mode N 20

Comparison between measured and calculated temperature data at points 1,2,10 and 11 for the investigated modes shown in Table 2. The results are in good agreement.

In the last row of Table 2 the relative error between the measured and the calculated temperature values are given according to the expression:

$$\varepsilon^j = \frac{1}{m}\sum_{k=1}^{m}\frac{\left|T^j_{k,calculated} - T^j_{k,measured}\right|}{T^j_{k,measured}} \tag{7}$$

Table 2. Measured and calculated temperature values at control points.

mode N	T, ^{0}C point 1 meas.	T, ^{0}C point 1 calc.	T, ^{0}C point 2 meas.	T, ^{0}C point 2 calc.	T, ^{0}C point 10 meas.	T, ^{0}C point 10 calc.	T, ^{0}C point 11 meas.	T, ^{0}C point 11 calc.
1	66.8	71.9	63.1	70.4	52.5	52.2	42.9	48.3
2	65.0	70.7	60.0	69.0	53.9	51.4	43.2	47.6
3	59.1	67.6	55.1	66.3	52.3	49.9	43.2	46.5
4	78.3	73.9	70.2	72.2	58.5	52.9	49.1	48.7
5	69.7	71.5	60.2	69.9	54.4	51.6	45.2	47.7
6	62.4	69.0	56.7	67.5	53.5	50.3	43.4	46.7
7	76.2	76.8	63.3	74.8	53.0	54.2	48.1	49.6
8	70.5	73.9	61.5	72.0	50.1	52.6	40.9	48.3
9	68.7	71.2	56.6	64.5	48.2	51.1	40.7	47.2
10	60.3	65.3	57.2	64.1	53.8	48.7	46.5	45.6
11	62.4	66.6	58.2	65.2	55.2	48.9	46.2	45.6
12	64.0	69.1	59.0	67.4	54.7	49.9	46.6	46.3
13	65.8	70.4	62.0	68.7	53.0	50.4	44.8	46.6
14	68.2	73.0	64.7	71.0	52.2	51.6	44.0	47.4
15	74.0	73.5	70.2	71.6	56.2	52.1	48.0	47.9
16	88.0	79.7	78.6	77.5	58.5	55.5	47.9	50.4
17	77.2	75.3	68.9	73.4	57.3	53.2	47.7	48.7
18	80.7	77.6	66.6	75.5	54.9	54.3	45.3	49.5
19	84.2	75.1	79.9	73.1	54.0	52.9	46.8	48.4
20	79.7	74.2	77.0	72.1	53.4	52.1	54.7	47.7
ε, %	6.50		11.56		5.23		6.51	

6. CONCLUSION

An identification problem - correct determination of the convection coefficients in the end zone of 200 MW turbogenerator has been solved using measured temperature data in control points on the surface of the flange and fingers.

Coupled field problem - electromagnetic and temperature field distribution in the end zone of the machine was studied using FEM for forward problem. Different modes (rated, at the leading power factor and at the lagging power factor) are considered. Secondary temperature field models have been obtained by combination of the Response Surface Methodology and the Design of Experiments. Each experiment was numerical FEM analysis experiment.

The identification problem has been solved as optimisation one. The optimisation criterion used was the quadratic sum of the differences of the computed and measured temperature data.

The calculation of the temperature at the control points using the obtained convection coefficients gives satisfactory accuracy.

REFERENCES

1. V. C. Silva, G. Meunier and A. Foggia, "A 3D Finite-Element Computation of Eddy Currents and Losses in the Stator End Laminations of Large Synchronous Machines", *IEEE Trans. Magn.*, vol. 32, No. 3, pp.1569-1572, 1996.
2. B. C. Mecrow, A. G. Jack and C. S. Cross. Electromagnetic design of turbogenerator stator end regions. *IEE Proceedings*, Part C, vol. 136, No. 6, pp. , Nov. 1989.
3. Schastlivyi G.G, G.M.Fedorenko,V.I.Vigovskyi, "Turbo and hydro generators under variable regimes", Kiev, Naukova dumka, 1985. (in Russian)
4. Krushinskyi A.G.,3D temperature field analysis using method of component's dividing., Tehn. Elektrodinamika, No1,1989, pp74-79. (in Russian)
5. Stancheva, R.D., I.I.Iatcheva, Investigation of electromagnetic field in the end regions of turbine-generator taking into account the anisotropy, Electrotehnic and Electronic,No.1, 1995, pp.6-10.(in Bulgarian)
6. Iatcheva I. I., R. D. Stancheva. Eddy Current Distribution in the End Region of a Large Turbine Generator Under Different Operating Conditions. Proc. 9th Int. Symp. IGTE 2000, Graz, Austria, 2000 , pp. 219-222.
7. Iatcheva I. I., Temperature field investigation in the end zone of large turbine generator under rated mode, Electrotehnic and Electronic, No.5-6, 2001, pp.34-38.(in Bulgarian)
8. Stancheva R., I. Iatcheva. Optimal Design of the End Zone of Powerful Turbogenerator. Proc. 5th Int. Conf. on Applied Electromagnetics PES 2001, Nis, Yugoslavia, 2001, pp. 147-150.

PERMANENT MAGNET MACHINE OPTIMIZATION
by using FEM and Sensitivity Analysis Techniques

Georgios D. Kalokiris, Antonios G. Kladas and John A. Tegopoulos
Laboratory of Electrical Machines, Electric Power Division
Department of Electrical & Computer Engineering, National Technical University of Athens
9, Iroon Polytechneiou Str.,15780 Athens, Greece.
E-mail: tegopoul@power.ece.ntua.gr, tel: +3-210-7723766, fax: +3-210-7723593

Abstract: The paper presents electrical machine design considerations introduced by exploiting new magnetic material characteristics. The materials considered are amorphous alloy ribbons as well as Neodymium alloy permanent magnets involving very low eddy current losses. Such materials enable electric machine operation at higher frequencies compared with the standard iron laminations used in the traditional magnetic circuit construction and provide better efficiency.

Key words: Finite element method, iron losses, permanent magnet machines, equivalent circuit parameters.

1. INTRODUCTION

Recently, technological advances in new materials are very promising for the development of new electrical machines structures exhibiting increased performance [3]. So, the permanent magnets have been extensively used to replace the excitation winding in synchronous machines with the advantages of simple rotor design without field windings, slip-rings and exciter generator, avoiding heat dissipation in the rotor and providing higher overall efficiency [5]. The rotor design can be distinguished in three main types according to the magnet position, namely the "interior" [4], the "peripheral" [2] and the "claw pole" type. In this paper the peripheral type machine structure has been adopted [1],[6].

M. Rudnicki and S. Wiak (eds.), Optimization and Inverse Problems in Electomagnetism, 287–294.

At the stator side, the materials considered are amorphous alloy ribbons involving very low eddy current losses. Such materials enable electric machine operation at higher frequencies compared with the standard iron laminations used in the traditional magnetic circuit construction and provide better efficiency.

2. NEW MATERIAL CHARACTERISTICS

Standard silicon iron laminations which have dominated magnetic circuits at low frequency applications (50 Hz) present disadvantages in high frequencies enabled by power electronic converter developments (a few kHz) due to the high iron losses associated. Such losses are mainly due to eddy currents as the lamination width is no more compatible with the respective skin effect depth.

In this range of frequencies ferrites and amorphous alloy ribbons seem to provide important advantages due to their low eddy current losses. Technological developments enabled production of such materials with high saturation induction approaching 1 T and very reduced coercive force providing the possibility of conception of high speed actuators with very attractive performance.

Moreover technological advances in permanent magnet materials enabled production of Neodymium alloy magnets with magnetization attaining 1.25 T and coercive force of 100.000 A/m.

Combination of such materials, that is high magnetization permanent magnets in the rotor parts and amorphous alloy ribbons in the stator parts, is expected to enable new machine designs, with light structure, high rotor speed and increased efficiency. Such machines will be supplied by convenient power electronic converters, enabling high frequency current generation. Moreover, sinusoidal waveforms are no more needed, and design procedure will tend to simplify the construction of the machine.

3. PERMANENT MAGNET MACHINE DESIGN

Under the above mentioned considerations a permanent magnet machine could be constructed by using simply concentric coils for stator winding constitution and U shaped amorphous ribbons for the stator magnetic circuit as shown in Fig. 1.

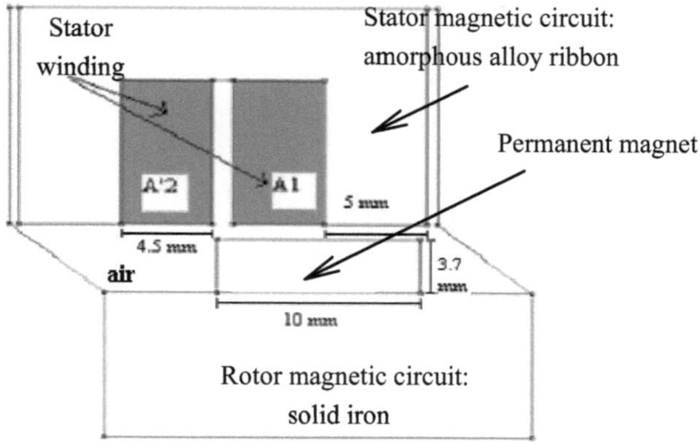

Figure 1. Geometry of one pole part of the permanent machine considered

A method for determination of the circuit parameters of the permanent magnet machine has been introduced by using the electromagnetic field calculation [7]. The field analysis has been performed by using a two dimensional finite element code. The geometry of the examined machine is illustrated in figure 1. It involves one pole part of a multipole permanent magnet synchronous machine structure. The total resistance of stator winding can be calculated by the following relations :

$$R = \rho \frac{l}{S} \tag{1}$$

where ρ is the electric resistivity of copper, l the conductor total length of each phase and S the conductor cross section. The winding length l can be estimated from relation (2):

$$l = 2*(l_{ax}+l_p)*\ N_w\ *P \tag{2}$$

where l_{ax} is the machine's axial length (m), l_p is the polar step length (m), N_w is the total number of series connected stator winding turns and P is the number of stator pole pairs.

In a first step, a simulation is used without the presence of the permanent magnet in order to calculate the magnetizing flux Φ_m and inductance L_m by means of the vector potential value A_{gap} in the middle of the air-gap as follows:

$$\Phi_m = \int_{S_1} \mathbf{B.dS} = \int_{C_1} \mathbf{A.dl} \cong (2\ A_{gap}).\, l_0 \tag{3}$$

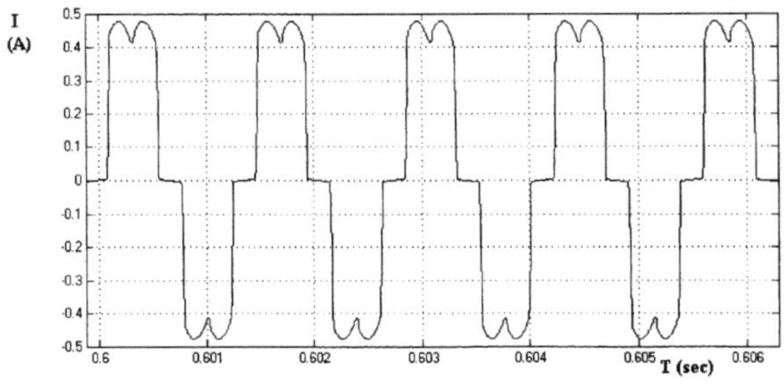

a

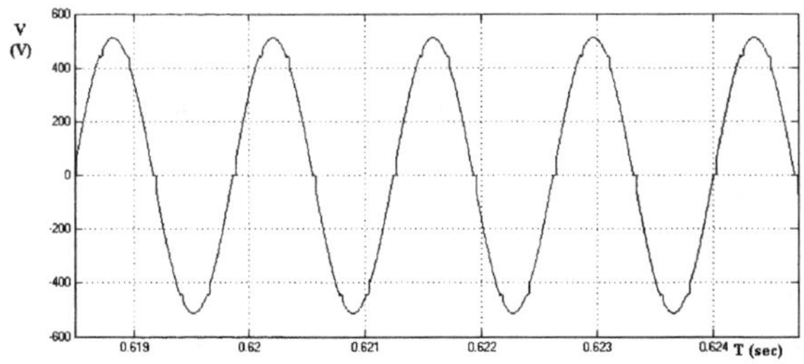

b

Figure 2. Permanent magnet machine supplying a resistive load through a rectifier
a: phase current time variation
b: phase voltage time variation

$$L_m = \frac{N_w^2 \Phi_m}{I_s} \cdot P \tag{4}$$

where l_0 is the length of the magnetic circuit and I_s the stator winding current.

The stator leakage flux $\Phi_{ls}$and leakage inductance terms L_{ls} are deduced by means of the vector potential value A_{slot} in the middle of the slot as follows:

$$\Phi_{ls} = \int_{S_2} \mathbf{B.dS} = \int_{C_2} \mathbf{A.dl} \cong \left(A_{slot} - A_{gap}\right) \cdot l_0 \tag{5}$$

$$L_{ls} = \frac{N_w^2 \Phi_{ls}}{I_s} \cdot P \tag{6}$$

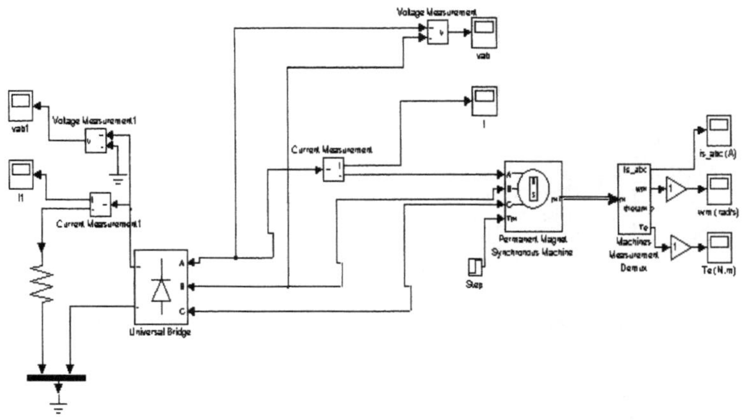

Figure 3. Configuration of the standard machine supplying a resistive load through a rectifier

The calculated parameter values are tabulated in table 1.

Table 1. Calculated equivalent circuit parameters for the 18-pole permanent magnet machine

L_l(mH)	$L_{,m}$(mH)	R (Ω)
0.081	0.123	0.37

As an example of application of such a machine, it has been considered the case of supply of a resistive load by means of a six pulse rectifier. The phase current and voltage time variations are shown in Fig. 2.

In this simulation sinusoidal electromotive force time variations have been considered by using the standard toolbox of Matlab/Simulink software shown in Fig. 3. This approximation is not realistic for all machine loadings as will be shown in the next paragraph.

4. RESULTS AND DISCUSSION

The permanent magnet machine design procedure described previously has been applied to construct a 2.5 kW, 26 rad/sec generator to be connected to a six pulse diode rectifier bridge. The air-gap width has been chosen 1 mm while a multipole "peripheral" machine structure has been adopted (figure 4). Figure 4a gives the geometry employed for the two dimensional finite element program of the machine involving approximately 2100 nodes 4000 triangular elements. The corresponding magnetic field distribution, at no load operating condition, is shown in Fig. 4b.

The electromotive force of the machine at no load predicted by the finite element method is shown in Fig. 5a while the measured one of the constructed prototype is shown in Fig. 5b, respectively.

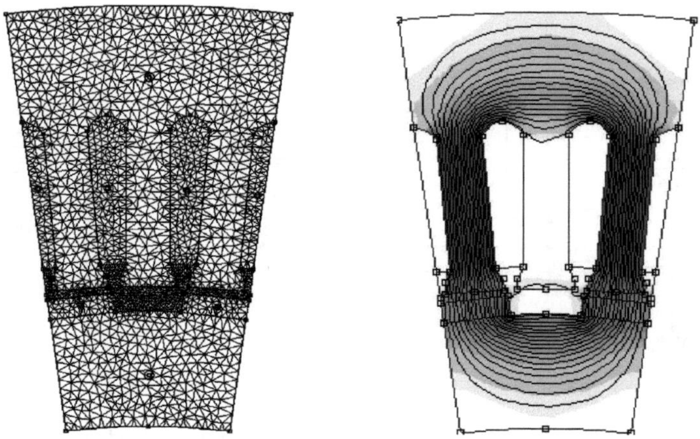

a b

Figure 4. One pole part of the permanent magnet machine constructed.
a: Employed triangular mesh
b: Field distribution at no load

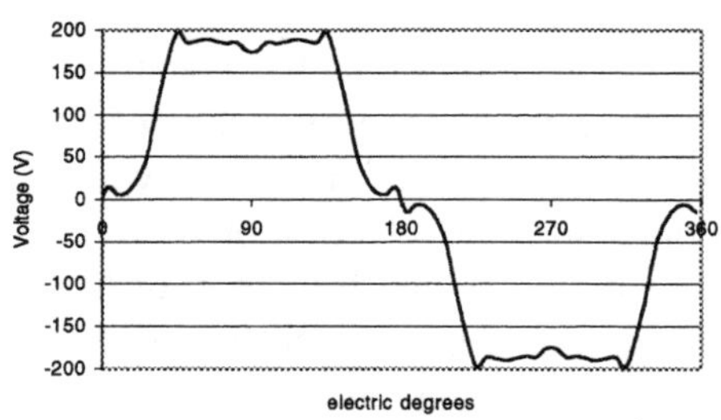

a

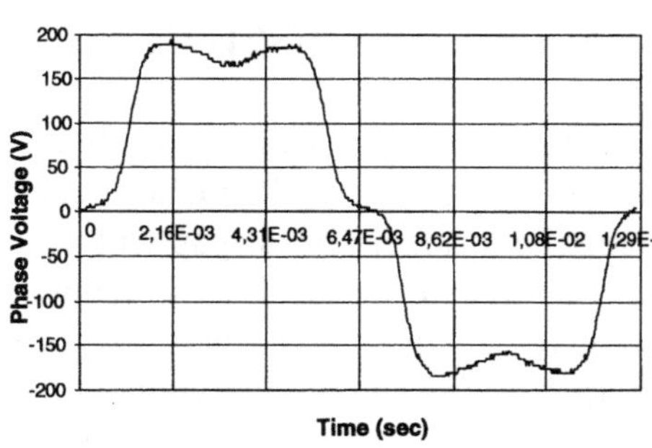

b

Figure 5. Voltage per phase (no load case)
a: simulated result
b: experimental result

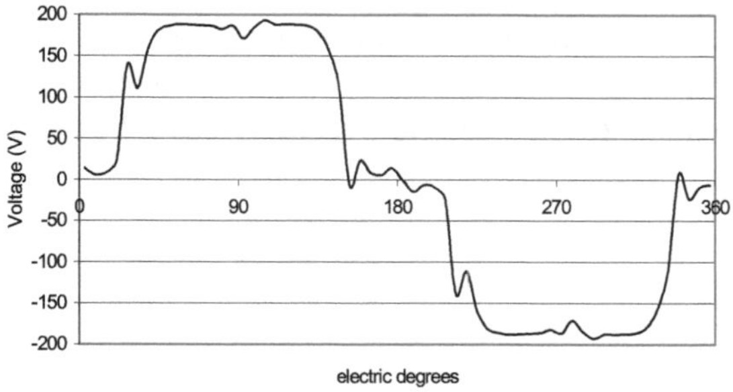

a

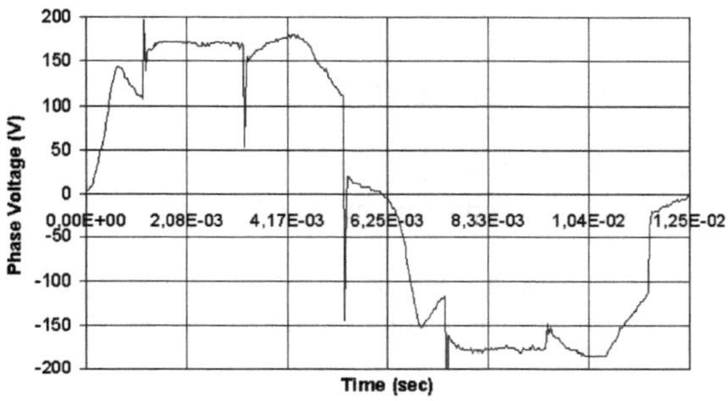

b

Figure 6. Voltage per phase (loaded machine case)
a: simulated result
b: experimental result

These figures illustrate that the simulated results are in good agreement with the measurements. Moreover, it may be noted that the sinusoidal electromotive force time variation assumed by the d-q axis transformation representation of the machine is not realistic.

The case of resistive loading of the machine through a six pulse diode rectifier has been considered. The simulation results for a load current of 1.5 A are shown in Fig. 6. Fig. 6a shows the simulation results for the phase voltage derived by the finite element method while Fig. 6b gives the measured results for the same conditions. Once again the good agreement between simulation and measurements can be observed.

5. CONCLUSION

The paper presented a design methodology for high efficiency machine design, by using new magnetic materials. Amorphous alloy ribbons and Neodymium alloy permanent magnets have been implemented involving very low eddy current losses.

An important advantage of such machines is that high speed operation can be introduced, avoiding risks of demagnetization of the permanent magnets. The proposed design involves standard formulae and finite element analysis. The results for a specific machine have been checked through measurements.

REFERENCES

1. Aliprantis D., Papathanassiou S., Papadopoulos M., Kladas A.: Modeling and control of a variable-speed wind turbine equipped with permanent magnet synchronous generator, Proceedings International Conference on Electrical Machines, Helsinki, Finland, pp. 558-562, 2000.
2. Higuchi T., Oyama J., Yamada E., Chiricozzi E., Parasiliti F. and Villani M.: Optimization procedure of surface permanent magnet synchronous motors, IEEE Transactions on Magnetics, Vol. 33, no 2, 1997, pp. 1943-46.
3. Jahns T.: Motion control with permanent magnet AC machines, IEEE Proceedings, Vol. 82, No 8, 1994, pp. 1241-52.
4. Marchand C., Ren Z., Razek A.: Torque optimization of a buried permanent magnet synchronous machine by geometric modification using FEM, Proceedings EMF'94, Leuven, pp. 53-56, 1994.
5. Lovatt H., Watterson P.: Energy stored in permanent Magnets, IEEE Trans. on Magnetics, Vol. 35, no 1, 1999, pp. 505-507.
6. Toba A., Lipo T.: Generic torque maximizing design methodology of surface permanent magnet Vernier machine, IEEE Transactions on Industry Applications, Vol. 36, no 6, 2000, pp. 1539-1546.
7. Tsekouras G., Kiartzis S., Kladas A., Tegopoulos J: Neural Network Approach compared to Sensitivity Analysis based on Finite Element Technique for Optimization of Permanent Magnet Generators, IEEE Transactions on Magnetics, Vol. 37, no 5/1, 2001, pp. 3618-3621.

OPTIMIZATION UNDER CONSTRAINTS OF STATIC CONVERTERS

Application to a flyback structure in PFC mode

C. Larouci, J.P.Ferrieux, L.Gerbaud, J.Roudet

Laboratoire d'Electrotechnique de Grenoble, CNRS UMR 5529 INPG/UJF, ENSIEG BP 46, 38402 St Martin d'Hères cedex, Grenoble, France, e-mail : Jean-Paul.Ferrieux@inpg.fr, Laurent.Gerbaud@inpg.fr, James.Roudet@inpg.fr

Abstract: The paper deals with an approach to size and optimize static converters using analytical models and a gradient optimization algorithm. The aim is to carry out a compromise between the computation rapidness and the result accuracy, which allows to treat more criteria in the same optimization process than a numerical or simulation approach. This method is applied to optimize the volume of a PFC (Power Factor Correction) flyback converter under EMC (electromagnetic compatibility), loss and thermal constraints. In this way, an analytical model is developed for each optimization aspect. However, these models are often described by discontinuous functions which are built from manufacturer data (transformer winding wire diameter, magnetic circuit volume...). To make easier the use of these models in an optimization approach with a gradient algorithm under constraints, specific tools are developed and integrated in an adequate optimization environment: Pro@design. The optimization results show that the optimal volume depends on both the converter parameters to be optimized (the input parameters) and the imposed constraints (the criteria). These optimization results are validated by experiment

Key words: Sizing, Optimization, Power Electronics, PFC

1. INTRODUCTION

The paper deals with the sizing of static converter with time multi-scale. Several aspects have to be taken into account: losses in the components – mainly in the transformer-, EMC and volume constraints. For that, a

M. Rudnicki and S. Wiak (eds.), Optimization and Inverse Problems in Electomagnetism, 295–302.

compromise has to be carried out between the computation rapidness and the result accuracy, to allow to treat numerous criteria in the same optimization process than a numerical or a simulation approaches.

In this way, gradient optimization processes under constraints have been chosen. To perform many criteria, an analytical model has to be developed as it will be presented in the first part of the paper.

Due to the optimization algorithm, the criteria have to be continuous as their derivative according to the model input, as detailed in the second part. However, this is not always possible, so, the paper proposes solution to carry out the discontinuities.

2. OPTIMIZATION PROCESS

The optimization process is represented on figure 1. From its input parameters (I_j), the sizing model has to give the sizing criteria (O_i) and their gradients (dO_i/dI_j), and also an objective function (Fobj) to be minimized, as shown in Figure 1. This model can be used by any optimization process for sizing, by constraining the I_j and O_i.

An input parameter has to be fixed or constrained in an interval of values. An output parameter has to be constrained in the same way than an input parameter or to be ignored (i.e. calculated but not considered by the optimization algorithm).

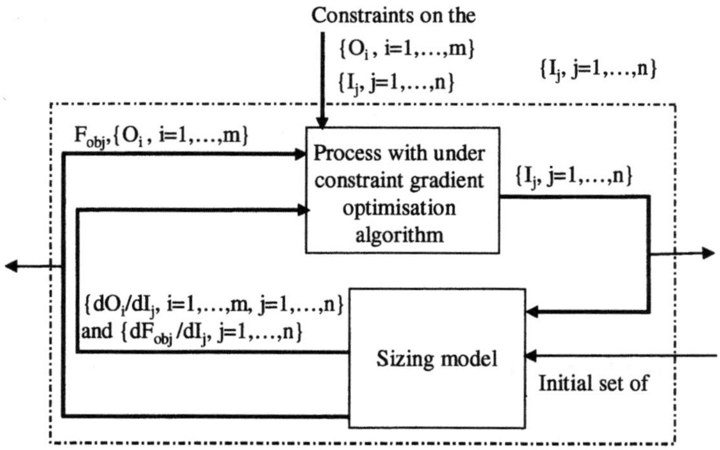

Figure 1 : Optimization and sizing model by using components with partial derivatives

To give the derivative, an other structuration of the model is possible using the differential, as proposed by an environment such Pro@design (Eden) [1].

3. FLYBACK ANALYTICAL MODELS

A flyback structure in PFC mode is selected as a significant support of our study because it contains the majority of aspects which interest the power converter designer (volume, losses, EMC and thermal performances...) [2, 3, 4].

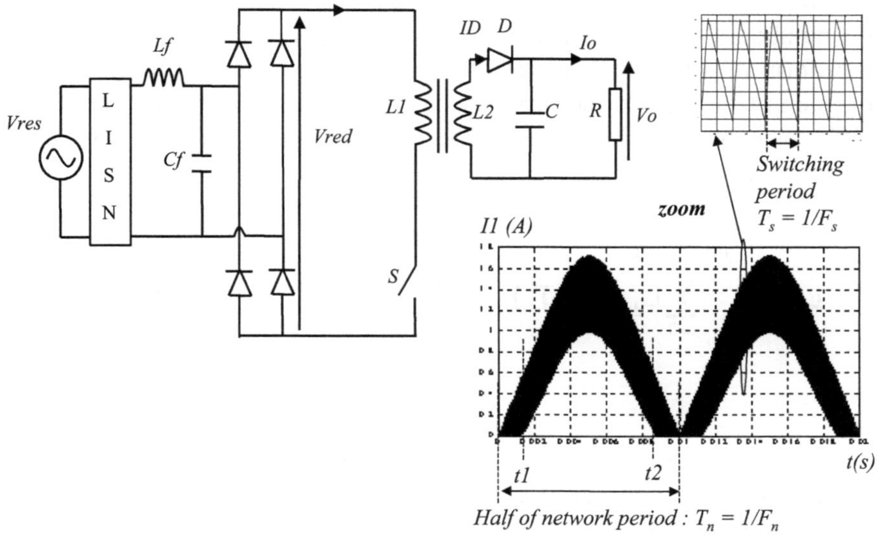

Figure 2 : Designed flyback converter

This static converter is presented in figure 2. It is characterized by an operating mode with several time scales as detailed in the following section.

The following analytical models are developed to take into account these different aspects in the optimization approach. These models are validated by numerical simulations and measures. Now, the paper details the specificity of this model.

3.1 The duty cycle analytical model

This model has been developed to carry out the evolving of the current, without using long simulation, as it should be with numerical simulation in tools like PSpice or Saber. Indeed, the operating of this static converter consider several time scales : the network frequency (Fn = 50Hz), the switching frequency (Fs =50KHz) and the EMC standard frequency (until 30MHz). For that, an analytical expression representing the control of switch S has been developed [5, 6, 7]. This allows to give the duration of every state of the converter. This one operates in both continuous mode and discontinuous mode.

3.2 The EMC spectrum analytical model

From the previous operating model, a symbolic model using Laplace transforms can be used to give the EMC model. It is made of series which the number of terms depends on the rate Fs/Fn [8]. For example, the Laplace transform of a variable expression such the transformer primary current is the sum of the transformer primary current on every switching period, with different expression according to the continuous and discontinuous part of the signal :

$$L(p) = \sum_{i=0}^{\max\left(\frac{t1}{T_s}\right)-1} L_{dc}(p,i\cdot T_s) + \sum_{i=\max\left(\frac{t1}{T_s}\right)}^{\max\left(\frac{t2}{T_s}\right)-1} L_{cc}(p,i\cdot T_s) + \sum_{i=\max\left(\frac{t2}{T_s}\right)}^{\frac{T_n}{2\cdot T_s}} L_{dc}(p,i\cdot T_s) \tag{1}$$

where t1, t2, Ts and Tn are defined in figure 1, and max(x) represent the nearest first higher integer of x. As Ts is a sizing parameter, the term number of each sum evolves discontinuously according to the switching period Ts, what is incompatible with the used gradient optimization algorithm.

3.3 The total volume analytical model

The volumes of the components are deduced from r.m.s. and maximal values estimated from the symbolic expressions given by the analytical model representing the evolution of the variables. This also uses abacus, so discontinuous expressions, with possible stair shape. Data tables also exist, and require one or two dimensions interpolation to be used in the sizing model. Such tables are used to give the volume of a capacitor according to its capacitance and its useful voltage, or the size of an inductor according to its inductance and nominal current.

3.4 The loss analytical model

Analytical models have been developed for the semiconductor losses, core and copper losses in the transformer [6]. These analytical models may use series and Bessel functions.

4. COMPUTING METHOD AND DEVELOPED TOOLS

4.1 Optimization environment

The model has been implemented in the Pro@design environment [11]. This environment proposes to translate an analytical model defined by symbolic equations, into a computation component called COB (see figure 3).

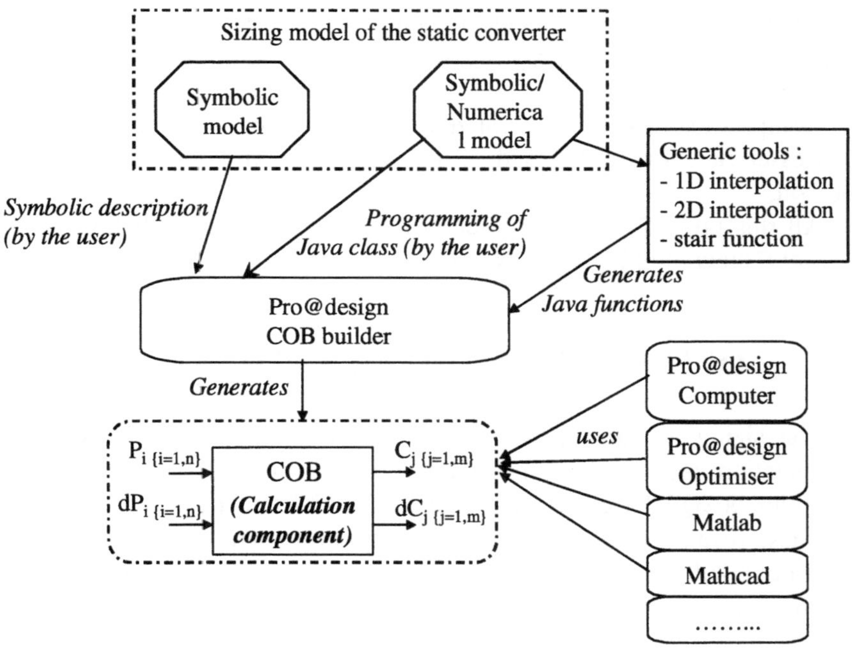

Figure 3 : Implementation of a sizing model in the Pro@design environment

4.2 Particular aspects of the models

The parts of the models that are not described by symbolic equations are defined by using Java classes.

In this way, to carry out tables in the sizing model, and to model parts described by stair shapes, specific tools for interpolation in one dimension and in two dimensions and to carry out stairs have been developed and integrated in the Pro@design optimization environment [1]. From a data table, they allow to generate the corresponding functions (in Java code) which allow to estimate, continuously, the function values and their differentials.

In addition, guides are proposed to help the designer to describe its sizing models in Java classes. This concerns series like (1), mainly to assure their continuity according to numbers of terms in the series, that are changes continuously by the optimization process but in fact cannot be considered as real.

Indeed, these terms are not necessarily integers due to the non synchronism of the network frequency and the switching frequency. The proposed solution is to consider the average value of the two sums given by the next lower and higher integer values of these terms, by changing the switching frequency [9]. This is also applied on the partial derivatives and gives good results.

Complex expressions are treated by separating imaginary and real parts.

Bessel functions (BER and KER) are approximated by an asymptotic method.

All this programming is presently made by the user, with Java classes, possibly integrating FORTRAN or C procedures.

5. OPTIMIZATION AND EXPERIMENT RESULTS

The sizing was made by using the previous models and tools in an optimization process (see the main result in table 1). The optimization algorithm is SQP, based on the gradients : VF13 from Harwell library [10].

These results are obtained around 1 minute for an optimization session. So, around 200 iterations of optimization (computing of the criteria and their differentials) are done.

The optimization results show that the optimal volume depends on both the converter parameter values and the constraints to be respected. These results are confirmed by experiments (as shown on figure 4 and figure 5).

Table 1. Optimization results

Parameters	Optimization results
L1(mH)	1.75
Switching frequency (kHz)	40
Transformer coefficient	0.34
Lf(mH)	0.4
Cf(μF)	0.56
Total volume (cm3)	44

Total losses (W)	2.99

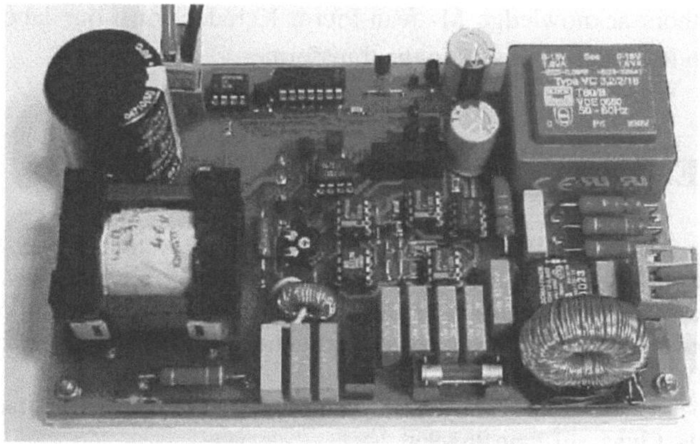

Figure 4 : Benchmark made from the optimization results

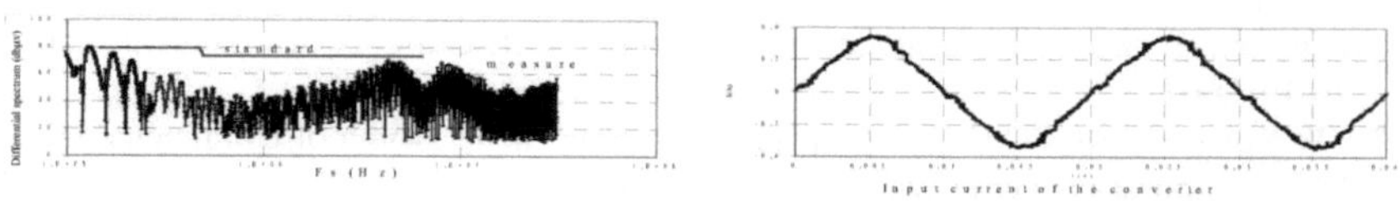

Figure 5 : Experimental results measured on the benchmark in figure 4

6. CONCLUSION

The paper has presented a way to size power electronics converters. Now, the modeling of some parts, like the CEM model and the losses expressions, may be automated. In this way, some procedures have to be developed to help the designer to build his models (see section 4). Optimization procedures to treat the discrete aspect have also to be combined with the present used deterministic optimization process. This will allow to treat the data or discrete size.

ACKNOWLEDGEMENTS

The authors acknowledge M. Jean-Pierre Keradec form our laboratory, for its contribution to model the losses in the transformer.

REFERENCES

[1] E.Atienza, J.Bigeon, F.Wurtz, B.Belhabib « Steps to an Electrical Design Environment », IEEE-IECON'99, San Jose, CA,USA, November 29th-December 3rd 1999, pp 815-820.

[2] L.A.Barcia, A.Fontan, A.Huertas, J.Gonzalez, J.Ara, A.Fernandez : Low power ac/dc on-board converters for rectified ac power distribution demonstrator for telecommunication, APEC'00, CD ROM proceeding.

[3] L.Yang, L.CQ : Variable duty-ratio control of flyback for power factor correction, IPEMC'94, China, vol 2, pp 963-968.

[4] B.V.Borges, V.Anunciada : A new single phase factor corrector converter, APEC'00, CD ROM proceeding

[5] R.Nalepa, N.Barry, P.Meaney : Primary side control circuit of a flyback converter, APEC'01, pp 542-547.

[6] C.Larouci, J.P.Keradec, J.P.Ferrieux, L.Gerbaud, J.Roudet, "Copper losses of Flyback transformer: Search for analytical expressions", IEEE-CEFC'2002 - Perugia, Italy, June 16-19, 2002

[7] JW.Kolar, H.Ertl, FC.Zach : A novel three-phase single-switch discontinuous-mode AC-DC buck-boost converter with high-quality input current waveforms and isolated output, IEEE-Transactions on Power Electronics, vol 9, 1994, pp 160-172.

[8] J.Lempinen, T.Suntio : Small-signal modeling for design of robust variable-frequency flyback battery chargers for portable device applications, APEC'01, vol1, pp 548-554.

[9] Chérif Larouci, Eric Atienza, Jean-Paul Ferrieux, Laurent Gerbaud, James Roudet, "Developed tools for an optimization approach. Application to optimise a flyback structure volume under EMC and loss constraints", IEEE-IECON'2001, Denver, Colorado, USA, Thursday, Nov 29 to Sunday, Dec 2, 2001

[10] VF17, Harwell library, http://www.cse.clrc.ac.uk/nag/hsl/

[11] http://www.designprocessing.com/

NONDESTRUCTIVE QUALITY TESTING OF HIGH TEMPERATURE SUPERCONDUCTING BULK MATERIAL USED IN ELECTRICAL MACHINES AND MAGNETIC BEARINGS

Ryszard Palka, Hardo May, Wolf-Rüdiger Canders
Institute of Electrical Machines, Traction and Drives, Technical University of Braunschweig
Hans Sommer-Str. 66, 38106 Braunschweig, Germany

Abstract: This paper deals with a non-destructive examination method for high temperature superconductors (HTSC's) based on the scanning of the magnetic field distribution which is trapped by a superconductor. The identification of the positions, dimensions and orientations of sub-domains within the HTSC –which are determining the quality– has been reduced to the determination of the critical current density (J_c) distribution within the HTSC using above mentioned measurements. This inverse field problem has been solved by an appropriate numerical algorithm based on extended finite element formulations.

Key words: quality control, high temperature superconductors, current density identification, electrical machines, magnetic bearings

1. APPLICATION OF HTSC'S IN ELECTRICAL MACHINES

The use of HTSC material to improve the performance of electrical machines has been described in many papers [1,2,4,5]. For the excitation coils of direct-current (DC) and synchronous (SM) machines as well as for the armature windings of all types of electrical machines the use of HTSC-conductors (strands) contribute to the reduction of the machine losses. Using their field trapping capability HTSC-bulks can be considered for the improvement of the excitation of DC-machines and SM's as well. Furthermore

M. Rudnicki and S. Wiak (eds.), Optimization and Inverse Problems in Electomagnetism, 303–312.

HTSC-clusters arrangements are of favourable utilisation for hysteresis and mixed-μ reluctance motors [5]. The excellent data of all types of these "Superconducting Machines" have been achieved by assuming ideal states of the HTSC's with infinite J_c's surrounding the surface of the HTSC-block in form of a current sheet (Meissner phase) or trapping the excitation field perfectly [1]. Under the conditions of real applications of HTSC's for energy converters, they are exposed high counter flux densities resulting in field dependent finite J_c's (Shubnikov phase). Furthermore the processing of the HTSC bulks suffers from the achievable sizes (cross section and thickness) and their inhomogeneity (multi-domains). Thus it has been of a great importance to develop methods and calculation codes capable to identify all these imperfections of today's manufactured HTSC-blocks. This is necessary for the development of the proposed energy converters with improved performances.

Following figures illustrate different applications of HTSC-bulks in electrical machines as for excitation (trapped field) or for parasitic field reduction (magnetic screens). *Fig. 1a* shows the classical rotating switched reluctance machine used for high speed applications e.g. To achieve remarkable force densities, this machine type has to be operated at very high flux densities (1.8-2.0T). A portion of this flux penetrates into the slot region (stray-flux) thus considerably reducing the force output. If an ideal HTSC is placed in the rotor slot (*Fig. 1b*), the screening effect will expel the field completely from this region and thus increasing the force and power.

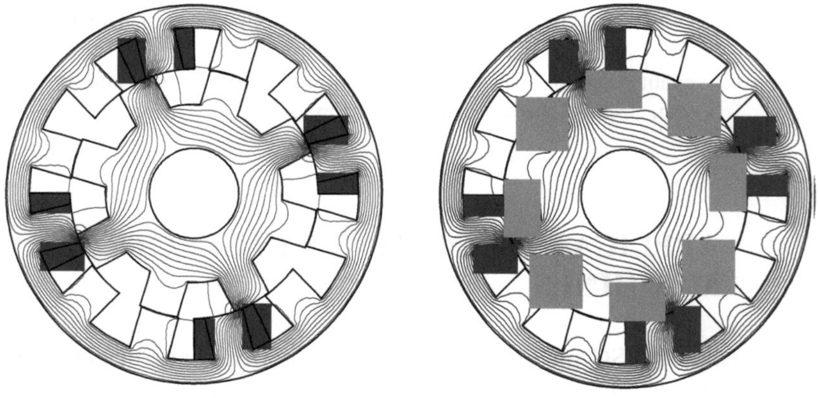

a) Rotor without HTSC screens resulting in increased stray field

b) Rotor with HTSC screens

Figure 1. Magnetic field distribution of a switched reluctance machine

The flux concentrating arrangement shown in *Fig. 2* is known as a further superconducting permanent magnet (SPM) excited SM design. The excitation system generates a horizontal directed flux, which is collected by the iron poles adjacent on both sides of the SPM and guided to the air gap of the machine. *Fig. 2a* shows the field plot of a SPM in the Meissner state (loaded case, one pole pitch). As a further improvement feasibility a HTSC bulk is placed as a magnetic screen in the centre of the iron-pole thus reducing the armature reaction. This simple measure leads to a remarkable increased power output (*Fig. 2b)*.

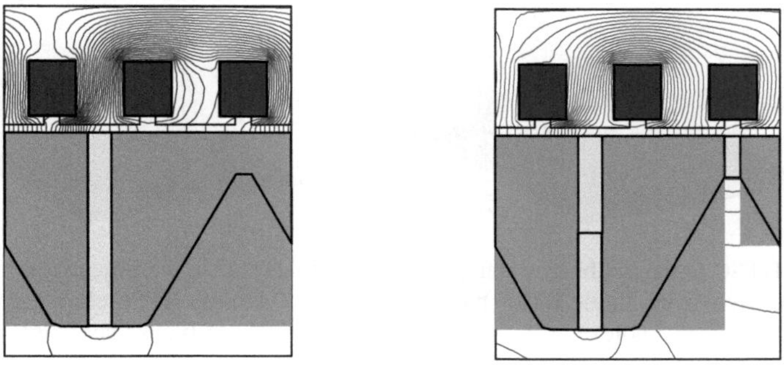

a) SPM in the Meissner state

b) SPM in the saturated Shubnikov state. Additional HTSC-screens in the centre of the iron-pole

Figure 2. Comparison of SM's with HTSC's in a flux concentrating arrangement (one pole-pitch, linear representation)

If a SPM in the trapped flux state is used as the field excitation for a synchronous machine in a gap oriented arrangement (*Fig. 3*), the effects of the saturation state (Shubnikov phase) on the field distribution influence the performance of the converter drastically. Furthermore it has to be noticed that only three dimensional field calculations are able to take the effects of the inhomogeneous flux distribution in the air-gap (plane A) into account (*Fig. 4*). Only in the Meissner state the excitation field distribution is fairly constant within the machine width (x-direction) and can thus be determined by a 2-D field calculation code [1] in plane B of *Fig. 3*.

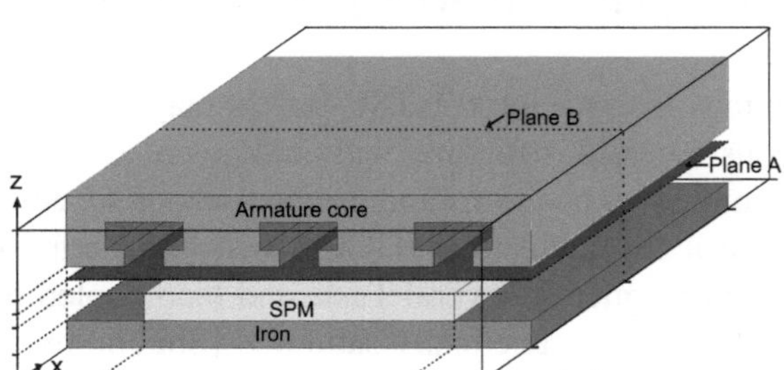

Figure 3. 3D-Calculation model for a HTSC-excited SM (linear representation, one pole-pitch). Gap oriented excitation by "Superconducting Permanent Magnets"

The results of the field calculation for two different HTSC arrangements are shown in *Fig. 4*.

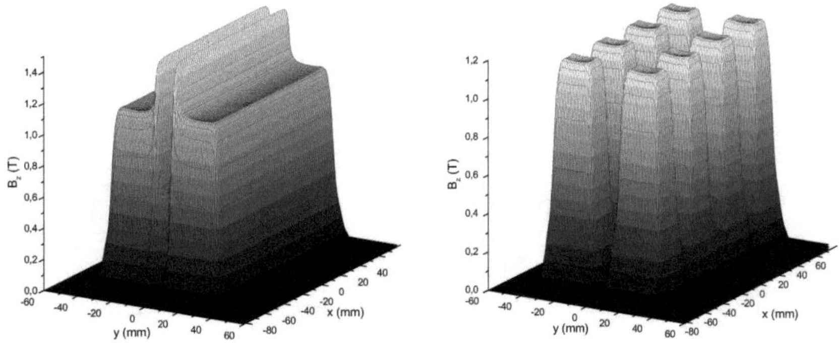

a) HTSC in the Shubnikov state, J_c= 100A/mm² (one single HTSC block per pole)

b) HTSC in the Shubnikov state, J_c= 100A/mm², HTSC cluster arrangement (8 blocks per pole)

Figure 4. Flux density distribution in plane A of *Fig. 3* (normal component)

Finally it may be concluded, that manufacturers of HTSC-bulk material are faced with two major demands. At first for SPM applications the critical current has to be increased as high as possible so that surfaces of adjacent HTSC-bulks with currents in an opposite direction compensate each other with the effect of a more or less homogenous excitation field distribution (neglecting slotting effects). A second demand on the HTSC-quality is related to the producible block-size without internal sub-domains. With respect to the total flux -that means the power of a machine- an increased block size with reduced current densities may be replaced by a cluster of several high quality SPM-blocks of minor size. The knowledge of the material parameters e.g. domain sizes and critical current densities are of a maximum importance for the design of above mentioned machines.

2. PROPERTIES OF REAL HTSC-BULKS

In large volumes manufactured HTSC-bulk's may exhibit a more or less non-homogeneous internal structure with unknown J_c's (distribution and value) and with unknown positions and orientations of superconducting sub-domains. To control the quality the HTSC's will be inserted into the warm bore of a high field magnet of several Tesla and then cooled below the critical temperature T_c. Thus magnetically saturated (Shubnikov state) the magnetic flux will be trapped within the HTSC by internal currents after the switch-off of the external field. To scan the trapped field distribution in the proximity of the HTSC surface, the measurement set-up shown in *Fig. 5* with a 3D-Hall probe has been applied. The values of all three flux density components will be used for the determination of the J_c's (distribution and value) and thus for the quality control of the HTSC-bulks.

Figure5. Magnetic field measurement set up (3D-co-ordinate table) with a 3D-Hall probe

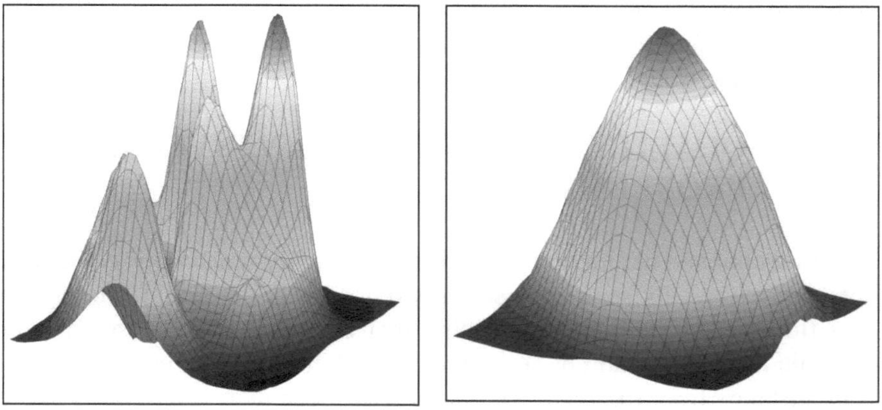

a) HTSC-bulk composed of several sub-domains

b) Single grain HTSC-bulk

Figure 6. Distribution of the magnetic field (normal component at 2 mm above the surface) trapped by HTSC-bulks of different quality in the saturated Shubnikov state

The two field distributions shown in *Figs. 6a* and *b* differ markedly as the left HTSC-bulk consists of several small sub-domains whereas the right one exhibits the trapped field of a perfect single domain high quality HTSC. Only these bulks can react on external magnetic fields in such a way as it is shown in *Figs. 7a* and *b* for an inherently stable magnetic bearing for high speed applications.

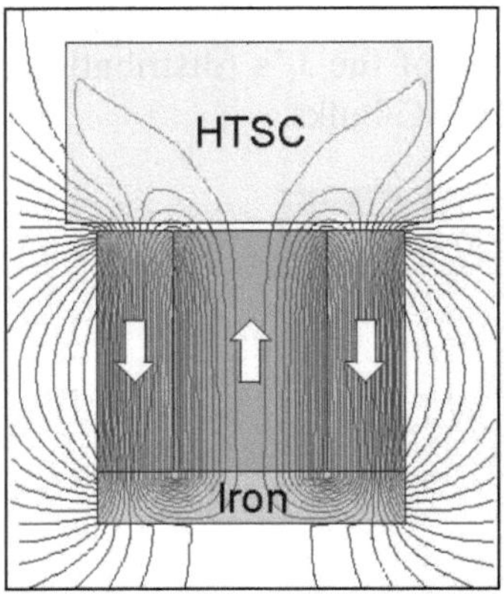

Figure 7a. Interaction between a multi-pole permanent magnet and a perfect single domain HTSC-bulk

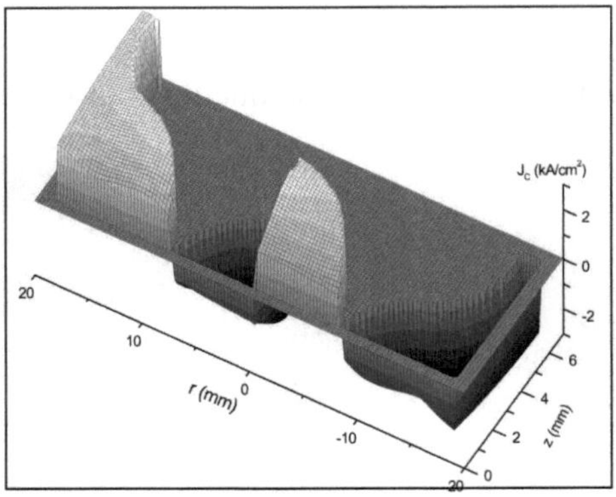

Figure 7b. Current density distribution within the HTSC from *Fig. 7a* (approach and retreat movement)

The goal of the proposed quality control is to distinguish the manufactured HTSC-bulks with respect to an increased reaction on external fields. This property is required by the optimal design of electrical machines (excitation magnets or magnetic screens *Figs. 1-4*), levitation systems (attractive or repulsive) and inherently stable contact-free magnetic bearings for high speed rotating shafts.

3. QUALITY ESTIMATION AS AN INVERSE FIELD PROBLEM

As mentioned before, the internal structure of commercially available superconducting bulk samples is strongly related with the manufacturing process (top-seeded melting), implicating that the orientation of the crystallographic a-b planes can not be ensured for arbitrarily large pellets. At a certain distance from the seed, the planes may loose their orientations and the material solidifies in disoriented directions of the original a-b-planes. As mentioned above the identification of the positions, dimensions and orientations of the superconducting domains together with the values of their internal J_c's is of a great importance. By the use of all three components of the external field in the close neighbourhood of the HTSC known from scanning measurements (*Fig. 6*), the current distribution within the sample can be identified. Such procedure constitutes a typical inverse problem and belongs to the class of the improperly posed tasks. That is, its solution may not be unique nor continuous to the input data thus the standard numerical field calculations can not be applied directly to the identification of the current density distribution within HTSC's.

To solve this problem the fundamental finite element equation set has to be extended by equation's describing the magnetic field in regions of measurements (*Fig. 8*). This leads to the following over-determined, linear and ill-conditioned set of equations with two unknowns: the vector potential values in the whole region and the current density values within the superconductor (in matrix representation):

$$\mathbf{MA} = \mathbf{U}. \tag{1}$$

Since the coefficient matrix **M** is of the size m×n (with m>n), the solution of the equation set exists in the sense of the least squares:

$$\mathbf{A_0} = \min\|\mathbf{U} - \mathbf{MA}\|^2. \tag{2}$$

This solution can be written as

$$\mathbf{A_0} = \mathbf{RU}, \tag{3}$$

where **R** denotes the Moore-Penrose pseudo-inverse [3,6]. If the matrix **M** is of the rank n, the least squares solution of Eq. 1 can be obtained by using the modified Gramm-Schmidt or Householder method. On the other hand, under the condition of rank **M**<n, a least squares solution exists, but it is not unique. In this situation it is possible to obtain the minimal least squares solution, i.e. the vector of minimum Euclidean length which minimises Eq. 2. This can be computed by the singular values decomposition of **M** that enables the calculation of the appropriate Moore-Penrose pseudo-inverse. The method consists of two steps: at first the sparse matrix **M** is reduced to the tri-diagonal form by the aid of Lanczos transformations. In the second step the singular values of this matrix are computed by the use of the QR algorithm [3]. This enables to determine the matrix **R** in the Eq. 3 and the minimal least squares solution $\mathbf{A_0}$ of the problem. It should be noted here, that - because of the possible numerical instabilities- the implementation of the above algorithm is very difficult. Therefore it is necessary do define several accuracy criterion to prevent convergence to solutions with unacceptable values. Additionally the matrix **M** should be scaled with the effect that the relative errors in its elements are all of comparable size. Such a scaling procedure helps to prevent the least squares problem from being unnecessarily sensitive to both measurement and numerical calculation inaccuracies.

At first –for testing of the proposed method– an arbitrary J_c distribution within the HTSC has been assumed and the external magnetic field has been calculated. *Fig. 8* shows the magnetised HTSC (radius 30 mm, height 15 mm) under the consideration, that the external flux densities can be measured (area of measurements $\mathbf{G_0}$ indicated in *Fig. 8*). Using these field values the current density distribution within the HTSC has been identified and compared with the assumed one. The number of nodes in the whole area (equal to the number of unknown vector potentials) has been equal to 14153. The discretisation of the HTSC into small rings with square cross-sections (0.25 x 0.25 mm) leads to additional 3600 unknown current density values. To obtain an over-determined equation set with advantageously much more equations than unknowns, it is necessary to take into account extended regions where all components of the magnetic field have to be measured.

The real superconductor is usually composed by some sub-domains which can carry the current of high magnitude and others with decayed properties with respect to their internal J_c's (drastically shown in *Fig. 6a*). As the knowledge of the continuous J_c distribution is not so essential for the force determination, thus from the practical point of view it is necessary to define the average current density values for some bigger sub-regions of the HTSC only. This simplifies the calculation process and gives the required information about the distribution of superconducting domains within the HTSC together with the individual J_c. Thus, according to the required accuracy, the superconducting region has been artificially divided into more or less separated parts (rings) with identical cross-sections and the individual mean J_c value for each individual part has been determined. By this approach the number of unknown J_c values has been advantageously reduced to a small number (e.g. 16, 25 or 36). This cuts down at the same time the extension of the measurement area leading finally to much smaller equation sets and stabilising the calculation process. The identification of the J_c distribution within the HTSC of *Fig. 8* has been performed for different sizes, positions and dimensions of the measurement region $\mathbf{G_0}$. For the above example the matrix $\mathbf{M}$ has finally the following dimensions: the number of equations m =48969 and –depending on the partitioning of the HTSC– the number of unknowns n=14169, 14178 or 14189 respectively. The following error function ρ acts as the convergence criterion for the calculations:

$$\rho = \|\mathbf{U} - \mathbf{MA}\| \leq tol\|\mathbf{M}\| * \|\mathbf{A}\|, \tag{4}$$

where tol denotes the required tolerance.

Additionally the difference between the defined (measured) A_m and calculated vector potentials A in the measurement area $\mathbf{G_0}$ was estimated:

$$\varepsilon = \iint_{G_0} (A_m - A)\,dz\,dr. \tag{5}$$

Typical values of the accuracy parameter for the above example have been: $tol=10^{-11}$, $\rho=10^{-8}$, ε=10-7 Vsm. It has been necessary to perform about of 75000 Lanczos iteration steps in order to obtain the above postulated accuracy. The field plot in the region under investigation is shown in *Fig. 8* and the J_c distributions in the cross section of the HTSC are shown in *Figs. 9a, b* and *c* respectively for different fragmentations.

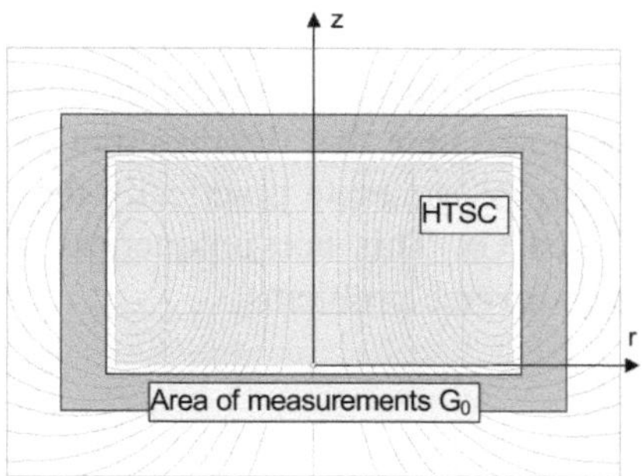

Figure 8. Field distribution of a cylindrical HTSC (HTSC subdivided into 4×4 rings)

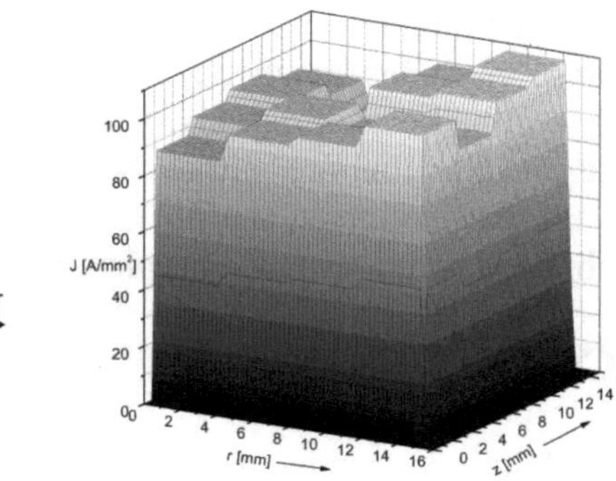

Figure 9a. Identified current density within one half of the HTSC ring. (J_c distribution in case of 4×4 subdivisions)

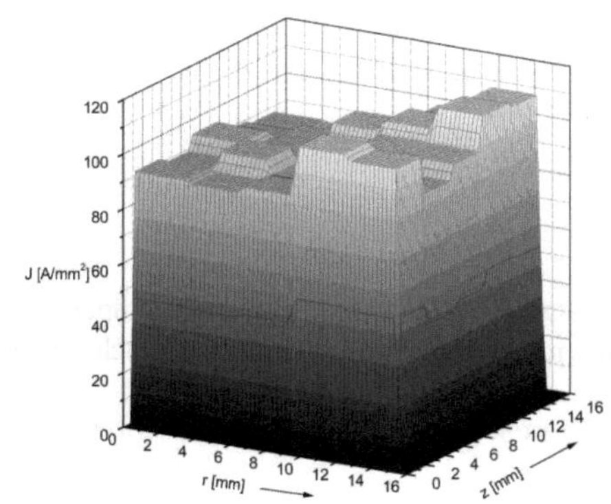

Figure 9b. J_c distribution in case of 5×5 subdivisions

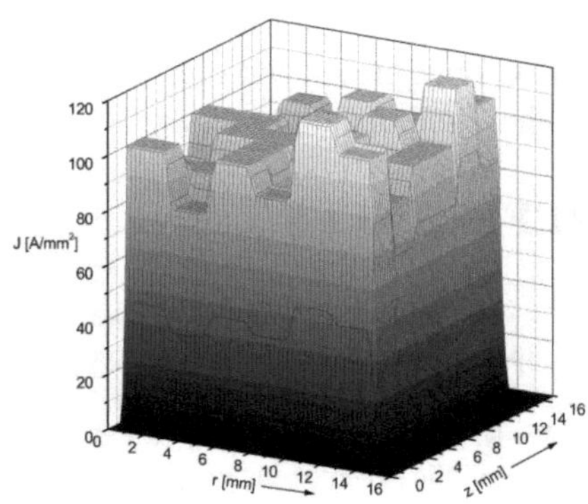

Figure 9c. J_c distribution in case of 6×6 subdivisions

As can be deduced from the J_c-distributions shown above, the proposed algorithm enables a stable and physically correct solution of the analysed inverse problem (determination of the current density distribution within the HTSC), thus it can be applied for the detection of flaws eventually existing within the HTSC's (regions with small J_c values). The solution of the problem differs strongly according to the number of assumed subdivisions of the HTSC region. For practical applications it is sufficient to examine the mean J_c distribution in maximum 36 HTSC rings.

The mean value of the determined J_c in each sub-region can be called the "engineering current density", as only this value is responsible for the force interaction between the HTSC and any external magnetic field. The final solution depends furthermore on the size and position of the measurement regions. It can be stated, that the measurements have to be performed at the points where the magnetic field possesses high values and gradients.

For the application of HTSC's as field excitation units within electrical machines the knowledge of the mean value of the J_c is only necessary in a reduced number of sub-domains, whereas for screening applications (reluctance machines, magnetic bearings) the current density distribution has to be determined more precisely. It should be noted here that the different current density distributions within the HTSC (*Fig. 9*) generate practically the same field distribution in the measurement area G_0. This non-uniqueness of the obtained solutions is very typical for such inverse problems.

CONCLUSION

A non-destructive quality control of HTSC-bulks based on measurements of the field distribution trapped by the HTSC has been developed. The proposed method is based on the determination of the current density within the HTSC. This identification code has been validated by many experiences and thus may serve as a non-destructive quality evaluation algorithm for single HTSC bulk pieces and clusters used for magnetic bearings and the excitation of electrical machines e.g.

REFERENCES

1. W.-R. Canders, H. May, R. Palka and E. Portabella, "Machines with high temperature superconducting bulk material in comparison with permanent magnet excited synchronous machines", ICEM'2000, Helsinki
2. W.-R. Canders, H. May and R. Palka, "Identification of the current density distribution of monolithic superconductors", ISTET'1997, Palermo
3. C.L. Lawson, R.J. Hanson, Solving least squares problems, Prentice-Hall 1974
4. H. May, R. Palka, E. Portabella and W.-R. Canders, "Calculation of the properties of electrical machines with high-temperature superconducting bulk material", ECCOMAS'2000, Barcelona
5. B. Oswald, T. Strasser, M. Krone, M. Soell, J. Oswald and K.J. Best, "Superconducting motors with HTS bulk material for medium power range", EUCAS'1999, Sitges
6. R. Palka, "Synthesis of magnetic fields by optimization of the shape of areas and source distributions", Archiv für Elektrotechnik 75, 1-7 (1991)

INCREASING OF OUTPUT POWER CAPABILITY IN A FLUX-WEAKENED PERMANENT MAGNET SYNCHRONOUS MOTOR WITH A THIRD HARMONIC CURRENT INJECTION

Bojan Štumberger, Anton Hamler, Mladen Trlep, Marko Jesenik and Viktor Goričan

Faculty of Electrical Engineering and Computer Science, University of Maribor, Smetanova 17, 2000 Maribor, Slovenia

Abstract: The paper presents a FEM-based analysis of output power capability (OPC) improvement in a six-phase flux-weakened permanent magnet synchronous motor with a third harmonic current injection. It is shown that the flux-weakened permanent magnet synchronous motor with the third harmonic current injection is capable of producing more output torque per RMS current flowing through the windings than the same motor supplied with sinusoidal currents. The influence of the iron losses on the OPC and the constant-power speed range (CPSR) is included in the analysis by the posterior iron loss calculation. The improvement of the machine OPC with a third harmonic current injection depends on the level of flux-weakening and it is higher at a high level of flux-weakening. The Results of the analysis are partially verified with the measurements.

Key words: permanent magnet synchronous motor, flux-weakening, third harmonic current injection

M. Rudnicki and S. Wiak (eds.), Optimization and Inverse Problems in Electomagnetism, 313–321.

1. INTRODUCTION

Flux-weakened Permanent Magnet Synchronous Motors (PMSMs) are often used in adjustable speed drive applications. The great disadvantage of such motor drives is the rapid decrease of motor output torque with the increase of the level of flux-weakening. A convenient measure of flux-weakening and output power capability (OPC) performance is the constant-power speed range (CPSR), which is the speed range at which the drive can maintain constant power with limited values of voltage and current.

It has been reported that the injection of the third current harmonics in the phase current waveforms can improve the output torque in reluctance and induction motors [1], [2].

The normally wound three-phase stator windings of interior permanent magnet synchronous motor (24 slots, 4 pole, skewing for one stator slot) [3] were replaced with two sets of three-phase electrically isolated windings, and one set was shifted from the other by 30^0 in space. In this way the six-phase stator windings were constructed. It was supposed that the motor is supplied from two sets of current controlled VSIs, and the phase currents in two three-phase groups are shifted 30^0 from each other in time. The injection of third harmonic current in the sinusoidal phase current waveforms in such a way constructed stator windings creates an additional rotating field with angular speed equal to the fundamental angular speed. The aim of the investigation is to reveal the influence of the proper adding of third harmonic current to the fundamental current component on the OPC of the interior PMSM. PMSM with complicated rotor geometry is analysed in this paper and Fig.1 shows its cross section.

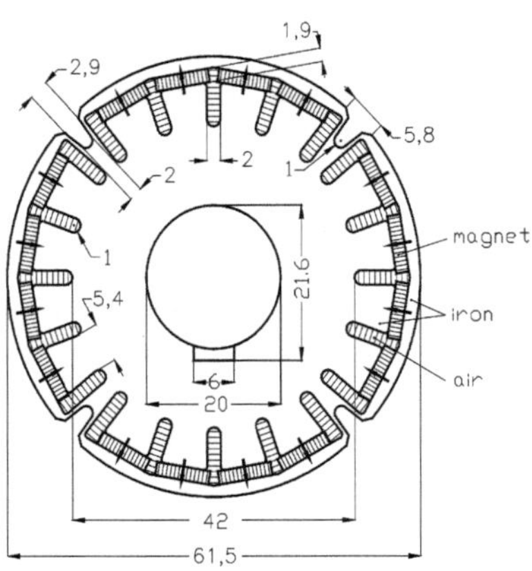

Figure 1. Rotor cross-section

The six-phase full-pitch windings with wye connection are placed in stator slots in the sequence of F, -B, -E, A, D, -C from 0 to 90 degrees in the mathematical positive direction. The phase currents in two three-phase groups are given by:

$$i_A = \sqrt{2}(I_1 \cos(\omega t + \varphi_1) + kI_1 \cos(3(\omega t + \varphi_1) + \varphi_3)) \quad (1)$$

$$i_D = \sqrt{2}(I_1 \cos(\omega t + \varphi_1 - 30) + kI_1 \cos(3(\omega t + \varphi_1 - 30) + \varphi_3)) \quad (2)$$

where $I_1(\varphi_1)$ and $I_3=k\ I_1(\varphi_3)$ are the RMS values (the phase angles) of fundamental and third harmonic current component.

In the remaining two phases B and C (E and F) the phase shift of currents is $\pm 120^0$ in the relation to the A (D) phase. In the analysis the total RMS value of phase current waveform is maintained the same with and without the third harmonic current injection.

2. ANALYSIS OF MAGNETIC CONDITIONS

To obtain the field distributions and the loci of local flux density vectors in the motor, a series of magneto-static field calculations for a complete cycle of field variation was computed by 2D FEM using the basic equation:

$$\mathrm{rot}(\nu\ \mathrm{rot}(\boldsymbol{A})) = \boldsymbol{J}_0 + \mathrm{rot}\ \boldsymbol{M} \quad (3)$$

where ν denotes the reluctivity, $\boldsymbol{A}$ is the magnetic vector potential, $\boldsymbol{J}_0$ is the current density and $\boldsymbol{M}$ is the magnetization of the permanent magnets. The non-linearity of used iron core material was accounted for with a single-valued *B-H* curve. The whole 2-D motor region was discretized into 15664 first order triangular elements. The magnetic conditions over a complete cycle of magnetic field variation were calculated in 121 discrete equidistant time steps by shifting the rotor position and simultaneously changing the stator excitation. The initial motor mesh was divided into rotational and stationary mesh areas. For simulation of rotor movement a lock step mesh generation scheme is used in the analysis of the magnetic field. In each step, the rotor position is at first rotated with respect to its previous position automatically, before the FEM calculation is performed. Between two sequential steps the rotor is moved for a fixed angle corresponding to the arc between two nodes at the inner rotor diameter. Skewing of the stator slots is accounted for with the multislice model.

3. RESULTS AND DISCUSSION

With the intention to determine the flux-weakening performance and the output power capability of the analysed permanent magnet motor, the output power characteristics versus speed were calculated for nominal current level and different current angles. In all cases it was supposed that maximum speed for a given current level and current angle (angle between the stator MMF and the direct axis in the case of sinusoidal current supply) is obtained when the line to line voltage limit (400 V) with increasing of speed is reached. The output power was determined by the power balance method by considering iron core loss, mechanical loss and ohmic loss component [3], [4]. The electromagnetic torque was calculated by using the Maxwell Stress Method. Increasing or decreasing of output power capability depends on the percentage and phase angle of added third harmonic current component in the phase currents waveforms. The highest output power capability in all cases was obtained when the simple rule $3\varphi_1+\varphi_3=90^0$ for determination of phase angle of added third harmonic current component in the phase currents waveforms was satisfied. The reduction in the terminal voltage and electromagnetic torque, as the current angle (angle between the stator MMF and the direct axis in case of sinusoidal current supply) is increased towards 180 degree, is presented in Fig. 2. The results of the electromagnetic power and the output power calculation versus speed are presented in Fig. 3.

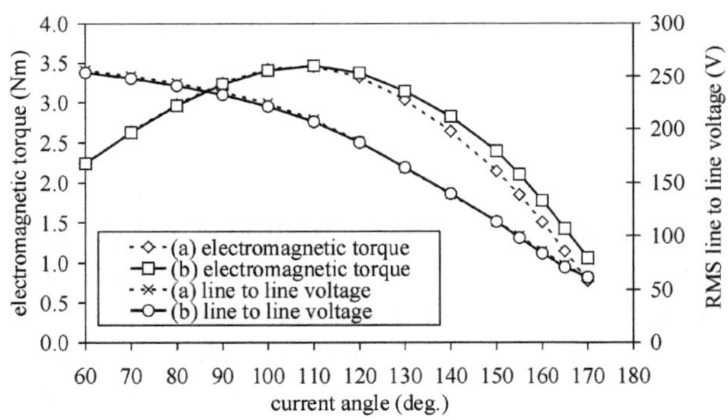

Figure 2. The highest average value of electromagnetic torque and RMS value of line to line voltage of six-phase machine at speed 1500 rpm and nominal current density: (a) with only fundamental current waveform applied; (b) with addition of 30% third harmonic current to the fundamental current waveform

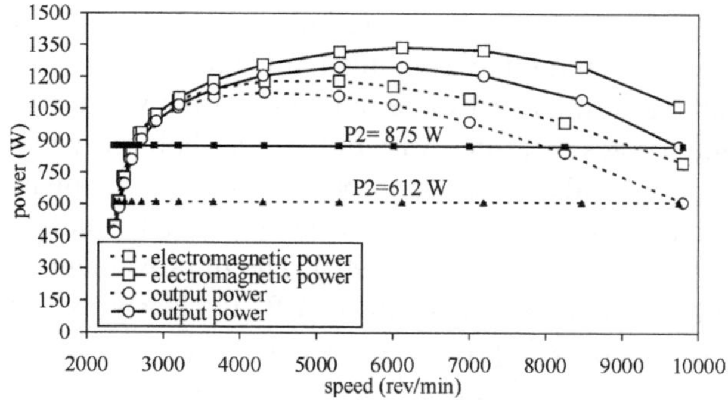

Figure 3. Calculated values of electromagnetic and output power versus speed: sinusoidal currents (dashed lines); with addition of 30 % third harmonic current to the fundamental current waveform (solid lines)

The improvement of the output power capability with a third harmonic current injection depends on the level of flux-weakening (FW) and it is higher at a high level of flux-weakening, which can be seen in Figs. 2-3. The electromagnetic torque ratio (electromagnetic torque with third harmonic current injection / electromagnetic torque without third harmonic current injection) is presented in Fig. 4, where a marked improvement in electromagnetic torque performance in flux-weakening region is clearly evident.

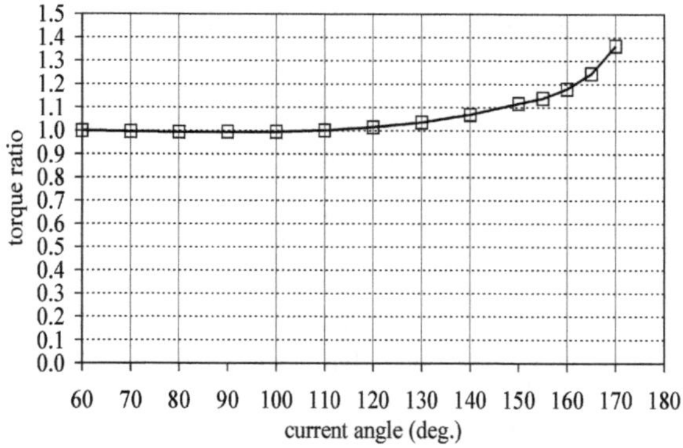

Figure 4. Calculated electromagnetic torque ratio

In order to compare the calculated torque capability of the motor under sinusoidal and combined fundamental and third harmonic excitation, some measurement were carried out. The shaft of the motor was locked using a locking device. Next, the dc values of the sinusoidal currents at one instant of time were applied. The motor shaft was moved incrementally, and the steady-state torque profile was measured using a torque transducer. Similar tests were carried out for the combined fundamental and third harmonic injection. In all cases, the RMS value of phase current was kept the same.

Fig. 5 demonstrates measured variation of the shaft torque with respect to rotor position for fundamental and for combined fundamental and third harmonic excitation with different percentage of third harmonic injection (k=0.0; 0.3; 0.5; 0.7; $\omega t=0$, $\varphi_1=0$, $\varphi_3=90^0$). The improvement of shaft torque capability with a combined excitation in FW is clearly evident. In Fig. 6, measured and calculated profile of shaft torque variation with only fundamental excitation (k=0.0, $\omega t=0$, $\varphi_1=0$) is presented.

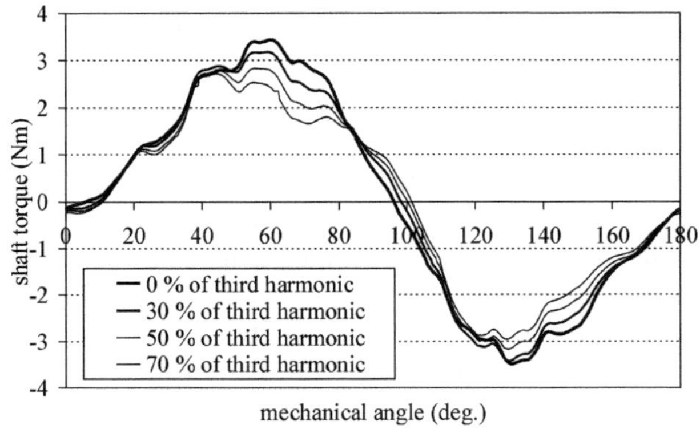

Figure 5. Measured shaft torque variation for fundamental and combined fundamental and third harmonic excitation

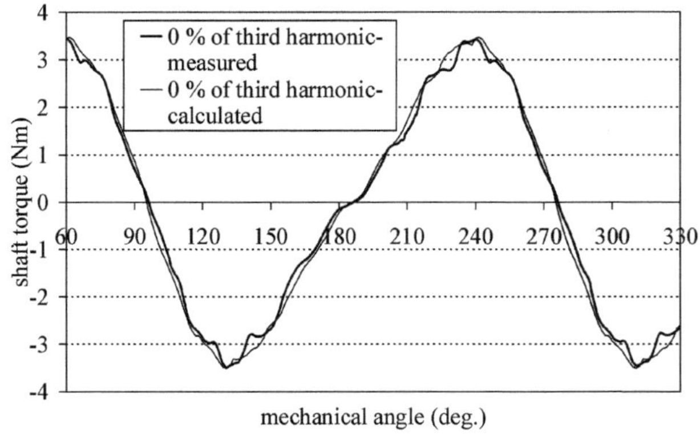

Figure 6. Measured and calculated shaft torque variation with only fundamental excitation

Measured shaft torque profile variations are dependent on the selected instant of time, which can be seen in Figs. 7-8. The measured and calculated shaft torque profiles in Fig. 9 are presented as an average value of two instant time moments (case1 and case 2). Measured and calculated torque ratio profile presented in Fig. 10 is in a good agreement with the torque ratio profile presented in Fig. 4.

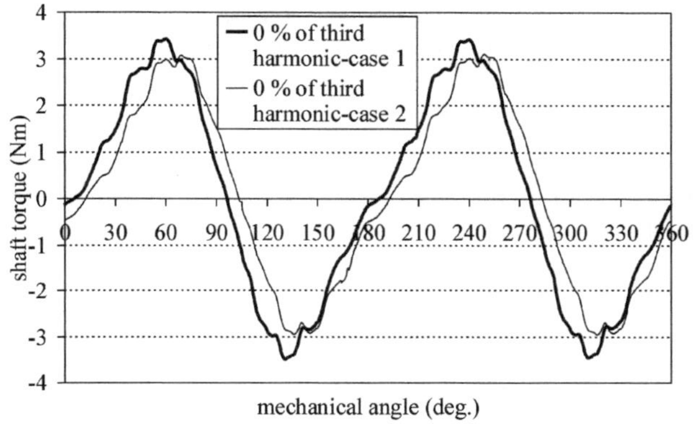

Figure 7. Measured shaft torque variation for fundamental excitation: case1 (k=0.0, ωt=0, φ_1=0,); case2 (k=0.0, ωt=15^0, φ_1=0)

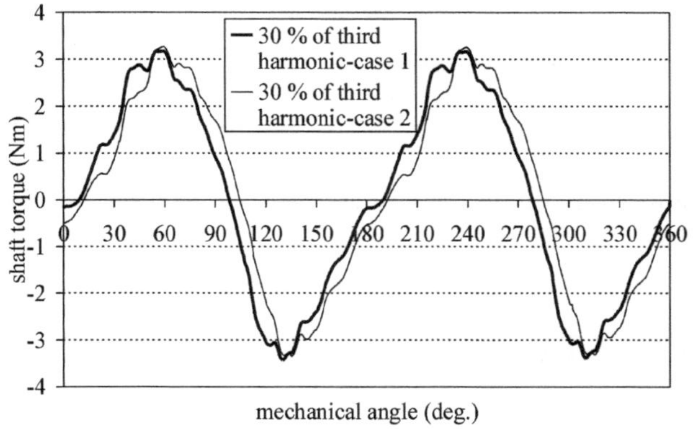

Figure 8. Measured shaft torque variation for combined fundamental and third harmonic excitation: case1 (k=0.3, ωt=0, φ_1=0, φ_3=90^0); case2 (k= 0.3, ωt=15^0, φ_1=0, φ_3=90^0)

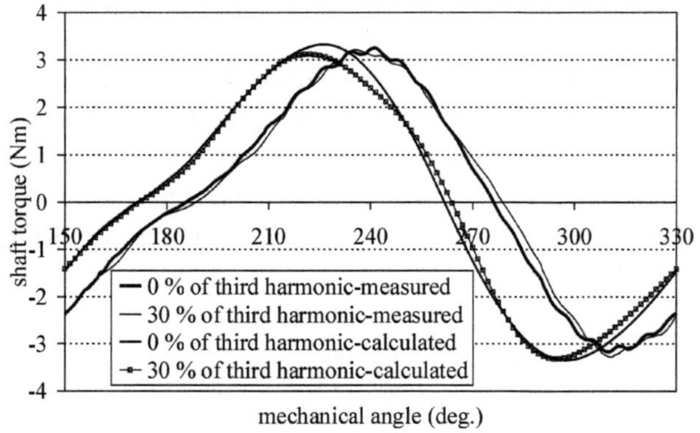

Figure 9. Measured shaft torque variation, determined as an average value of two instant time moments (case 1 and case 2), calculated shaft torque profiles are shifted from the measured ones for 15 mechanical degrees

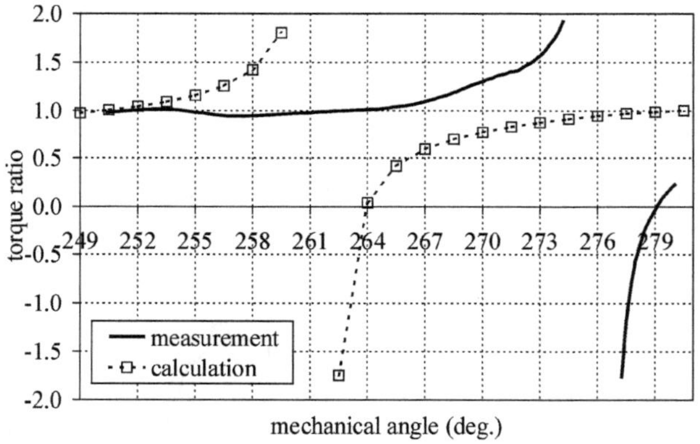

Figure 10. Measured and calculated values of torque ratio calculated from torque profiles presented in Fig. 9

4. CONCLUSION

Presented measured and calculated results validate the theoretical analysis and motivate additional investigation on this system. The improvement of the output power capability with a third harmonic current injection depends on the level of flux-weakening (FW) and it is higher at a high level of flux-weakening.

REFERENCES

1. H. A. Toliyat, L. Xu and T. A. Lipo, "A five-phase reluctance motor with high specific torque," IEEE Transaction on Industry Applications, vol. 28, No. 3, pp. 659-667, May/June 1992.
2. R. O. Lyra and T. A. Lipo, "Torque density improvement in a six-phase induction motor with third harmonic current injection", in Proc. of the 36th IAS Annual Meeting, Illinois, USA, Sept. 30 – Oct. 4, 2001, pp. 1779-1786
3. B. Štumberger, A. Hamler and B. Hribernik, "Analysis of iron loss in interior permanent magnet synchronous motor over a wide-speed range of constant output power operation," IEEE Transaction on Magnetics, vol. 36, No. 4, pp. 1846-1849, July 2000.
4. B. Štumberger and B. Hribernik, "Prediction of permanent magnet synchronous motor characteristics using finite elements," COMPEL, vol. 19, No. 2, pp. 517-523, 2000.

A METHOD OF SHORTENING BREAKING TIME OF THE MAGNETIC CONTACTORS BY THE EXTERNAL MAGNETIC FORCES

Hyun-Kil Cho*, Eun-Woong Lee**

*LG Industrial systems. Ltd. Co. ** Dept. of Electrical Engineering, Chung-nam Nat'l Univ*

Abstract: In this paper, the contact surface of moving and fixed contactors of 3 different arc extinction unit used for DC 100A magnetic contactor has been computed with Holm's equation . Also, electromagnetic blowout force acting on the arc and arc velocity are analyzed using 3D FEM and the breaking time for power source DC current 100A, voltage 100V has been measured and compared computing value with measuring value.

Key words: DC magnetic contactor, arc extinction, Blowout force, Contacts

1. INTRODUCTION

The electrical life of the magnetic contactor depends on the arc extinction performance during the interrupting current.[1]-[3] The main subjects of magnetic contactors are shortening the arcing time for the interruption of currents and the reduction of contacts wear. The arc extinction time of a DC magnetic contactor is generally longer than that of an AC magnetic contactor. So, it is important for improving the breaking performance especially for DC magnetic contactor. We have developed new chamber of arc extinction unit that can shorten the arcing time and reduce contact wear.

Holm's equation [1] is applied to magnetic contactors each of 3 different arc extinction units model to compute the appropriate contact surface between moving and fixed contacts and the electromagnetic blowout force

M. Rudnicki and S. Wiak (eds.), Optimization and Inverse Problems in Electomagnetism, 323–334.

analysis using 3D FEM. A chamber for arc extinction experiment was made to investigate the relationship of electromagnetic force acting on an arc, drag force and the arc breaking time. As well, the actual arc breaking time was measured. The electromagnetic force acting on an arc is presented as the electromagnetic factor influencing the arc extinction performance through the relative comparison the arc force obtained by FEM and the breaking time obtained by experiments. Also the method of anticipating arc extinction performance with electromagnetic field analysis is suggested.

2. ARC EXTINCTION

2.1 3 types of arc extinction unit

3 kinds of analysis model are showed in Fig. 1 (a), (b), (c) along with the vector direction of arc current I, permanent magnet. The arc extinction model 1 of Fig. 1(a) employs a U-shaped grid, which is the same as the one used in the arc quenching room of an AC magnetic contactor. And the model 2 of Fig. 1(b) has an U-shaped grid and permanent magnet to increase the magnetic flux crossing the arc current. The model 3 has divided grid and permanent magnet to increase the magnetic flux crossing the arc current. The volume of the arc extinction unit and the size of the permanent magnet are shown in Fig. 1(d). And all dimensions of 3 different arc extinction units are the same size. The associated materials are a SCP1 steel plate and an Alnico5-series magnet. The arc quenching force is generated by the generation of the driving force F directing forward and backward when the magnetic flux of permanent magnet flows left and right through direction B and the arc current flows through direction I.

2.2 Electromagnetic blowout force

Fig. 2(a) shows the arc extinguish chamber of electric contactors having grids only for model 1. The magnetic flux generated by arc current in the U-shaped grid flows through grid (magnetic plate) and the open part of the U-shape. If the arc current flows toward the z direction column on the origin of the x-y coordinate, the vector of magnetic flux by the arc column agrees with Ampere's law. Accordingly, the magnetic flux density of Y direction B_Y is lower closing place of the U-shaped grid than opening place of the U-shaped grid. As shown in the Fig. 2(b), the magnetic flux by arc current flows along grid and the open part of the U-shaped grid. The most magnetic flux of permanent magnet concentrates the grid but the leakage flux of permanent magnet intersects the arc column. The leakage flux affects to reduce the magnetic flux of arc current on the left side of Yaxis but to increase on the other side. Fig. 2(c) shows the divided U-shaped grids model. The amount of the permanent magnet flux flows along grid and goes across in the air between divided grids. If it fades away from the permanent magnet to the distance, the magnetic flux density in the air between divided grids reduces gradually. As the results, the magnetic flux on left side of Y axis reduces resultant flux due to opposite face. On the right side, it increases resultant flux owing to face the same direction. Resultantly, the blowout force of arc column is proportional to the difference of the magnetic flux density between on the left side and on the other side.

The electromagnetic force F acting on the arc can be obtained by Eq. (1)

$$F_e = \int_{V_a} \bar{J}_a \times \bar{B}_a dV \qquad (1)$$

where $\bar{J}_a$: current density, V_a : arc area , $\bar{B}_a$: magnetic flux density

2.3 Drag Force and arc motion

Arc flow by the blowout forces is prevented drag force by frictions of the air viscosity on the front edge of the arc column. On the rear edge of the arc column, an in ward suction force to suck the air from the surrounding area is generated. The force acting on the arc column by the surrounding air is named the drag force. [2]

An arc is a current path, which is a stream of ions and electrons. And the electromagnetic force and the drag force deform arc freely. Because it is very difficult to obtain the motion of all particles in the arc, arc model is simplified as follows. The center of an arc contains strongly ionized gas at a high temperature and the boundary of arc column is not clear from surrounding gas. But, in order to calculate drag force, we assume the arc region is clear from surrounding gas.

The drag force for an arc column is defined by

$$F_d = \frac{1}{2} C_D \rho S u^2 \qquad (2)$$

Where S is the frontal area of the column exposed to the flow and u is the moving velocity of the arc column. C_D is the drag coefficient

$$C_D = 0.62\left(\frac{L}{d}\right)^{0.12}.$$

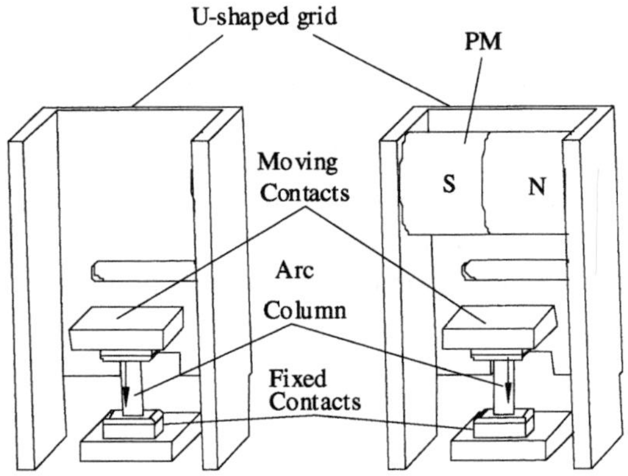

(a) Model 1 (U-shaped grid only)

(b) Model2 (U-shaped grid and permanent magnet)

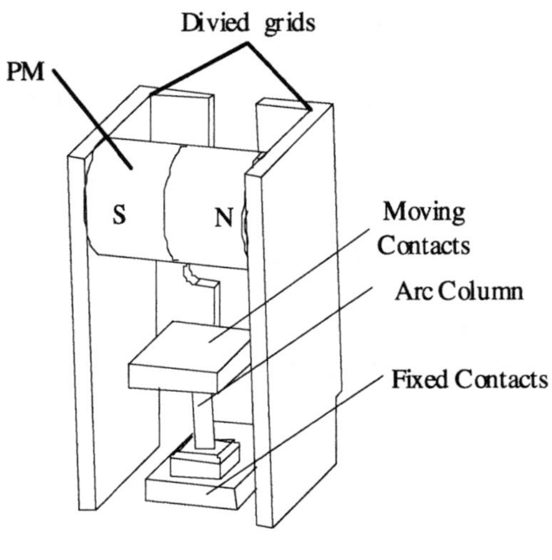

(c) Model 3(divided grid and permanent magnet)

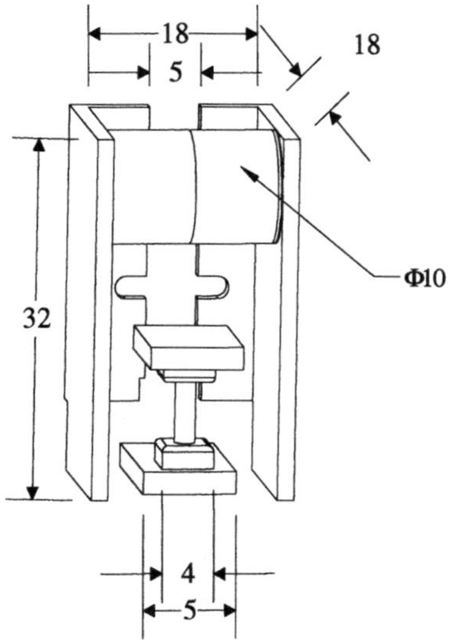

(d) dimension of arc extinction unit.

Fig.1 3 magnetic contactors of different arc extinction unit.

The equation of arc motion is given by

$$F_e - F_d = m\frac{du}{dt} \qquad (3)$$

F_d is the frictional force of the arc, m is the mass of the arc.

On the assumption that the arc consists of electrons and singly ionized silver atoms, the mass of the arc is

$$m = \left(\frac{M}{N_0}\right)\times\left(\frac{\pi}{4}D^2L\right)\times\left(\frac{P}{2kT}\right) \qquad (4)$$

Where M is the atomic weight of ionized atoms, No is Avogadro's number, k is Boltzmann's constant, T is the arc temperature, and p is the arc pressure.

The contact radius, r, is obtained by Eq. (5) from Holm's equation for the modeling of arc, where P is contacting pressure [N], H_d is the Brinell hardness and ζ is a constant determined by the condition of the contact shape.[1][4][5]

$$r = \sqrt{\frac{P}{\pi \varsigma H_d}} \qquad (5)$$

TABLE 1 The actual values of contacts model

Material of contacts	H_d	Z	P[kgf]	r[mm]
AgCdO	90	0.5	0.7	0.25

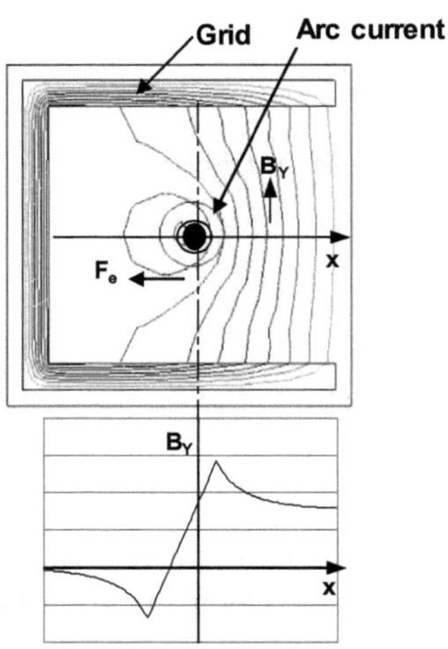

(a) Model 1

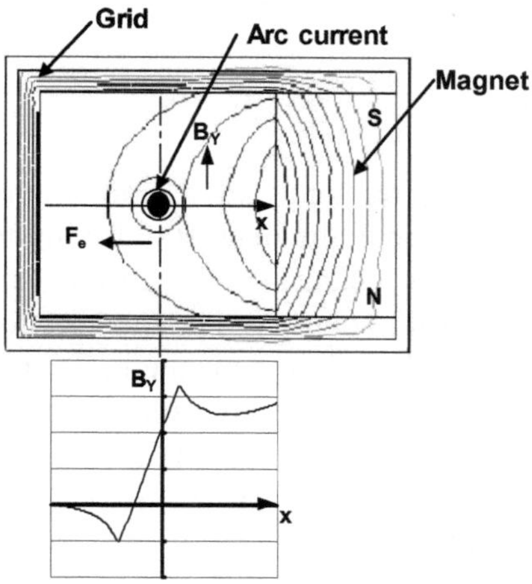

(b) Model 2

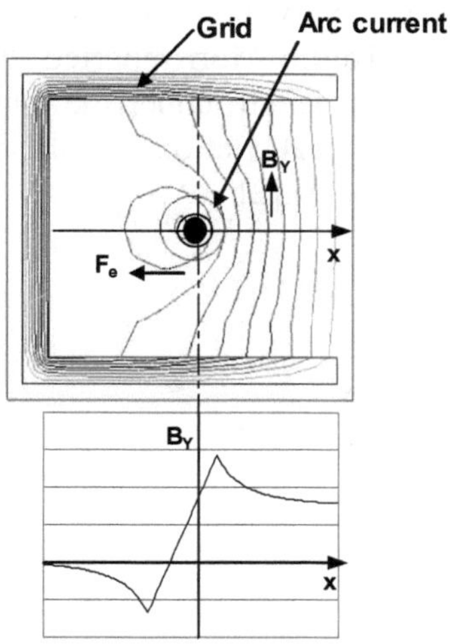

(c) Model 3

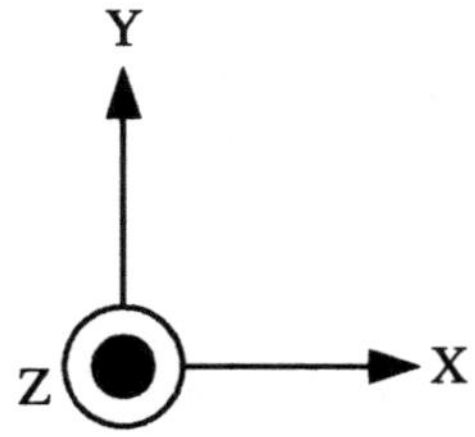

(d) Vector coordination

3. COMPARATION OF SIMULATIONS AND TESTS

3.1 Result of simulation

Fig. 3(a) shows driving force Fe acting on arc as a function of current on each model. Provided that the current is same, the magnetic driving force of model 3 changes rapidly according to the change of the current since the electrical force depends on the magnetic flux density. The model 1 to which the external magnetic field is not applied has slower magnetic driving force change. Magnetically driving forces acting on the arc column are 0.0017, 0.0089 and 0.029 N for I=100A respectively model 1, 2 and 3

Fig.3 Results of Analysis; (a) Driving force for arc as a function of current, (b) Blowout Velocity of each Model

Fig. 3(b) shows driving velocity acting on arc as a function of current on each models of silver vapor arc. The arc velocity depends on magnetically driving force and drag force acting on arc column for each model. Driving velocities of arc are 9.3, 22 and 36.9 m/s respectively model 1, 2 and 3 when contacts are completely opened for I=100A. Teijiro Mori reported that the calculated velocities of a 100A arc with a 5mm length under the magnetic flux densities of 10mT were 25m/s. Our calculation result agrees approximately with value calculated by Teijiro Mori.[2]

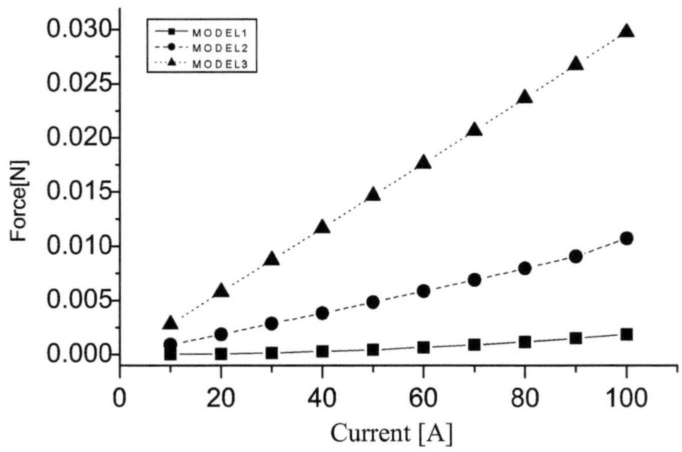

(a) Driving force for arc as a function of current

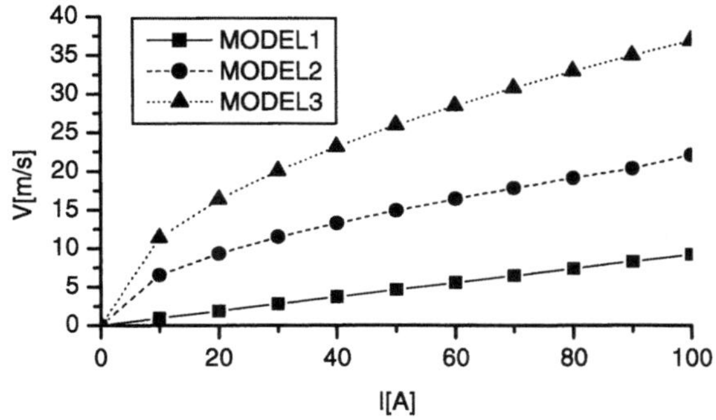

(b) Blowout Velocity of each Model

Fig.3 Results of Analysis

3.2 Test equipment and result

The experiment for each model was conducted by applying DC 100A to the contact area with the time constant of 10.2 [ms] on DC load. The increase of arc blowout force results in the improvement of arc extinction performance and the shortening of arc dwelling time in an arc extinction chamber. Hence, the shape of a current waveform was measured at the precise moment of being turned off the load. Fig.4 show oscillographs of arc current and arc voltage during breaking load current at DC 110V. The arcing time of models 1,2 and 3 are 20.9[ms], 10.1[ms] and 5.2[ms], respectively. The arc voltage rises in the proportion to the variation of arc current and arcing time.

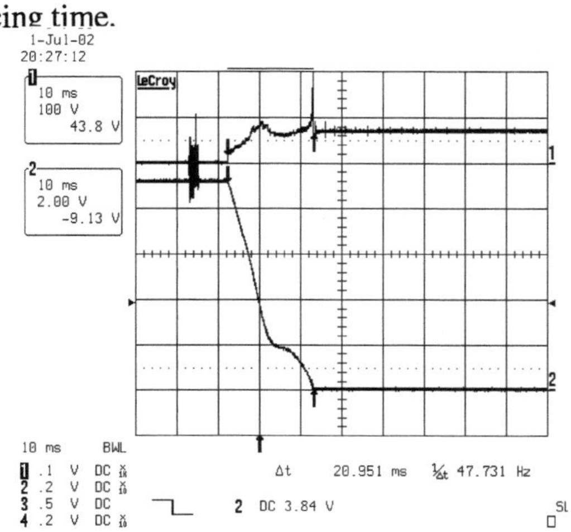

(a) Model 1

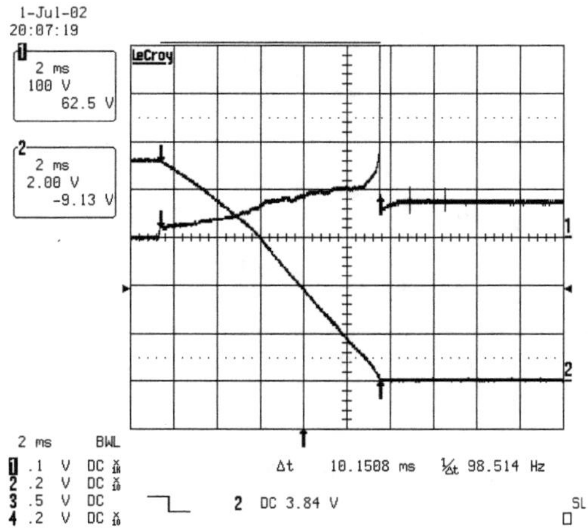

(b) Model 2

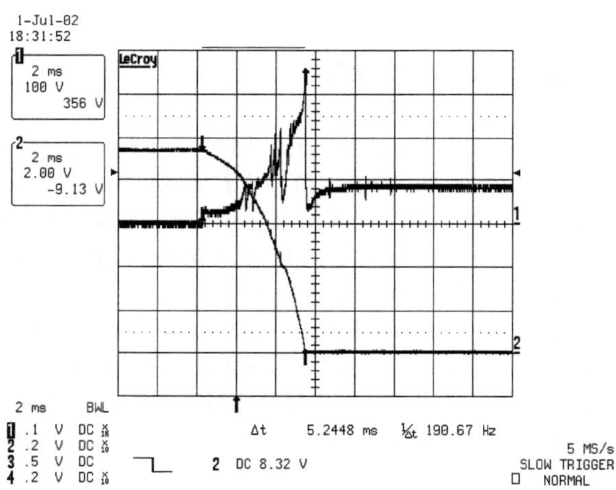

(c) Model 3

Fig .4 *Measurement of arcing time and arc voltage*
(1 : arc voltage, 2: current)

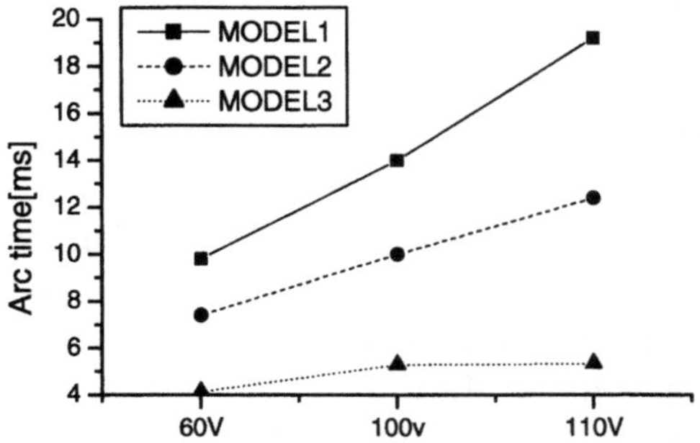

(a) Measurement of Arcing time

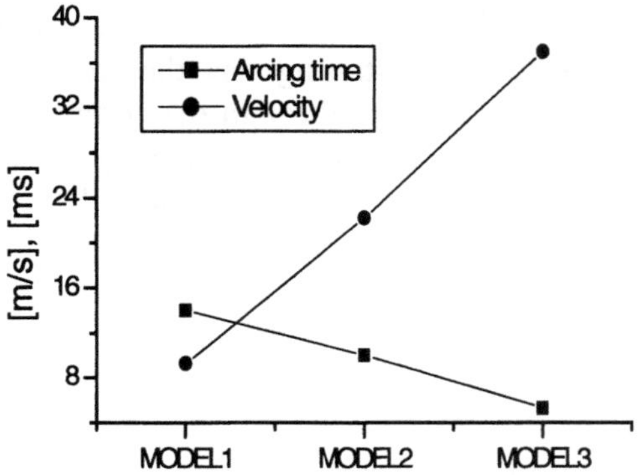

(b) Comparing calculated velocities and measured arcing time (I=100A)

Fig. 5 Measurement of Arcing time

Measured results of arc time are showed by Fig. 5(a) . Interrupting tests were made in dc circuits having line voltages of 60V, 100V and 110V. Interrupting tests were repeated 30 times for each model and each voltage levels. There was significant difference of arcing time in the 3 type models. For the model 1, average arcing time was increased largely with line voltage. On the other hand, for the model 3, average arcing time was increased gradually with line voltage. At he voltage of 100V, average arcing time of model 3 was about 38% of that the model 1. Fig. 5(b) showed estimating method of comparing results of analysis with measured results. Velocities of moving arc for each model were carried out simulation at the current 100A. Arc extinction time and velocity of arc driving force can be compared relatively because of same arc quenching room. The result shows that bigger arc blowout forces lead to shorter arc extinction time, which coincides with the FEM analysis and drag force.

4. CONCLUSION

In this research, 3 different models of DC arc extinction structure have been formed and the appropriate contact area between the moving contacts and fixed contacts were estimated to implement 3 different types of arc model and to analyze them with 3D FEM.

The arc blowout force acting on the arc stream and the arc velocity were computed accordingly and the arc breaking time was measured employing an arc extinction chamber which was equivalent to the one used in FEM analysis.

This paper showed that arc blowout force is a main factor for arc velocity and the arc breaking performance by demonstrating that the arc breaking time is inversely proportional to the arc velocity through the comparison of arc velocity with the arc breaking time obtained by experiment. Also, the model for improving the performance of DC contactor is suggested

REFERENCES

1. R.Holm, "Electric Contacts", 4th ed., springer-verlag,(1967)
2. Teijiro Mori, Kenichi Koyama, "A new interruption Method for low-voltage, small-capacity, air-break contactors", IEEE Industry Application, Vol. 27. No. 1. PP.161-165. Jan. 1991.
3. Teijiro Mori, Kenichi Koyama, "A new interruption Method for low-voltage, small-capacity, air-break contactors", IEEE Holm Conference on Electrical Contacts . PP.2308-2313, 1989
4. John. Shea, Yun-Ko Chien, " Blow-open forces on double-break contacts" , IEEE Holm Conference on Electrical Contacts . PP.103-109. 1993
5. Shokichiro Ito and Yoshiaki Takato, " Numerical Analysis of Electromagnetic Forces in Low Voltage Circuit Breakers using 3-D Finite Element Method into account eddy Currents", IEEE Transaction on Magnetics, Vol.34, No.5, pp.2597-2600, Sept. 1998

MIX
Papier aus verantwortungsvollen Quellen
Paper from responsible sources
FSC® C105338

If you have any concerns about our products,
you can contact us on
ProductSafety@springernature.com

In case Publisher is established outside the EU,
the EU authorized representative is:
Springer Nature Customer Service Center GmbH
Europaplatz 3, 69115 Heidelberg, Germany

Printed by Libri Plureos GmbH
in Hamburg, Germany